MW00527886

Women at War in the Classical World

Dedication

For the women and girls inside Syria and those who have fled the conflict for safety in Lebanon, Jordan, Turkey and Iraq. Many have been subject to sexual and gender-based violence, coerced into early marriages, overwhelmed by economic strife, and psychologically scarred by loss in a war that seemingly has no end. Women and girls affected by conflict must be regarded as more than victims of brutality; they are agents of change who, if given the opportunity, can transform their societies.

Ambassador Melanne Verveer
Executive Director
Georgetown Institute for Women, Peace and Security
https://www.rescue.org/sites/default/files/resource-file/IRC_WomenInSyria_Report_WEB.pdf

Women at War in the Classical World

Paul Chrystal

Pen & Sword
MILITARY

First published in Great Britain in 2017 by
PEN & SWORD MILITARY
An imprint of
Pen & Sword Books Ltd
47 Church Street
Barnsley
South Yorkshire
S70 2AS

ISBN 978-1-47385-660-8

A CIP catalogue record for this book is available from the British Library.

Typeset by Concept, Huddersfield HD4 5JL.
Printed and bound in England by TJ International Ltd, Padstow, Cornwall.

Pen & Sword Books Ltd incorporates the imprints of Pen & Sword Archaeology, Atlas, Aviation, Battleground, Discovery, Family History, History, Maritime, Military, Naval, Politics, Railways, Select, Social History, Transport, True Crime, and Claymore Press, Frontline Books, Leo Cooper, Praetorian Press, Remember When, Seaforth Publishing and Wharncliffe.

For a complete list of Pen & Sword titles please contact
PEN & SWORD BOOKS LIMITED
47 Church Street, Barnsley, South Yorkshire, S70 2AS, England
E-mail: enquiries@pen-and-sword.co.uk
Website: www.pen-and-sword.co.uk

Contents

List of Plates vii
Acknowledgements ix
Abbreviations x
Introduction xi
Women and War in Earlier Ancient Civilisations xiv

Part One: Greece 1

1. Goddesses and War in Greek Mythology 3
2. Warlike Women in Homer 7
3. *Teichoskopeia*: A Woman's View from the Walls 16
4. The Amazons 19
5. Women and War in Greek Tragedy 28
6. Women and War in Greek Comedy 36
7. Women and War in Greek History and Philosophy 38
8. Women Warriors Catalogued 60
9. Spartan Women: Vital Cogs in a Well-Oiled War Machine 82
10. Macedonian Women at War: Pawns and Power-Players 89

Part Two: Women as Victims of War 99

11. War Rape and Other Atrocities in the Classical World 101

Part Three: Rome 121

12. Military Women in Roman Legend 123
13. Military Women in Roman History 126
14. Foreign Women Fighters 154
15. Women and War in Roman Epic 183
16. Women and War and *Militia Amoris* 194
17. Military Tendencies in Women in Seneca's *Troades* 199

Part Four: Warrior Women in the Arts and Entertainment 201

18. Military Women in the Visual Arts 205
19. Women as Gladiators 205

Epilogue 211
Notes 213
Bibliography 226
General Index 244
Index of Women 247

List of Plates

The skeleton of a young woman from Cemetery 2, Kurgan 8, Burial 4 at Pokrova, Russia.

Andromache, Astyanax and Hector in a touching scene from the *Iliad*, Apulian red-figure column-crater, *c.*370–360 BCE.

Menelaus goes to strike Helen, but stunned by her beauty, he drops his sword, detail of an Attic red-figure crater, *c.*450–440 BCE.

A scene showing Achilles fighting with the Amazon Penthesilea; on a sarcophagus dated *c.*250 CE.

Papyrus fragment with a drawing of the abduction of Briseis by Talthybius and Eurybates in the *Iliad* (Book 1, lines 330–48), fourth century BCE.

Amazonomachy marble, sarcophagus panel, *c.*160–70 CE.

Amazon mosaic from Paphos found in the Orpheus House, 3rd century CE.

Amazonomachy scene on a *lekythos*, *c.*420 BCE.

Jacques-Louis David (1748–1825), *The Intervention of the Sabine Women*, 1799.

Jean Jacques François Lebarbier, *A Spartan Woman Giving a Shield to Her Son*, 1805. (© The Portland Art Museum, Ohio)

The rampart walk on the East Wall in Jerusalem with the Dome of the Rock in the background. (Thanks to Professor Tod Bolen at Bibleplaces.com, Santa Clarita, CA)

Woodcut illustration of Veturia and Volumnia confronting Coriolanus, from an incunable German translation by Heinrich Steinhöwel of Giovanni Boccaccio's *De mulieribus claris*, *c.*1474.

A detail from the frieze of the Basilica Aemilia depicting the punishment of Tarpeia, *c.*14 BCE.

Vincenzo Camuccini, *Roman Women Offering their Jewellery in Defence of the State*, *c.*1825. (© Glasgow Museums)

Paul Jamin, *Brennus et sa part de butin – Brennus and his Share of the Spoils*, 1873.

Statue of a mourning barbarian woman, a victim of war, which stands in the Loggia dei Lanzi, Florence.

The terrible scene that is the end of the Roman Empire, with women pursued in the chaos – Thomas Cole, *The Course of Empire 4: Destruction* (1836). (Thanks to The Schiller Inc, Washington DC, http://www.theathenaeum.org/art/list.php?m=a&s=du&aid=375)

Pavel Svedonsky (1849–1904), *Fulvia with the Head of Cicero*.

Agrippina I on the Rhine fortifications. (Thanks to Jasper Burns for permission to use this, originally published in his *Great Women of Imperial Rome*)

Peter Froste (b.1935), *The Temple of Claudius* in Camelodunum, going up in flames at the hands of Boudica in 60 CE. (Courtesy of Colchester and Ipswich Museum Service)

A detail from the Column of Marcus Aurelius showing captured women and children.

Woodcut illustration showing Artemisia II of Caria drinking the ashes of her husband Mausolus, from an incunable German translation by Heinrich Steinhöwel of Giovanni Boccaccio's *De mulieribus claris*, *c.*1474.

Marble relief from Halicarnassus showing two female gladiators fighting, first or second century CE.

Acknowledgements

My thanks to Professor Amy C. Smith at the Ure Museum of Greek Archaeology in the Department of Classics, University of Reading for permission to use a number of their images.

Abbreviations

AC	*L'Antiquite Classique*
AJAH	*Americal Journal of Ancient History*
AJPh	*American Journal of Philology*
AncW	*Ancient World*
AZ	*Archaeologische Zeitung* (Berlin, 1848)
CEG	P.A. Hansen, *Carmina Epigraphica Graeca* (Berlin, 1983)
CJ	*Classical Journal*
CIG	*Corpus Inscriptionum Graecarum* (Berlin, 1825–77)
CIL	*Corpus Inscriptionum Latinarum*, (Berlin, 1863)
CP	*Classical Philology*
CQ	*Classical Quarterly*
CW	*Classical World*
EMC	*Echos du Monde Classique*
FGrH	F. Jacoby, *Die Fragmente der Griechischen Historiker* (Leiden, 1923–57)
G&R	*Greece and Rome*
GRBS	*Greek, Roman and Byzantine Studies*
HN	Pliny the Elder, *Historia Naturalis*
HSCP	*Harvard Studies in Classical Philology*
IG	*Inscriptiones Graecae*
JRS	*Journal of Roman Studies*
P. Oxy	*The Oxyrhynchus Papyrus*
PCG	*Poetae Comici Graici*
PMG	*Poetae Melici Graeci* (Oxford, 1962)
RhM	*Rheinisches Museum fur Philologie*
SO	*Symbolae Osloensis*
TAPA	*Transactions of the Proceedings of the American Philological Association*
WS	*Wiener Studien*

Introduction

This is the first single-author book to address and analyse the role of women in classical warfare – both as agents of and participants in war, and as victims of war. It is often assumed, beginning with Hector's remarks in the *Iliad*, that conflict was the exclusive preserve of men in the battles and wars fought by the ancient Greeks and the Romans. After all, Alexander the Great, Scipio Africanus and Julius Caesar are much more familiar than, for example, Fulvia who embroiled herself in the conflicts of Mark Antony, her husband, during the Perusine War, or Agrippina the Elder who got involved with the troubled legions in Germanicus' campaigns in Germania. At least Cleopatra and Boudica are much better known. These military women are just four of many who took on exceptional and significant roles in the wars of the ancient Greeks and Romans from Homer to the end of the Roman Empire: in those 1,200 or so years there is a large number of women who had a role in the causation, direction or conduct of wars and battles. Guile, military intelligence, diplomacy, tactical excellence, courage and ferocity are just a few of the qualities exhibited by the women featured here. This may still surprise some readers today, but, when we observe that many of the books and journals published in recent years on Greek and Roman warfare ignores the role of women, then I remain unsuprised that they are surprised.

The female of the species, of course, features prominently in the Greek and Roman pantheons and in mythical representations of war: Andromache, Athena and the Amazons are examples; women are present in epic poetry in Helen of Troy or Briseis, and in drama in the shape of the vengeful or victimised women of the tragedies, or as the 'revolting' women in the *Lysistrata*. In the real world, or what was imagined to be the real world, she populates the strange foreign countries described with some incredulity by, for example, Herodotus – take Queen Tomyris, Artemisia and Pheretima; she emerges even as a poet warrior in Telesilla.

But behind the celebrities we know that the everyday women in the classical period who married soldiers – and there must have been countless thousands of them down the years – were typical army wives, forever providing the routine support that army wives have always provided – not least, in the extended absences of warrior husbands: holding the household, the *oikos*, together, running the farmstead, raising the children and schooling the next generation of soldiers. There were camp followers foraging for, selling and cooking rations, working the wool, making and repairing clothing, organising worship in the field, nursing casualties and burying the dead, and selling sex. They were all essential back-up for the soldiery but these activities were indicative of a dire and determined need among women to assist, subsist and survive in war-torn environments, or to exploit the system, working the black markets and profiting from war, often just to scrape the most basic living.

So much for all the background activity that the women of Greece and Rome got on with, quietly and relentlessly over something like a millennium and a half that is characterised by virtual non-stop war. This was only part of the picture though: given the relatively subdued profile of women generally in Greek and Roman societies and the social, civic and political unobtrusiveness that was fostered, it is perhaps surprising to learn that women were not totally excluded from military strategy-making nor were they completely absent from combat situations; in sieges and street-fighting women sometimes did their bit.

The aim of this book is to trace and analyse the direct, as well as indirect, involvement of women in the war machines of Greece and Rome, and to demonstrate the important part women played in classical military history, both as participants in war and battle, and as agents of conflict and victims of war. In so doing, the book redresses the balance between men and women in ancient warfare and accords military women their rightful place in the annals of Greek and Roman military and social history. The military, political and social consequences of their bellicose actions are fully discussed in the context of women's place and role in their societies.

Warfare did not, of course, start with classical antiquity – far from it. To give context to the subject, the first chapter of the book provides a survey of women's activity in warfare as waged by earlier civilisations relatively close in time and place to ancient Greece and Rome. In ancient Greece we examine the role and purpose of mythical and fictional military women and deities in Homer, the tragedians and the comic playwrights; a woman's view from the ramparts is given and the warlike Amazons are then discussed. Belligerent women in Herodotus (for example, Tomyris and Telesilla) and the comparative reticence of Thucydides regarding women at war, are explored; the potential role of military women in Plato's ideal states is covered next. The obscure *Tractatus de Mulieribus* is examined in the context of catalogues of the woman-warrior, along with similar works by Plutarch and Polyaenus. We then follow the relatively prominent role of Spartan women in the military, and Macedonian women in the dynastic wars and those women who associated with Alexander the Great in the fourth century BCE.

So far the book deals mainly with women as agents of military activity and war: a pivotal chapter analyses women as victims of war in Greek and Roman military conflict, covering displacement, gender based violence, rape and enslavement.

Women feature prominently in the legends surrounding the very foundation of Rome and constitutional reform, starting with the abduction of the Sabine women and taking in Lucretia and Verginia. We follow women's involvement in, for example, the terrible aftermath of the Punic Wars, and the influential role played by Cleopatra and then Fulvia in consort with Mark Antony. When foreign women fighters are explored we meet, among many others, the British queens Cartimandua and Boudica – one an ally, the other a formidable enemy; war in Roman literature forms the basis of a number of important chapters, including the relationship with the *miles amoris*, the soldier of love, and women in Roman epic: leaders, battlers and visionary, life-changing witches. The official Roman attitude to women and their situation in the army environment – to accompany or not to accompany – is discussed.

In the early Empire we meet the elder Agrippina who helped save the Roman legions, and her husband's career, from disaster on the German frontier. The shuttle diplomat Octavia and the haughty younger Agrippina are discussed and then a number of prominent and influential warlike women in the later Empire. The book concludes with warring women as depicted in the visual arts and with combative female gladiators in the arena.

Any book tackling a subject such as this, involving women – ordinary people in the classical world – is plagued with the usual problems: men wrote the histories of Greece and Rome and, as members of the elite in their societies, tended to write about elite activity and elite men; women did get an occasional look-in but usually only when and where they impinged on men and on the activities of men. So, to some extent, the ordinary Greek and Roman man and woman is often written out of the history books, leaving us to build a picture composed from what little we have in the way of descriptions of the ordinary man and woman in other literature, from the visual arts and from epigraphical evidence. Scant though it is, there is, nevertheless, more than enough material to enable us to provide a detailed and accurate picture of women at war in Greece and Rome.

Women and War in Earlier Ancient Civilisations

Ever since man first picked up a weapon and assaulted his fellow man, it is probably true to say that women were involved and implicated in some way, either as causes of disputes and wars or as victims of those disputes or wars. If nothing else, early man was, by nature, anxious to preserve his life and his livelihood: to that end he used conflict as a means of defending his home, his land, his crops and his livestock. Women played a key biological and societal role in survival, the extension of his family line and the preservation of his homestead: early man, therefore, also defended early woman.

It is now reliably established that the earliest known use of projectile weapons capable of killing at a distance was about 400,000 years ago. Wooden spears resembling modern javelins have been recovered from a site near the Schöningen brown coal mine in Germany. Found with stone tools and butchered remains of more than fifteen horses, the seven spears 'strongly suggest that systematic hunting, involving foresight, planning, and appropriate technology, was part of the behavioural repertoire of premodern hominids'.[1] Such spears may even have been used 1 million years ago: big game was a sizeable element of the human diet during this period, and we can assume that spears were used to kill game and to drive off scavengers. According to Kelly:

> the origin of war ... facilitated the mobilization of all adult male group members and their participation in preplanned dawn raids on settlements in which the tactical advantages of surprise and numerical superiority could be brought to bear ... the location of combat shifts from the border zone to the sleeping quarters at the core of a group's territory. At the same time, the intrinsic military advantage shifts from defenders to attackers. All of the attackers are combatants, whereas less than half of those under attack are armed. Attackers characteristically inflict numerous casualties while suffering few or none.[2]

Bows and arrows probably came into use around 60,000 years ago. Cave paintings in Spain between 20,000 and 12,000 years old depict battle scenes with serried ranks of archers.

The Talheim Death Pit, or Massaker von Talheim, was discovered in 1983 and is a mass grave in a Linear Pottery Culture settlement – Linearbandkeramik (LBK) culture – dating back to about 5000 BCE. In the pit were found the remains of thirty-four bodies, the first known evidence of orchestrated violence and genocide in Early Neolithic Europe. There were sixteen children, nine adult males, seven adult women and two more adults of indeterminate sex. Several skeletons showed signs of repeated and healed trauma, suggesting that violence was an habitual part of daily life and took place over time; all of the skeletons exhibited significant trauma which was the likely cause of death. Eighteen skulls showed wounds indicating contact with the sharp edge of adzes; fourteen were with wounds produced from the blunt edge of adzes; and two, maybe three, had entry wounds caused by arrows. There was no evidence of defensive wounds, perhaps indicating that the victims were in flight when killed. Reasons for the atrocity are speculative but would include reprisal attacks, conflicts over land and resources, poaching, assertion of superiority, kidnapping slaves and abducting women, or fodder for ritual cannibalism of the victims in LBK culture.

A mass grave near Schletz, north of Vienna, Austria, dating back about 7,500 years is more proof of genocide in Early Neolithic Europe among LBK tribes. The site is still to be fully excavated, but it is estimated that the grave could contain up to 300 bodies, probably members of other LBK tribes. Proportionately fewer young women have been found than men at Schletz suggesting that other women may have been kidnapped by the attackers.

Another Early Neolithic mass grave has been unearthed at Herxheim in Germany and gives proof of ritual cannibalism. The grave contained 173 skulls and skull-plates, the scattered remains of over 450 individuals as well as two complete skeletons located inside the inner ditch. The crania from these bodies were excavated at regular intervals in the two defensive ditches surrounding the site. Victims had been decapitated and their heads were either thrown into the ditch or stuck on top of posts that later subsided inside the ditch. The heads exhibited signs of trauma from axes and another weapon. The organised situation of the skulls suggests a repeated ritual act; the claim for cannibalism is supported by the discovery of numerous high-quality pottery artefacts and animal bones with the human remains.

Men, of course, have usually been the protagonists in the waging of war, but frequently women have, down the centuries, assumed a bellicose role as *casus belli*, combatant, strategist or foreign-policy maker.[3] As the rest of this chapter will show, women have also always been victims of war – as slaves or prisoners, as rape victims or as the casualties of devastation, defeat, displacement and destitution. Tragically, in the early twenty-first century CE all of this remains true: while arguments relating to the active role of women in war as combatants are becoming increasingly tense, rape, mass rape and sexual mutilation all are constants – often with the connivance of commanders and dissimulating politicians.[4]

A *stela* at Karnak dating from the sixteenth century BCE gives us our first evidence of a woman proving influential in a military sphere: on it, **Ahhotep I** (*c.*1560–1530 BCE) is described as 'having pulled Egypt together, having cared for its army, having guarded it, having brought back those who fled, gathering up its deserters, having pacified the South, subduing those who defy her'.[5] The tomb of **Ahhotep II** contained her now-destroyed mummy and gold and silver jewellery as well as daggers and an inscribed ceremonial axe blade made of copper, gold, electrum and wood; three golden flies were found too: these were usually awarded to people who served bravely in the army.[6] In the late fifteenth century BCE female pharaoh **Hatchepsut** may well have led from the front in campaigns against Nubia and Canaan.

The Bible tells us that in the thirteenth century BCE – **Deborah**, prophet and only ever female Judge of Israel, led her ill-provisioned Israelite army on a military campaign in Qedesh where she spearheaded a successful counter-attack against the superior forces of Jabin, King of Canaan and his general Sisera, enemy of the Israelites. Deborah duped the over-confident enemy into driving their iron-wheeled chariots onto marshy land where they became bogged down. The Israelite slingsmen and archers then picked them off one by one in a wholesale rout. Sisera fled from the battlefield to the camp of Jael the Kenite. *The Song of Deborah*, the earliest example of Hebrew poetry dating from about 1125 BCE, in which this exploit is narrated, remains one of history's earliest passages that describes fighting women; courageous, if ruthless, tent-maker Jael assassinated Sisera by hammering a tent peg through his temple as he slept, and delivered Israel from the army of King Jabin.[7] The capacity for extreme violence in belligerent women had thus been well and truly established. 'Extolled above women be Jael ... She stretched forth her hand to the nail, Her right hand to the workman's hammer, And she smote Sisera; she crushed his head, She crashed through and transfixed his temples.'

At Judges 4:5–9 we read how Deborah sent for Barak and said to him:

> 'The Lord, the God of Israel, commands you: Go, take with you ten thousand men of Naphtali and Zebulun and lead the way to Mount Tabor. I will lure Sisera, the commander of Jabin's army, with his chariots and his troops to the Kishon River and deliver him into your hands.' Barak replied 'If you go with me, I will go; but if you don't go with me, I won't go.' 'Very well,' Deborah said, 'I will go with you. But because of how you are going about this, the honour will not be yours, for the Lord will hand Sisera over to a woman.'

The Book of Judith tells how **Judith** was a brave and beautiful widow who berated her Jewish compatriots for not trusting God to deliver them from the

Assyrians, their foreign conquerors. Judith took matters into her own hands and went with her maid to the camp of the enemy general, Holofernes, into whose trust she gradually insinuated herself, promising him secret intelligence about the Israelites. She is allowed into his tent one night as he lies in a drunken stupor; here she decapitates him and takes his head back to her anxious countrymen. The Assyrians, bereft of their leader, disperse, and Israel is saved. More extreme female violence, mixed with guile, bravery, good use of miltary intelligence and a healthy impatience with dilatory or drunken men.

Cemetery 117 was discovered in 1964 – a Nubian cemetery near the present-day town of Jebel Sahaba close to the northern border of Sudan; the three cemeteries here have offered up copious remains radiocarbon-dated between 13,140 and 14,340 years old in a UNESCO rescue dig when the level of the Aswan Dam was raised.[8] Fifty-nine bodies were exhumed: twenty-four females and nineteen males over 19 years of age, along with thirteen children ranging in age from infancy to 15 years old. Three additional bodies were discovered of indeterminable age and sex. About 40 per cent of them died from violent and traumatic wounds: pointed stone projectiles were found in their skeletons which suggest the bodies had been pierced by spears or arrows. The wounds were generally found around the sternum, abdomen, back, and skull through the mandible or neck. The absence of bony calluses, which occur naturally in healing around these types of wounds, would suggest that the attacks were instantly or very soon fatal.

In the Vedic period (*c.*1200–1000 BCE) the *Rigveda* refers to a female warrior called Vishpala.[9] She lost a leg in battle, had an iron prosthesis made when the wound had healed and returned to the fray, thus making her not only one of the earliest women warriors but one of the first recorded casualties of life-threatening trauma and a pioneer patient in battlefield medicine.[10]

In Britain, the legendary Queen Gwendolen is described by Geoffrey of Monmouth in his pseudohistorical work *Historia Regum Britanniae* (*c.*1138) as having defeated and killed her husband in battle at the River Stour and then assumed the leadership of the Britons – their first recorded queen regnant. After drowning both Estrildis, her late father's mistress, and Habren their bastard daughter, she reigned peacefully for fifteen years, then abdicated in favour of her son. Gwendolen was one of the first queens to demonstrate the power women, especially royal women, could have, deploying her gender to advantage in defiance of her male counterparts. Gwendolen is an early example of a queen willing to go to any lengths to protect her kingdom, resorting to violence and invasion as necessary.

Geoffrey of Monmouth also records Queen Cordelia – the youngest daughter of Leir (Shakespeare's *King Lear*) and the second ruling queen of pre-Roman Britain. Cordelia was Leir's favourite daughter and the younger sister of Goneril and Regan. When Leir decided to divide his kingdom between his daughters and their husbands, Cordelia would not accept this so Leir refused her any land in Britain or a husband. However, when Aganippus, the King of the Franks, showed an interest in Cordelia, Leir allowed the marriage but denied him any dowry. Cordelia moved to Gaul; Leir was later exiled from Britain and fled to Cordelia, intent on restoring his throne which had been seized by his other daughters' husbands. Cordelia raised an army and invaded Britain, defeating the ruling dukes and restoring her father. After Leir's death three years later, Aganippus died and Cordelia returned to Britain and was crowned Queen. She ruled peacefully for five years until her sisters' sons, Cunedagius and Marganus, came of age; these were the dukes of Cornwall and Albany who despised the rule of a woman. They raised armies and fought against Cordelia, who participated in the battles. She was eventually captured, imprisoned and committed suicide.

Shammuramat was a wife of King Shamshi-Adad V and, on his death in 811 BCE, ruled the Neo-Assyrian Empire as its regent for five years until her son Adad-nirari III came of age. She was empress regnant of Assyria between 811 and 808 BCE; as one of the first known women to rule an empire, it can be assumed that she wielded military power in the execution of that rule. The empire was extensive, extending from the Caucasus Mountains in the north to the Arabian Peninsula in the south, and from western Iran in the east to Cyprus in the west. Shammuramat is often associated with **Semiramis**, the legendary wife of King Ninus, succeeding him to the throne of Assyria. The account of her life by Diodorus Siculus writing in the first century BCE, reveals her as being one of the first women to be used as a pawn in the political machinations of ruling men-folk and as a cause of conflict at the highest level.[11] Semiramis married Onnes, one of Ninus' generals, and fought with him at the capture of Bactria. We need not pay too much credence to Diodorus' report that Ninus' army 'numbered, as Ctesias has stated in his history, one million seven hundred thousand foot-soldiers, two hundred and ten thousand cavalry, and slightly less than ten thousand six hundred scythe-bearing chariots'. Nevertheless, Ninus was so taken with Semiramis' bravery there that he compelled Onnes to 'willingly give her to him, offering in return for this favour, his own daughter Sonanê as wife'. Onnes was not interested – so Ninus 'threatened to put out his eyes unless he immediately complied with his commands'. Onnes was terrified, 'fell into a kind of frenzy and madness' and hanged himself. Ninus then married Semiramis; she bore him a son called Ninyas.

What exactly was it that attracted both Onnes and Ninus to Semiramis? Diodorus describes her as 'endowed ... with understanding, daring, and all the other qualities which contribute to distinction' – qualities that she was able to apply in a military context:

> When Semiramis arrived in Bactria and observed how the siege was going, she noted that all the attacks were being made on the plains and at vulnerable positions, but that no one ever assaulted the acropolis because of its strong position, and that its defenders had left their posts there to reinforce those who were under pressure on the walls below. Consequently, taking with her such soldiers as were trained in scaling rocky heights, and making her way with them up through a difficult ravine, she seized part of the acropolis and gave a signal to that effect to those who were besieging the wall down on the plain. The defenders of the city, terrified at the seizure of the acropolis, deserted the walls and gave up all hope of saving themselves.

Ninus conquered Asia but was fatally wounded by an arrow. Semiramis then masqueraded as her son and fooled her late husband's army into following her because they believed they were following Ninyas; she went on to reign as queen regnant for forty-two years, conquering much of Asia in that time. Her duping of the Assyrian army is one of the first examples we have of a woman apparently acting like a man to achieve masculine power and authority, as perfected later by many others, not least Joan of Arc. Often, as we shall see, the illusion of masculinity was thrust upon women by incredulous men struggling to equate or reconcile what they believed to be exclusively manlike achievements with a woman: Agrippina the Elder is a later example, described as exhibiting masculine qualities in her adroit marshalling of her husband's, Germanicus', legions on a troublesome German frontier.

Semiramis was hungry for more: she took on the monumental task of founding the city of Babylon: 'and after securing the architects of all the [known] world and skilled artisans and making all the other necessary preparations, she gathered together from her whole kingdom two million men to complete the work', and reinforced it with a high brick wall surrounding the city.[12]

She built several palaces in Persia, along the Euphrates and Tigris rivers including Ecbatana, and annexed Libya and Aethiopia to her empire. Somewhat bored with peace, Semiramis declared war on King Stabrobates of India, 'since she had great forces and had been at peace for some time she became eager to achieve some brilliant exploit in war'.[13] Well aware that she was at a strategic disadvantage as a result of the absence of elephants in her army, she inventively and ingeniously had her engineers create a herd of faux elephants to deceive the Indians into thinking she was deploying the real thing. Camels and river boats were also used to good effect. She laughed off Stabrobates' slander denouncing her as a whore – another common theme we will meet many more times when men call into question women's sexual behaviour in order to discredit them – and his threats to crucify her when he had defeated her. Semiramis' strategy worked, but she was wounded in the counterattack and her army retreated west of the Indus. Even in retreat she remained resourceful, causing the slaughter of many Indians on an overcrowded pontoon bridge: 'she cut the fastenings which held the bridge together; ... the pontoon bridge, having been broken apart at many points and bearing great numbers of pursuing Indians, was brought down in chaos by the violence of the current and caused the death of many of the Indians'.[14]

The sexual innuendo referred to above persisted in other ways. Semiramis gets the credit for inventing the chastity belt while the Roman historian Ammianus Marcellinus records her as the first person to castrate a male youth and create a eunuch.[15] The Armenians portayed her as a whore and a homewrecker. One of their legends involved King Ara the Beautiful after whom Semiramis allegedly lusted: she asked Ara to marry her, but he refused. Indignant at this regal snub, she assembled her armies and invaded Armenia; during the battle she slew Ara. The tabloid reports got even worse when Pliny (*Natural History* 8.155) and Hyginus (*Fabulae* 243.8) both record that Semiramis' sexual voracity extended into bestiality when she had sex with a horse. Sexual slurring apart, Semiramis also earned the reputation for being a witch – another frequent and effective way of disparaging a prominent and successful woman. Making the most of this, though, she responded by pretending to raise Ara's body from the dead. When the enraged Armenians attacked to avenge their dead leader, she disguised one of her lovers as Ara and convincingly spread the rumour that the gods had brought Ara back to life. This ended the war.

In the eighth century BCE an Arabian woman called **Samsi** reigned as queen; she bravely rebelled against Tiglath-Pileser III, the King of Assyria, who became the first foreign ruler to subdue the Arabs when he attacked and defeated Samsi; he forced her surrender and imposed on her a tribute to enable her to remain in power as a puppet, which she did for the next twenty years. That tribute included gold, silver, male and female camels, and all types of spices. Her defeat, however, was significant: the Assyrians took from her numerous prisoners of war, 30,000 camels and more than 20,000 oxen as booty. An inscription tells us that 9,400 of her soldiers were killed, while 5,000 bags of various lucrative spices, religious icons and armaments and her estates were seized. When she fled to the desert, Tiglath-Pileser set fire to her remaining tents and she was said to have departed the battlefield like a 'wild she-ass of the desert'. Profound as this defeat was, it shows clearly the military power and wealth she formerly enjoyed and the magnitude of the armed forces she commanded.

Babylon and neighbouring civilisations had thriving pantheons which included a number of war goddesses: Belus, Babylonian god of war; Inanna, Sumerian goddess of physical love, fertility and warfare; Ishtar, Assyrian and Babylonian equivalents of Inanna; Nergal, Babylonian god of war, fire, the Underworld and pestilence; Pap-nigin-gara, Akkadian and Babylonian god of war; Sebitti, minor Akkadian and Babylonian war gods; Shala, Akkadian and Babylonian goddess of war and grain; Shara, Sumerian god of war; and Shulmanu, god of the Underworld, fertility and war. Shaushka was a Hittite goddess of fertility, war and healing, while Wurrukatte was the god of war.

In Egypt, war was just as prevalent in the portfolios of the gods: Bast was a cat-headed goddess linked with war, the sun, perfumes, ointments and embalming among other things; Menhit was goddess of war, 'she who slaughters'; Neith, goddess of war, hunting and wisdom; Pakhet, goddess of war; Satis, deity of the floods of the Nile River and a war, hunting and fertility goddess; Sekhmet was goddess of warfare, pestilence and the desert; Set, god of chaos in war; Sobek was god of the Nile, the army, military, fertility and of crocodiles.

These deities no doubt underscore the importance to, and ubiquity of, war in these cultures, and the comparatively prominent part played by women. As we shall see, the Greeks and Romans also had numerous female deities looking after many aspects of life, including war and warlike behaviour.

PART ONE

GREECE

Chapter 1

Goddesses and War in Greek Mythology

Greek goddesses play a significant role in the prosecution of war among mortals. Enyo, Athena, Hera and Thetis all have a major responsibility for aspects of warfare. Victory too, Nike, was a female goddess. However, despite the pre-eminence of female divinities, the father of the gods himself, Zeus, had some patronising advice for Aphrodite (*Iliad* 5.330–430) when she emerged injured from the battlefield wounded by Diomedes: war is not for you, my child: you stick to the marriage bed – Ares and Athena will look after military affairs.

Enyo was goddess of war and calamitous destruction, partner of the war god Ares. She is also his sister Eris, and daughter of Zeus and Hera.[1] She is the mother of the war god Enyalius, god of soldiers and warriors, by Ares.[2] The wholesale destruction of cities was Enyo's speciality; she is 'supreme in war' and a frequent fighter alongside Ares.[3] She was particularly active during the fall of Troy, where she handed down terror and carnage along with her dreadful stable mates Eris (Strife), Phobos (Fear) and Deimos (Dread), the latter two being two sons of Ares. Eris, and the two sons of Ares, are depicted on Achilles' shield.

Enyo was implicated in the Seven Against Thebes and in Dionysus' war with the Indians.[4] When she elected not to take sides in the battle between Zeus and the monster Typhon, it was because this decision would extend the duration of the conflict, so much so did she delight in war: 'Eris (Strife) was Typhon's partner in the melee, Nike (Victory) led Zeus into battle … impartial Enyo held equal balance between the two sides, between Zeus and Typhon, while the thunderbolts with booming shots danced like dancers in the sky'.[5] Enyo was one of the three Graiae – three grotesque-looking sisters who shared one eye and one tooth between them.[6] Aeschylus describes them best in a horrid vignette:

> the Gorgonean plains of Kisthene where the daughters of Phorkys dwell, ancient maids (δηναιαι κοραι), three in number, shaped like swans (κυκνηορφοι), possessing one eye among them and a single tooth; neither does the sun beam down on them, never the nightly moon. And near them are their three winged sisters, the serpent-haired Gorgons, hated by mankind: no mortal will look at them and live to tell the tale.[7]

Athena was something of a paradox: her divine portfolio included wisdom, inspiration, enlightenment, law and justice, mathematics, the arts, crafts and skill – all peaceful, all constructive and civilising. She was patroness of weaving, that badge of the ideal Greek wife, mother and homemaker. However, she was also goddess of military strategy, and she takes credit, as Athena Hippeia, as inventor of the chariot, and was patroness of the metals used for weaponry, thus adding a bellicose aspect to her currulum vitae. Indeed, when Athena was born from Zeus' head, not only was she already fully grown, but she emerged wearing a full suit of armour. As Athena Promachos she was Athena 'who fights from the front'. When her attempts at diplomacy to prevent the Trojan War failed, she was on the side of the Greeks giving sound advice and encouragement, especially to Achilles; she saved Menelaos from the arrow of Pandaros, and diverted the spear of Diomedes to injure Ares. The mighty bronze Athena Promachos statue fashioned by Pheidias from the Persian spoils at the Battle of Marathon showed Athena standing with her shield resting upright against her leg, and a spear in her right hand; it towered between the Propylaea and the Parthenon on the Acropolis, a highly visible symbol of female belligerence. In contrast, however, with her more violent, blood-crazed brother, Ares, Athena showed a more considered and analytical attitude to war in her role as military strategist. Conflict and war were last options for Athena, deployed only when all diplomacy had proved unsuccessful.

Hera too played a major part in the Trojan War and in its depiction by Homer in the *Iliad*; she loathed the Trojans with a vengeance after Paris decided that Aphrodite, and not her, was the most beautiful goddess; accordingly, she spent the next ten years doing her best to disadvantage the Trojan forces and support the Greeks during the war. She persuaded Athena to side with the Greeks and in Book 5, conspires with Athena to harm Ares, who was assisting the Trojans. Book 8 sees Hera attempting to enlist Poseidon's support for the Greeks but he refuses. Hera's machinations continue in Book 21 when she orders Hephaestus to prevent the river from harming Achilles. Hephaestus sets the battlefield on fire; the river implores Hera to stop the attack with the promise to help the Trojans.

Thetis was a Nereid, but our interest in her is as the divine mother of Achilles, and how she influenced the actions of her son in the Trojan War. Achilles, in Book 1 (1.400f.) of the *Iliad* describes Thetis' credentials as a warlike goddess when she defended Zeus in an attempted coup by three Olympians, namely Hera, Poseidon and Pallas Athene, by installing the terrifying Giant 'monster of the hundred arms whom the gods call Briareus' as a guard. She was, of course, implicated in the cause of the Trojan War since it was her wedding to which Eris was not invited which led to the fateful Judgement of Paris and the Trojan War itself. Thetis plays a major and significant role advising Achilles during the final year of the war, particularly after the death of Patroclus. She advises patience and caution to which Achilles adheres, even when urged by Hera, via Iris, to rejoin the fray. Thetis is also instrumental in having a splendid suit of armour made for Achilles by Hephaestus – greaves, shield, helmet and breastplate – armour that helps him to slay Hector. She was there for him when he was utterly demoralised by the loss of Briseis, the love of his life. She was there for him when he fretted about the decomposition of Patroclus' body and embalmed the corpse with preserving ambrose and nectar. She was there for him when he risked annoying the gods by retaining Hector's corpse – intending to mutilate it, only releasing it when Thetis, on the bidding of the gods, advises him to do so.

Enyo, Athena, Hera and Thetis were not the only female divinities associated with war. As in many other aspects of ancient Greek life, a whole host of goddesses and female spirits were on hand to oversee and patronise the minutiae of war and conflict; a selection of these are described below.

Alala, spirit of the war cry from the onomatopoeic ἀλαλή (alalē) and the verb ἀλαλάζω (alalazō), 'to raise the war-cry', reputed to be from the eerie sound owls make. Pindar references her: 'Listen! O Alala, daughter of Polemos! Prelude of spears! To whom soldiers are sacrificed for their city's sake in the holy sacrifice of death.'[8] She is the daughter of Polemos and niece to Enyo; her uncle was Ares.

The **Androktasiai** were the female spirits of battlefield slaughter; Hesiod in the *Theogony* gives their mother as Eris, Strife, and some fairly unsavoury siblings as Ponos (Hardship), Lethe (Forgetfulness), Limos (Starvation), Algae (Pains), Hysminai (Battles), Makhai (Wars), Phonoi (Murders), Neikea (Quarrels), Pseudea (Lies), Logoi (Tales), Amphillogiai (Disputes), Dysnomia (Anarchy), Ate (Ruin) and Horkos (Oath).

Bia was the spirit of force and violent compulsion. She and her sister and brothers, Nike, Kratos (Strength) and Zelos (Rivalry), were constant companions of Zeus, an honour accorded after they helped Zeus in the war against the Titans.[9]

Erida was twin sister of Ares and goddess of hatred and blood-lust.

Eris, goddess of discord, chaos and strife, notably in battle – one of two goddesses of that name, the other being much more benign than ours, whom Homer in Book 4, 440 of the *Iliad* describes as: 'Strife whose wrath is relentless, is the sister and companion of murdering Ares, she who is only a little thing at first, but then grows until she strides the earth with her head striking heaven. She then hurls down bitterness equally between both sides as she walks through the slaughter making men's pain heavier. She also has a son whom she named Strife.'

The **Hysminai** were female spirits of fighting and combat descended from Eris. Quintus Smyrnaeus described them vividly in his *Fall of Troy*: 'Around them hovered the relentless Fates; beside them Battle incarnate pressed forward yelling, and from their limbs streamed blood and sweat'.[10]

The Keres were female spirits of violent or cruel death, including death in battle and death by accident, murder or ravaging disease. They are characterised as dark beings with gnashing teeth and claws and with a thirst for human blood, hovering over the battlefield in search of dying and wounded men. A description of the Keres is in Hesiod's *Shield of Heracles*: 'The black Dooms – gnashing their white teeth, grim-eyed, fierce, bloody, terrifying – fought over the dying men: they were all longing to drink dark blood. As soon as they caught a man who had fallen or one newly wounded, one of them clasped her great claws around him and his soul went down to Hades, to cold Tartarus. And when they had sated their hearts with human blood, they would toss that one behind them and rush back again into the battle and the tumult.'[11]

There may be a link between Keres and various Celtic battlefield deities and the Norse Valkyries.[12] The **Makhai** were spirits of fighting and combat, sons or daughters of Eris.[13] **Nike** was a goddess who personified victory and was the divine charioteer, flying round battlefields rewarding the victors with glory and fame, in the shape of a wreath of laurel leaves. **Otrera**, wife of Ares, was goddess of violence and chaos, mother of the Amazons, daughter of Eurus, the east wind. Palioxis was the spirit of flight, and retreat from battle while **Proioxis** was spirit of battlefield pursuit.[14]

Chapter 2

Warlike Women in Homer

The eighth century brings us to Homer and his description of the Trojan War as fought out 500 years earlier in the thirteenth century BCE.

It is with Homer that we first meet the widely held axiom of the classical Greek and Roman worlds that war is man's work while wool-working is the preserve of women: the two lie at opposite poles of Greek and Roman societal and gender convention and, according to most Greeks and Romans, never should the two meet or be confused. The one informs men, the other women; war is a badge of maleness, wool an emblem of the good wife and homemaker. Hector and Telemachus vocalise it quite clearly in Homer, and ever since it has echoed down through classical life and literature as a mantra to normal life and the much desired status quo. At the same time, though, it came under attack from a momentous gender role reversal in which some women went off warring and, occasionally, men are left holding the bobbins and shuttles. Like the stereotypes, it first manifests in Homer, specifically with Penthesilea the Amazon fighting Achilles.

The world, as we know, is not always normal. Herodotus is staggered to report that Egyptians are all crazy: 'the women go to market and men stay at home and weave [the exact opposite to Greek practice]. Women even urinate standing up and men sitting down.'[1] Aristophanes' *Lysistrata* attests that war is as much the responsibility of women as it is of men, she turns the state upside down when she dresses up the *proboulos* (deputy) in women's clothes and teaches him the ways of wool while she and her comrades take over the running of the city, the complicated affairs of which include the Peloponnesian War, a conflict that can be disentangled as a ball of wool (*Lysistrata*, 567ff). Elsewhere, an armoured Athena, female goddess, protects Athens while effeminate Cleisthenes has his shuttle; Diodorus (3.53) reports that the Amazon men of Libya stay at home, weave and look after the children while the women go out fighting wars; Pindar's Cyrene eschews the loom and prefers to slay wild beasts with her sword (*Pythian*, 9.19–22); Euripides' Ague goes one better, getting self-fulfilment by killing animals with her bare hands (*Bacchae*, 1236); the Bacchants too have deserted the loom for a life much more challenging; when he observes Artemisia's military excellence and belligerence, a bemused Xerxes reflects that his men are acting like women and she like a man. The warlike Amazons consign their men to a life of woolworking. Throughout Greek and Roman history exceptional women are described as exhibiting *andreia* or *virtus* (bravery), with all the connotations of manliness and bravery.

It is this gulf between the norm and the reality, this 'world-on-its-headness', that provides the essence of this book: women who do war.

Men, of course, were the protagonists in the Trojan War: Homer's *Iliad* tells of the ten-year conflict with Achilles, Patroclus, Hector, Agamemnon, Ajax, Aeneas and Deiphobus among the many alpha male warriors. But there would not have been a Trojan War without the involvement of women, both as the very *casus belli* itself and as characters who influenced the action and direction of the war. Helen, Queen of Sparta must take responsibility for causing the war when she allowed herself to be abducted by Paris, while Briseis and Chryseis both played a role far more influential than their status as spoils of war would suggest. Hector's wife, later war widow, Andromache, tried to influence her husband's actions and strategy, while ever-patient, faithful army wife Penelope, in the *Odyssey*, endured virtual widowhood for twenty years as she waited, and waited, in the hope that Odysseus would come back to her and dispel the repellent suitors circling around her shark-like, with an eye on her virtue. Achilles' mother, Thetis, was, as we have seen, a significant influence in her son's agonising decision – to fight or not to fight in the war against Troy.

Women, though, have no active role on this battlefield, with the notable exception of the Amazon, Penthesilea. However, Athena can be found there supporting her favourite heroes and Thetis it was who gave Achilles back his weapons to enable him to resume fighting. After Penthesilea, the nearest we come to active participation is Hector's wife, Andromache, who looks after his horses for him (*Iliad*, 8.185–90) and offers some sensible strategic advice before his final departure for the battlefield.

Homer's women do, nevertheless, have a vital role in narrating the tragedy that befalls Troy – and, as we might expect, in funerals and in mourning dead heroes – for example, *Iliad*, 24.723–46, 24.748–60, 24.761–76. With the exception of Helen and of Homer himself, the destruction of Troy is always described from the perspective of Trojan women which gives them an important role as exclusive narrators of the Trojan catastrophe. This is best illustrated when the women of Troy mob Hector at the Scaean Gate, desperate for news of their fathers, husbands, sons and friends: Hector is evasive and advises prayer: Homer, however, comes straight to the point when he interjects into the action the doom-laden words: 'grief – *kedea* – awaits many' (*Iliad*, 6.241). Hector's mother, Hecuba, takes control of the situation and advises Hector to sacrifice to Athena in order to enlist her protection of the Trojan wives and children from the marauding Diomedes.

Homer's women remain proud of their fighting men; Hecuba urges the Trojan women to remember Hector as god-like in their lamentations, the greatest glory (*Iliad*, 22.430). Despite their physical absence from the field of battle, Homer's female characters remain at the heart of the ten-year conflict.

Helen of Troy – *Casus Belli*

Helen became embroiled in the Trojan War when Paris decided that Aphrodite (and not Hera or Athena) was the fairest goddess of them all, thus winning for himself the prize of the most beautiful woman in the world. Helen was no stranger to abduction and her role as *casus belli* goes back much further than the Judgement of Paris. Indeed, the true origins of the Trojan War can be found in her abduction by Theseus after which, as a young girl, she became the target for a number of suitors. Coincidentally, Helen's abduction here gave rise to a conflict when her brothers, Castor and Pollux, attacked Athens, captured Theseus' mother, Aethra, in retaliation and returned their sister to Sparta.[2] The suitor who won the day was King Menelaus of Sparta who duly married Helen, beating off such peerless contenders as Odysseus, Menestheus, Ajax the Great, Patroclus and Idomeneus. The victory came after the drawing of straws and a deal in which Odysseus secured not just an agreement in which all the suitors swore a solemn oath to defend the successful husband in any future dispute – the Oath of Tyndareus – but also the hand of Penelope.

But it was not a happy marriage: while Menelaus was out of town attending a funeral, Paris and Helen fled to Troy, despite brother Hector's protestations, causing the Oath of Tyndareus to be invoked and a fleet of a thousand ships to be launched in retaliation. The Trojan War had begun.

Helen assumes a prominent role in the progress of the war when, at the Scaean Gate, the very epicentre of the battle, she discusses how things are going, with Priam and some other wise elders. Priam absolves her of any blame for the conflict and asks her to provide military intelligence regarding some of her ex-kinsmen, notably Agamemnon. She obliges, but not before wishing herself dead for causing so much trouble. One of the elders, Antenor, confirms the accuracy of her information, so establishing her as a viable and reliable source of intelligence. Death, however, eluded Helen so she spent her time 'pining away, weeping'. In the *Iliad* we first meet Helen weaving a tapestry, like Penelope in the *Odyssey*. Helen's tapestry charts the course of the war, and indicates that she is the weaver, or cause, of the war.

During the war, Menelaus and Paris fought a duel which Menelaus won easily; however, Aphrodite spirited Paris away to the safety of Troy before he could be killed. Homer shows Helen's take on life as a military woman when he makes it clear that Helen, with Menelaus, had been trapped in a loveless marriage and that Paris presented an opportunity for her to live a more fulfilling married life. She explains to Paris that, while appreciative of Menelaus' prowess as a warrior, she had little interest in army life or in being an army wife – no more heroes for Helen: Helen wanted to grow old with Paris; every day she spent with bellicose Menelaus made her want to walk into the sea and end it all. Moreover, she confides in Hector that 'I can't ask anyone to fight for me; I am no longer queen of Sparta', thus parading the loss of self-esteem and guilt she now feels for being the one person responsible for the bloody regional conflict that was the Trojan War.

Helen comes out of the fall of Troy very badly: both Homer, and Virgil after him, paint a picture of a woman, now partnered with a Trojan warrior, duplicitously and cruelly assisting the Greek military offensive. In the *Odyssey*, Homer describes how Helen circled the Trojan Horse three times, cruelly imitating the voices of the Greek war wives left waiting anxiously at home and putting the men inside, which included Odysseus and Menelaus, through emotional agony as they envisaged their loved ones. Helen effectively drove them to the brink of death: they had to be restrained by Odysseus from bursting out of the horse to certain massacre. In Virgil's *Aeneid*, Deiphobus recounts Helen's treachery: he tells a shocked Aeneas how, when the Trojan Horse was admitted into the city, she was at the head of a chorus of Trojan women feigning Bacchic rites, and brandishing a torch from the citadel – the signal the Greeks were waiting for to launch their deadly attack.[3]

Paris died later in the war and Deiphobus, his younger brother, married Helen, only to be slain by Menelaus during the sack of Troy when Helen, again acting treacherously as a fifth columnist, hid his sword and rendered him easy prey.[4] Virgil's graphic description of Deiphobus' mutilation serves only to magnify the enormity of Helen's crime, both as wife to husband and as belligerent traitor to country: 'Priam's son, his body slashed to bits, can now be seen – his face mangled, his face and hands covered in blood, his head shockingly shorn of ears and nose. Aeneas could barely recognise this shivering shade as Deiphobus; it struggled to hide its face and the scar of shame.'[5] Indeed, Deiphobus is unequivocal about Helen's crime, in words replete with and redolent of deceit and dishonour:

> This was my fate (*fata*), and that Spartan woman's murderous crime (*scelus exitiale*) to mire me in this mess (*his mersere malis*) – these are the souvenirs she has left me with. We spent that final night in false (*falsa*) joy . . . when the deadly (*fatalis*) horse leaped into impregnable (*ardua*) Pergamon, pregnant with infantry armed to the teeth; she led the Phrygian women in choral dance and false (*simulans*) Bacchanalean song.[6]

In the aftermath of the fall of Troy, the once double-dealing Helen is reduced to a cowering wreck, desperate to save her skin, deploying her most potent weapons – her beauty and sexuality. Sources differ on the exact details surrounding the reunion of Helen and Menelaus after the fall: one version has it that an angry Menelaus resolved to kill Helen himself but when he finds her and raises his sword he takes pity on her as she weeps, begging for her life. Menelaus' fury evaporates and he takes her back as wife. Another has Menelaus, again intent on slaying her, captivated by her beauty and sparing her life; he takes her back to his ship 'to punish her at Sparta'.[7]

The *Bibliotheca* (pseudo-Apollodorus) tells us that Menelaus raised his sword in front of the temple in Troy but his anger subsided when she tore her clothes and revealed her breasts.[8] Stesichorus (*c.*640–555 BCE), in his *The Sack of Troy*, reports that Menelaus gives her up to his soldiers so that they could stone her to death; however, so stunned were they by her beauty when she ripped open her clothes, that the warriors dropped the stones from their hands, all agog.[9]

Whatever happened to her, Helen remains for us a key player in the cause, progress and aftermath of the Trojan War. Indeed, things could have been much worse even than events at Troy – she could have shared responsibility for igniting another much more serious war, the war to end all wars: according to Hesiod, Zeus planned to exterminate the race of men and heroes and was going to use the Trojan War – as triggered by the elopement of Helen – as the catalyst for this apocalyptic cataclysm.[10]

Briseis and Chryseis – More Than Just Spoils of War

Like Helen, both Briseis and Chryseis are driving forces behind the Trojan war; they form a basis for Homer's description of the military action at the gates of Troy and the eventual raising of the siege there. Briseis first encountered Achilles at the wrong end of his sword when he ruthlessly slaughtered her father, mother, three brothers and husband during a Greek assault on Troy.[11] A bereft Briseis was awarded to Achilles as war booty with a life of concubinage to look forward to; as such, she is our first real individual female victim of war.

Initially, all went well: the relationship between Briseis and Achilles blossomed into mutual love with the promise of marriage after the war assured by Patroclus. However, Apollo and Agamemnon spoilt the party when the king was required to give up his own concubine, Chryseis. A petulant and selfish Agamemnon then insisted that Achilles hand over Briseis in recompense. Achilles did not react well: he withdrew his troops from the Greek force with dramatic strategic consequences and retired to sulk in his tent, mortified and wounded by his loss. From Achilles' point of view, if the cuckolding of Menelaus could start a war, then how should he react to Agamemnon for taking Briseis from him? Answer: by withdrawing and thereby compromising any successful outcome to the war. It took the death of Patroclus and the return of Briseis – both whom he loved in different ways – to spur Achilles back into action and save the Greek cause. Predictably,

Agamemnon swore that he never laid a hand on Briseis – well, he would wouldn't he?[12]

Chryseis has the honour of participating in the opening scene of the *Iliad*. She, like Briseis, was war booty, given up to Agamemnon. He, with breath-taking insensitivity, tactlessly described her as a better woman than his own wife, Clytemnestra, with, as we know, fatal consequences when he finally got home from Troy. Agamemnon stubbornly held onto Chryseis when her father, Chryses, a priest of Apollo at Chryse, attempted to ransom her: 'I would not accept that marvellous ransom for the girl, the daughter of Chryses, since I much prefer to keep her in my home. For sure, I prefer her to Clytemnestra, my wedded wife, since she is just as good as her, in terms of beauty or in stature, or in mind, or in any handiwork.'[13] The priest's lack of success here, and Agamemnon's oafish, hubristic disrespect towards him angered the gods to such an extent that they unleashed a 'loathsome pestilence' on the Greek armies.[14] If he wanted an army to command, Agamemnon had no choice but to renounce Chryseis and send Odysseus to return 'fair-cheeked' Chryseis back to her father. Agamemnon then follows one military disaster with another when he, with supreme military myopia, selfishly compensates himself by taking 'fair-cheeked' Briseis from a weeping Achilles – with menaces, should he refuse to relinquish her.[15]

Pawns as they are in the machinations of Agamemnon and Achilles, both Briseis and Chryseis, for all that, lie at the heart of the progress, or lack of progress, in the Greek war against Troy. It is the spat between Achilles and Agamemnon over whose bed Briseis warms which stalls the conflict and compromises the Greek war effort. Only when Briseis is returned does Achilles marshal his armies again with renewed vigour to avenge the death of Patroclus and win the war for the Greek alliance.

Andromache – Devoted Wife, and Military Advisor

Andromache's fame lies not just in being the devoted wife of Hector and loving mother of their son, Astyanax. She shrewdly uses a lesson in military strategy in an attempt to protect Hector from the worst dangers of the war raging around them.

She, like Penelope, was one of the thousands of army wives who waited patiently for her returning hero, buoyed up by hope and by an enduring love. Tragically, Andromache was about to welcome her hero home, ensconsed in domestic security, when she was floored by the dreaded news that he had just been killed: 'She was at work in an inner room of the lofty palace, weaving a double-width purple tapestry, with a multi-coloured pattern of flowers. Totally unaware [of Hector's fate] she had asked her ladies-in-waiting to set a great cauldron on the fire so that Hector would have hot water for a bath when he got back ...'[16] Andromache's tragic story began when Thebes, her home city, was sacked by Achilles, and her father and seven brothers died in the ensuing carnage. Her mother then succumbed to an illness and Andromache became just another piece of collateral damage and one of the countless spoils of war. But Hector had come along to provide stability and love in a life rent asunder by conflict.[17] All the more bitter then is the tragedy when Andromache's life is ripped apart again at Troy when Achilles kills Hector – leaving her once again bereft, rootless and alone in the world, a displaced person relegated to the margins of society and a pathetic emblem of the fate that awaits conquered women in ancient warfare.[18] Hector had foreseen all of this, the writing was on the walls for Troy: he bewailed the certain fact that Andromache would be forced into slavery – weaving at another woman's whim, fetching the water as a slave and wailing in captivity – for a second time.

Just before they part for the last time, the couple hold a conversation on the exposed and dangerous ramparts of Troy, 'the great wall of Ilios', a place generally inimical to a woman and alien to a distraught Andromache. In contrast, though, to the martial environment, domestic harmony prevails as Hector describes a typical scene with women weaving and men warring: in real life this is a metaphor for the one partner protecting the *oikos*, the homestead, the other defending the *polis*, the city-state – both central and essential to the survival and continuation of normal Greek life. The pathos is palpable. Hector:

> laid his child in his dear wife's arms, and she held him to her fragrant breast, smiling through her tears; Hector was moved with pity when he looked at her, and he caressed her with his hand, and said: 'My dear wife, do not be too upset ... go home and busy yourself with your own tasks, the loom and the distaff, and tell your maids to get on with their work: war is for us men'.

Many have seen this as a sexist statement, firmly putting women in their place and telling women bluntly to keep out of men's business. But it signifies much more than that: if we go back to the protection just mentioned that Andromache gives to the *oikos*, it surely is a confirmation of her invaluable role, and the role of all other married women and mothers, in preserving and extending the homestead as a safe and nurturing environment in which to raise new generations and new citizens for the *polis*. Without a thriving *oikos* there was no *polis* and nothing for the Hectors of the world to defend. Women like Andromache, then, were performing an important, albeit indirect, military role by providing something worth defending and fighting for.

Homer reinforces this important message by repeating the phrase on two occasions in the *Odyssey*, with slight variations.[19] The first comes at the beginning of the poem where Odysseus' precocious son, Telemachus, asserts his authority in the Odyssean household: he dismisses the grieving Penelope and tells her, and the slaves, to get on with the wool-working, making it abundantly clear, not least to his mother, that he is in charge and that he will do the talking to the suitors, for talking is man's work. The second echoes the first when Telemachus again rebukes Penelope, reminding her that he is the master of the house and that the bows and arrows are his province; Penelope is again ordered to go back to the wool work; she, somewhat meekly, retires to grieve over ever-absent Odysseus. Just as Hector was fighting to save his city, so Telemachus is talking, and shooting arrows, to save his household, his *oikos*, and his mother from the circling suitors.

Even more unconventional and extraordinary is the gender role-reversal implicit when Andromache offers Hector military advice. On one level, this is a loving but astute wife's ploy to detain Hector with a lesson in military strategy, thus keeping him in the relative safety of the ramparts and out of the much more hazardous open fighting taking place down below; on another level, it is nothing short of sound, calm-headed advice based on a woman's observation in a sea of confusion and panic: 'Come on, show some pity, and stay here on the wall, in case you orphan your child and widow your wife. Post your army by that wild fig-tree, where the wall is most vulnerable to a scaled assault, and where the city is exposed. ... [the Greeks] have tried to get in there three times already.'[20]

Andromache is, like Penelope, the caring, loving mother and wife. She weaves a cloak for Hector in the domestic seclusion and safety of the interior rooms of the house and she runs that bath for him just before he is due home, sadly in vain. She is, though, at the same time, a wise and perceptive woman who has obviously observed the military activity raging around her and feels confident enough to offer a strategy based on her informed observations in the male world of war. Indeed, the reason she is out on the walls in the first place was because she heard that the Trojans were on the defensive, and that victory was in the grasp of the Greeks. Most civilian Trojans would surely have fled in the opposite direction; not so steadfast, military-minded, 'white-armed' Andromache.[21] Andromache shares with her mother-in-law **Hecuba**, who detains her son in the relative safety within the walls of Troy, the distinction of being able to understand and expatiate on war and its consequences. Hecuba helps Hector – the warrior and son – in his prosecution of the battle: she facilitates his prayers to Zeus and his libation, and she has him drink a restorative cup of wine, an energy drink, before returning to the fray.

More generally, Homer's women in the *Iliad* are clearly shown as appreciating the fact that their men are obliged to fight – not least because of the inviolable treaty invoked at the snatching of Helen. The women never demand an end to hostilities, unpalatable as those hostilities are. Helen herself is never criticised by the Trojan women because by criticising her the

Trojan women would be calling into question the validity of the war. The best they can do is encourage their warrior husbands to adopt a less dangerous approach to the conflict ... and pray to Athena.

Three heroic mortal women who deserve mention here are Epipole, Messene and Aglauros, the first for her exploits in the Trojan war. Although not mentioned by Homer, **Epipole** was a daughter of Trachion, of Carystus in Euboea. She so much wanted to fight with the Greeks against Troy that she bravely dressed up as a man and inveigled her way into the massed armies – a very early case of classical cross-dressing. Unfortunately, when Palamedes discovered her sex, she was stoned to death by the Greek army.[22] It is, of course, impossible to know how many, if any, Greek or Roman women ever dressed up as men in order to fight in the lines. Unlikely as it may seem, we should nevertheless not forget, from different eras, Hua Mulan in China, Joan of Arc, the conquistador Catalina Erauso in sixteenth-century Mexico and Deborah Sampson wounded in the American Civil War, and one of a thousand or so women who fought in that war dressed as men.

Messene was the daughter of Triopas, King of Argos; she was married to Polycaon, son of King Lelex, of Laconia. Messene was a very ambitious woman. After her father-in-law died, her husband's brother Myles assumed the throne of Laconia, thus overlooking Polycaon; but it was not in Messene's script to be married to a nonentity, so she set about raising an army from Argos and Laconia and invaded and occupied nearby territory. This territory was then named Messenia in her honour. The couple then went on to found the city of Andania, where they built their palace.[23]

Aglauros was a brave virgin girl who stepped up to the plate when Athens, her city, needed someone to bring a protracted war to a successful conclusion. An oracle had pronounced that Athens would win the war if a citizen committed suicide for the sake of the city. Aglauros jumped off a cliff to her death; her reward was everlasting fame and a temple on the Acropolis built in her honour. It then became the custom for young Athenians, when they donned their first suit of armour, to swear an oath that they would always defend their country to the last.[24]

Chapter 3

Teichoskopeia: A Woman's View from the Walls

The *teichoskopeia*, or teichoscopy, is an epic device in which observers on the city walls describe the battle scenes raging below them; since the people watching from the walls are non-combatants, they are usually women and eldery men. Apart from providing commentary on the action below, the teichoscopy also gives us an insight into the effects of battle on women or on a particular woman, especially in relation to combatants who may well be husband, brother, father or son, or someone else close to them.

Probably the most famous of, and certainly one of the earliest, examples of teichoscopy is Helen's in Homer's *Iliad*.[1] Helen is busy at her loom when she is approached in her bedchamber by Iris, disguised as her sister-in-law Laodice, the daughter of Priam. Helen is taken to the walls where Priam asks her to point out the Achaean heroes she sees on the Trojan plain where the Greek and Trojan armies are preparing for the duel between Menelaus and Paris, former and current husbands of Helen. Helen thus becomes a vital source of military intelligence for the Trojans. In a parade of heroes, she points out Agamemnon, Odysseus, Telamonian Ajax and Idomeneus; she praises both the Greek and the Trojan armies: an indication perhaps of the paradoxical situation she finds herself in – she, a Greek woman, is now firmly ensconsed in the Trojan camp. The stakes are high, not only because whoever wins the duel will win Helen, but more crucially, he will also ultimately win the war for either Greece or Troy. Helen watches Menelaus defeat Paris – and then witnesses Aphrodite saving Paris, divine interference which serves to extend the war and results in many more casualties. What she sees of the duel turns Helen against Paris – she wishes him dead, praises the might of Menelaus and challenges Paris to resume the duel, sure that he will be slain. Paris is unimpressed and can think only of bedding Helen while Menelaus desperately tries to find Paris in order to finish him off.

Hesiod's *Shield of Heracles* features a teichoscopy on the shield and is notable not just for the striking image of women wailing and tearing their cheeks – which must have been typical female behaviour on the walls – but also the fact that it was there on the shield in the first place, indicating that, for women, watching proceedings from the walls was itself not unusual behaviour.[2]

A fragment from Ennius' *Annals* describes women crowding the walls, eager to look on the battle scene: *matronae moeros complent spectare fauentes*, suggesting again the normality of such behaviour and the desperation of the women forming the audience for sight of their loved ones.[3]

Virgil's manning of the walls in Book XI of the *Aeneid* by the women and children of Latium is for defensive reasons as well as for supplication and spectating: 'the women and children, in a motley cordon, defend the ramparts; the decisive hour calls for a task for everyone. On the citadel a line of matrons, with the doleful Queen [Amata], heads toward Pallas' temple'.[4]

Horace, in a lyric poem more famous for its *dulce et decorum est pro patria mori*, describes a Roman youth who is encouraged to terrify 'wild Parthians' with his spear under the walls – a scene watched both by the Parthian king's mother or wife, and, with some trepidation, by a *virgo* – the bride of the king, or of a Parthian prince. This bride is anxious that her prince (or king) does not provoke the Roman into violent action, likened as he is to a ferocious lion, capable of extreme carnage. She sees caution as the best option that is open to her Parthian: the lesson to be learnt, though, according to Horace, is that 'it is sweet and fitting to die for one's country' – although there is no escaping death anyway, even when a soldier flees the battlefield.[5]

Scylla as featured in the epyllion *Ciris*, composed in the first century CE, is very different from the standard classical woman. Not for her the loom, music or gold; rather, she can be found, driven crazy (*demens*) for love of Minos, on the city walls during Minos' siege of Megara, looking out for Minos below in the camp burning in a conflagration. Ovid gives us more detail, illustrating well the kind of detailed military intelligence that a woman with an interest and involvement can glean by regular observation from city walls: 'so, when the war began, she [Scylla] often viewed the dreadful conflict from that height; until, while the enemy camp remained, she got to know the names, and knew the habits, horses and the arms of many a chief, and could see the signs on their Cydonean quivers'.[6]

As in the *Ciris*, Scylla is overcome with desire for Minos and, in her frenzy, contemplates going down from her vantage point to meet him through the hostile enemy lines, or else flinging herself from the walls, or even, Tarpeia-like in a sexually charged metaphor, unbarring the city gates to let Minos and his armies into her city. Indeed, she is ready to do anything that Minos asked of her. As she gazes at the enemy's white tents below, Scylla agonises over the dichotomy her view from the walls has given her: should she grieve over the war raging below, or should she rejoice? The fact that Minos is her enemy should elicit grief, but without the war there would be no Minos and that makes the war a cause for joy. She concludes that a solution to the problem might be for Minos to take her as a hostage, a concubine and a peace pledge, in exchange for a cessation of hostilities. Scylla then envisages herself flying through the air bird-like to deliver this proposal, intent on doing whatever Minos required except, that is, in a reversal of her earlier thoughts of treachery, surrendering her city to him.

In Statius' epic the *Thebaid*, the poet redrafts a scene from Euripides' *Phoenissae* in which Antigone witnesses the fight to the death between her two brothers, Eteocles and Polynices, as reported to her by her female companion Phorbas.[7] This view from the walls is reprised in Book 11: here Antigone extends the usual visible action implicit in a teichoscopy to include the audible and vocal when she shouts down to Polynices in desperation and pleads with him to suspend his attack, all reinforced with the threat that she will come down and join in the action. When Polynices looks up at her on the tower, he reverses the typical action of a teichoscopy by looking up at the audience looking down at him.

Valerius Flaccus, in the *Argonautica*, has a pre-infanticide Medea ordered up onto the walls by a disguised (as Chalciope) Juno to watch the conflict between the Colchians and Perses whose allies are Jason and his Argonauts. The pretext is that Medea must know who her father's allies are, thus making the teichoscopy a means to acquire military intelligence.[8] In this respect it recalls Homer's teichoscopy with Helen. When the two princesses reach the walls they are instantly horrified and petrified, as frightened birds in a storm. The portrait of a Medea here as an innocent girl, untainted by the experience of death and destruction, is in stark contrast to the Medea we, and Valerius' audience, know will end up slaughtering her own children. The view from the walls affords her a grandstand from which to watch the battle, including, most crucially, sight of Jason; she is smitten and the whole sorry tale of Medea's subsequent tragic life begins to unravel. By the end of Jason's *aristeia*, his finest hour in battle, she is convulsed with passion for the hero.

Medea gets closest to the action below when she leans over the parapet, ill-advised (*improba*) as this is, to get a closer look at the carnage. She is soon inured to death, insensible to it through the monomania caused by her love for Jason, desperate that he survives, even though he must slaughter others to do so.[9] This teichoscopy gives us not only a female perspective of the fighting but also something that is distinctly disturbing: war's easy brutalisation of a young innocent girl who, before she ascended the walls, had no experience at all of the horrors of war. Her new, ready acceptance that to win in battle involves killing lays the foundations for her later infanticide.

Chapter 4

The Amazons

'We are armed with the bow and javelin and we ride horses. We know nothing at all about women's work'

Herodotus (4.114.3) citing the Amazons' mission statement

Mention fighting women in the classical world and many people immediately think of the Amazons: they are the best known female wagers of war in classical history and mythology. The only sure thing that can be said about the Amazons is that nothing much about them can be said for sure. Much to do with the Amazons remains mired in controversy or shrouded in speculation, but the following chapter shows what we can deduce with some confidence, mindful at the same time that Amazons were not the classical world's only woman-warriors.

In Greek mythology, the Amazons (Ἀμαζόνες) were redoubtable woman-warriors who commanded respect. Herodotus believed them to be related to the Scythians and located them vaguely between Scythia and Sarmatia, roughly modern Ukraine. Others would put them in Pontus, Anatolia on the River Don or Libya. The Greeks called the River Don 'Tanais'; but the Scythians called it the 'Amazon', a nod to the fact that the women fighters inhabited the area. Pliny the Elder suggests the valley of the Terme River may have been their home and mentions a mountain named after them (the modern Mason Dagi), as well as a settlement called Amazonium; Herodotus first mentions their capital Themiscyra, which Pliny locates near the Terme.[1] Philostratus places the Amazons in the Taurus Mountains; Ammianus locates them east of Tanais, as neighbours of the Alans. Wherever they were or wherever they came from, the important fact remains, in the words of Peter Walcott, 'Wherever the Amazons are located by the Greeks ... it is always beyond the confines of the civilized world. The Amazons exist outside the range of normal human experience'.[2] They are then, strange, liminal and not normal. This is Hippocrates' description of the origins of the Amazons:

> And in Europe is a Scythian race, living around Lake Maeotis, which is different from the other races. Their name is Sauromatae. Their women, when they are still virgins, ride, shoot, throw the javelin on horseback, and fight with their enemies. They do not surrender their virginity until

> they have killed three of their enemies, and they do not marry before they have performed the traditional sacred rites. A woman who takes a husband no longer rides, unless she is compelled to do so by a general expedition.[3]

How this came about is explained by Herodotus when he tells that the Sarmatians were descendants of Amazons and Scythians, and that their wives observed their ancient maternal customs, 'frequently hunting on horseback with their husbands; in war taking the field; and wearing the very same dress as the men'. He goes on to say how a group of Amazons was blown across the Maeotian Lake (Sea of Azov) into Scythia (today's southeastern Crimea). They mastered the Scythian language and agreed to marry Scythian men, on the condition that they would not be required to follow the customs of Scythian women.[4] Domesticity springs to mind. They then moved northeast, settling beyond the Tanais (Don) River, and became the ancestors of the Sauromatians. Herodotus adds that the Sarmatians fought with the Scythians against Darius the Great in the fifth century BCE.

Despite their pugnacity, Scythian women were indubitably human, with human needs. Herodotus (4.1–4) tells us how the Sythian men came home after a long campaign fighting the Cimmerians and the 'empire of the Medes'; but when they arrived home, they were confronted by another Scythian 'army'. Due to the lengthy absence of their men, the wives had been having sex with the slaves; a battle between the offspring of the slaves and the original Scythians ensued in which the Scythians cleverly exchanged their weapons for whips: the slaves fled. Later in the same book, Herodotus explains how the Sauromates people came about (4.110–11): the Amazons escaped from the Athenians and ended up in Scythian territory where the Scythians mistook the Amazons for men and attacked them; it was only when they were examining the corpses that they realised that their enemy had, in fact, been women. They decided then not to fight them; sexual relations began and the Sauromates race was born.

Lysias, the fifth-century Attic orator, describes the Amazons as the only women to wear iron armour and the first to ride horses; he adds 'they were seen as males because of their courage rather than females because of their nature' – an early instance of courage being seen as a male virtue. In Roman times Julius Caesar reminded the Senate of the conquest of large parts of Asia by Semiramis and the Amazons. Strabo compares successful Amazon raids against Lycia and Cilicia with resistance by Lydian cavalry against the invaders.[5] Gnaeus Pompeius Trogus reports how the Amazons originated from a Cappadocian colony of two Scythian princes, Ylinos and Scolopetos.

The most famous Amazons were undoubtedly Penthesilea, who served in the Trojan War, and her sister Hippolyta, whose magic belt, given to her by her father, Ares god of war, constituted the ninth of the labours that exercised Hercules. Hippolyta was the founding queen of the race of Amazons; aptly for women very much at home on horseback, her name means 'unbridled mare'. So prominent were the Amazons in the Greek psyche, and powerful on the field of battle, that the Greeks had a word for conflict with Amazons – *amazonomachy*; their name has become a byword for female warriors in general. Their obvious battlefield skills apart, Amazons were great colonists and civilisers: they are said to have founded many cities including Smyrna, Paphos, Ephesus and Magnesia; as natural horse-born fighters they are also credited with inventing the cavalry.

The etymology of the word Amazon is disputed to this day: it may derive from a Greek word meaning 'without men or husbands'; alternatively, it is from ἀ- and μαζός, 'without breast', reflecting an aetiological tradition that Amazons cut off their right breast to facilitate their archery. Greek art does not support this, as Amazons are always depicted with both breasts intact. Hippocrates, the influential medical scientist, though, is adamant: 'They have no right breasts ... for while they are still babies their mothers make red-hot a bronze instrument constructed for this very purpose and apply it to the right breast and cauterize it, so that it stops growing, and all its strength and mass are diverted up to the right shoulder and right arm'.[6] Some 800 years later Justinus still agrees: 'They exercised the virgins on weapon-wielding, horse-riding and hunting, and burned the children's right breasts, so that arrow-throwing wouldn't be impeded; and for such reason, they were called Amazons.'[7] The 'right mastectomy theory' may have just been made up to deter women from taking up archery; whatever, women are perfectly capable of loosing an arrow with both breasts intact. The word may well be derived from *hamazakaran*, 'to make war' in Persian.

If the Amazons spurned men and they were 'really killers of men' (*androktones*), as Herodotus would have us believe, how then did their race survive? Apparently, once a year, the Amazons called on the neighbouring Gargareans and had sex with their men, for purposes of procreation. According to Hellanicus, any resulting male children were – in a reversal of Greek practice – either killed, sent back to their fathers or exposed; the girls, however, were retained and raised by their mothers with training in agriculture, hunting and combat. Others explain the survival of the race by asserting that when the Amazons went to war they would spare some of the men they defeated and take them as sex slaves, having sex with them to produce the girls they needed to sustain the race.[8]

Penthesilea

When Penthesilea, daughter of war god Ares, accidentally killed Hippolyta in a deer hunting accident, it foreshadowed Penthesilea's own death at the hands of Achilles.[9] Penthesilea was so distraught at this unintended sororicide that she wished only for death, but she had a problem. Being an Amazon, the only fitting way to die was to do so with honour and in battle. According to Quintus Smyrnaeus' *Posthomerica*, she saw the Trojan War as an opportunity to achieve this end and enthusiastically joined in on the side of Troy, vowing, somewhat overconfidently perhaps, to slay Achilles. Her military exploits were widely famous, not least because of her appearance on the doors of the Temple of Juno: Aeneas was as transfixed by her fierce combat skills, equal to any man's, as he was by her protruding naked breast; Penthesilea, like Camilla, was a *bellatrix*, a woman of war. She arrived in Troy with twelve Amazon comrades, but the omens were not good: Priam saw an eagle holding a dove, a sure sign that Penthesilea would soon die; Theano, priestess of Athena, argued that to fight would be suicidal but Penthesilea carried on regardless.[10] She slew Podarces, one of the former suitors of Helen, and a number of other Greeks; her fight with a mocking Telamonian Ajax was inconclusive. Achilles, however, killed Penthesilea by impaling her and her horse with a single blow to her breastplate, knocking her to the ground; she begged for mercy but Achilles, unmoved, finished her off and scoffed over her corpse – until, that is, he removed her helmet and gazed on her beauty. It was at this point that he felt remorse for what he had done and realised that rather than kill her he should have married her.

The conflict between Achilles and Penthesilea is covered in the fragmentary *Aethiopis*, written in the seventh century BCE. The death of Penthesilea was, according to Diodorus Siculus, the beginning of the end for the Amazons: 'Now they say that Penthesileia was the last of the Amazons to win distinction for bravery and that in the future the race diminished more and more and then expired; consequently in later times, whenever any writers recount their prowess, men consider the ancient stories about the Amazons to be pure fiction.'[11]

Pseudo-Apollodorus, in the *Bibliotheca*, says that Achilles, 'fell in love with the Amazon after her death and slew Thersites for mocking him about it'.[12] In the early Roman Empire Propertius, the elegiac poet, supports this notion.[13] So does Pausanias, who describes the throne of Zeus at Olympia on which Panaenus' painted image shows the dying Penthesilea expiring and being supported by Achilles.[14] Thersites' cousin Diomedes was furious at Achilles so he harnessed Penthesilea's corpse behind his chariot, dragged it and dumped it into the River Scamander. Someone, either Achilles or the Trojans, retrieved the body and gave it a proper burial.

Robert Graves has an interesting take on the incident in his '*Penthesileia*':

> Penthesileia, dead of profuse wounds, was despoiled of her arms by Prince Achilles who, for love of that fierce white naked corpse, necrophily

> on her committed in the public view. Some gasped, some groaned, some bawled their indignation, Achilles nothing cared, distraught by grief, but suddenly caught Thersites' obscene snigger and with one vengeful buffet to the jaw dashed out his life. This was a fury few might understand, yet Penthesileia, hailed by Prince Achilles on the Elysian plain, paused to thank him for avenging her insulted womanhood with sacrifice.[15]

According to a lost poem by Stesichorus, Penthesilea killed Hector.

Hippolyta

Hippolyta, as we have already noted, was accidentally killed by her sister Penthesilea in a hunting accident. The magic belt she wore and which caused Hercules so much trouble was a badge of office, denoting her status as an Amazon queen.[16] Hippolyta was apparently so taken with Hercules that she surrendered the belt without a fuss while visiting him on his ship. According to Pseudo-Apollodorus, however, this did not take into account Hera who, masquerading as one of the Amazons, spread a malicious rumour among them that Hercules and his crew were intent on abducting their queen; the Amazons then attacked the ship, Hercules slew Hippolyta and relieved her of the belt, beat off the attackers and sailed away.

Another Hippolyta myth involves Theseus, some versions of which say he abducted her, and others that she fell in love with Theseus and betrayed the Amazons by absconding with him. In any event, she ended up in Athens where she was to marry Theseus – the only Amazon ever to marry – causing the other angry Amazons to attack Athens in what became known as the Attic War, a conflict that they lost to Athenian forces under Theseus or Hercules. Plutarch assures us, though, that 'the invasion of Attica would seem to have been no slight or womanish enterprise'.[17] Other versions have it that Theseus rejected Hippolyta in favour of Phaedra, so Hippolyta rallied her Amazons to attack the wedding ceremony and was killed in the ensuing fracas. The Attic War is commemorated as an *amazonomachy* in marble bas-reliefs on the Parthenon, in the sculptures of the Mausoleum of Halicarnassus, reliefs from the frieze of the Temple of Apollo at Bassae, now in the British Museum, on the shield of the statue of Athena Parthenos and on wall-paintings in the Theseum and in the Stoa Poikile. Pliny the Elder records five bronze statues of Amazons in the Artemision of Ephesus.[18]

Hippolyta married Theseus and despite having thereby renounced her amazonian status and nullified her credentials as an Amazon, was keen to accompany him as a warrior on his expedition against Thebes. Her 'swelling womb', however, prevented this and she was encouraged by Theseus, in a fine euphemism combining the erotic with the martial, to discard the 'cares of Mars and dedicate her retired quivers to the bedroom'.[19]

Myrina

Another queen of the Amazons, Myrina not only epitomises the military capabilities of the Amazons but demonstates for us what great builders, town planners and civilisers the Amazons were. Her tomb in Troad is mentioned in the *Iliad* and according to Diodorus Siculus she led a military expedition in Libya and was victorious over the Atlantians, laying waste their city, Cerne.[20] Myrina was not so successful fighting the nearby Gorgons, failing to burn down their forests. She struck a peace treaty with Horus of Egypt, conquered the Syrians, the Arabians and the Cilicians. She subdued Greater Phrygia, from the Taurus Mountains to the Caicus River, and several Aegean islands, including Lesbos; she discovered a previously uninhabited island which she named Samothrace, building a temple there and went on to found the cities of Myrina in Lemnos, another Myrina in Mysia, Mytilene, Cyme, Pitane and Priene.[21] Myrina died when her army was eventually defeated by Mopsus the Thracian and Sipylus the Scythian.

* * *

Jordanes' *Getica* (written *c.*560) is the earliest surviving history of the Goths. In it Jordanes tells how the ancestors of the Goths, descendants of Magog, originally inhabited Scythia, on the Sea of Azov between the Rivers Dnieper and Don and how, having repulsed a raid by a neighbouring tribe, while the menfolk were away fighting Pharaoh Vesosis, the women formed their own army under **Marpesia** and crossed the Don to invade Asia; Marpesia's sister, Lampedo, remained in Europe to guard the homeland. The invading women were in fact descended from the Amazons: they procreated with men once a year, conquered Armenia, Syria and all of Asia Minor as far as Ionia and Aeolia. Jordanes also relates how they fought with Hercules, and in the Trojan War, how a contingent of them lived in the Caucasus Mountains until the time of Alexander. He mentions by name the Queens Menalippe, Hippolyta and Penthesilea.

There are at least eighty-two Amazons known to us; some of the more important ones are detailed below.

Aello was the first to attack Hercules when he came for Hippolyta's belt. Unfortunately, Hercules was now clad in the lion skin acquired from his first labour, making him invulnerable. Aello, therefore, was killed. Her name means 'Whirlwind'.

Andromache was an Amazon queen. Her name translates as 'Man Fighter'.

Antiope was an Amazon queen defeated in battle by Theseus; undeterred, she married Theseus and bore his son, Hippolytus. One version has it that Antiope was betrayed by Theseus when he married another woman; she led an attack with her Amazons on the day of the wedding, planning to massacre the guests. Theseus and Hercules killed her. Her name means 'Confronting Moon'. A similar story is ascribed to Hippolyta, as we have seen.

Lysippe, another Amazon queen, settled her Amazons near the Black Sea. Lysippe established the policies and rules that Amazons lived by. She was a highly intelligent woman, an excellent general and founder of the city named Themiscrya. She was killed in battle; her name means 'She Who Lets Loose the Horses'.

Melanippe was the sister of Antiope. When Hercules came to get Hippolyta's girdle, Melanippe was taken prisoner. While in captivity, she launched a successful mutiny among the crew on one of Hercules' ships enabling her to escape with some other captive Amazons. They commandeered the ship, killing the Greeks and dumped their bodies overboard. Unfortunately, the seafaring skills of the Amazons were somewhat inadequate and they were blown to the shores of Scythia. Here they stole horses and became career horse rustlers. The name Melanippe means 'Black Mare'.

Pantariste pursued Hercules' officers when they fled; two Greek soldiers attacked her, but she killed them, one by grasping his throat until he suffocated. She threw her spear at Tiamides, who blocked it with his shield, but the force knocked him to the ground. She then unleashed her *labrys*, a double-headed axe, and beheaded him.

Thalestris was an Amazon queen and mistress of Alexander the Great. They went lion hunting together but, more interestingly, spent thirteen nights in lovemaking – thirteen is significant: it is a sacred fertility number for moon worshippers and relates to the number of moons in a year. Despite this torrid sex, Thalestris died without issue, dashing her hopes of a child by Alexander.

Valasca (or Dlasta) was a tyrannical and cruel Amazon warrior queen, who cut off the right eye and thumbs of all males, to render them ineffective in battle. Her aim was to start a new era for the Amazons; only when she died 'the nation resumed its normal order'.

The Amazons were not averse to intermarrying. Herodotus tells us how they settled down with the Scythians.[22] In the *Iliad*, the Amazons are referred to as *Antianeirai* – 'those who fight like men'. They invaded Lycia only to be defeated by Bellerophon, who was sent against them by King Iobates who was hoping in vain that the Amazons might kill him.[23] They attacked the Phrygians, who were assisted by Priam, then a young man.[24]

The Amazons featured prominently in religion. According to Plutarch in his *Theseus*, and Pausanias, Amazon tombs are to be found scattered throughout Greek lands, including Megara, Athens, Chaeronea, Chalcis, Thessaly at Skotousa and in Cynoscephalae; statues of Amazons crop up all over Greece. Plutarch reports that there was an Amazoneum, or shrine of Amazons, at both Chalcis and Athens; on the day before the Theseia (held on the eighth of the month Pyanopsion (October); the eighth day of every month was also sacred to Theseus; at Athens there were annual sacrifices to the Amazons. In historical times, Greek maidens of Ephesus performed an annual circular dance with weapons and shields that had originally been established by Hippolyta.

By and large, the Amazons received a good press when it came to assessments of their martial capabilities: when no less an authority than Priam relates his earlier war experiences to Helen of Troy he concedes that the Amazons are the equal of men: 'Once before I visited Phrygia of the vineyards ... and I myself, a helper in war, was marshalled among them on that day when the Amazon women came, men's equals'.

The military exploits of the hero Bellerophon elicit a similar assessment: 'but third he slaughtered the Amazons, who fight men in battle'.[25]

The first-century Greek historian Diodorus describes their revolutionary *modus vivendi* as follows:

> The sovereignty was in the hands of a people among whom the women held the supreme power, and its women performed the services of war just as did the men. Of these women one [Penthesilea], who possessed the royal authority, was remarkable for her prowess in war and her bodily strength, and gathering together an army of women she drilled it in the use of arms and subdued in war some of the neighbouring peoples. And since her valour and fame increased, she made war upon people after people of neighbouring lands.[26]

What few men and boys there were in the Amazon world had a distinctly hard time:

> But to the men she assigned the spinning of wool and such other domestic duties as belong to women. Laws were also established by her, by virtue of which she led forth the women to the contests of war, but upon the men she fastened humiliation and slavery. And as for their children, they mutilated both the legs and the arms of the males, incapacitating them in this way for the demands of war.[27]

Diodorus too subscribes to the 'right mastectomy' etymology: 'and in the case of the females they seared the right breast that it might not project when their bodies matured and be in the way; and it is for this reason that the nation of the Amazons received the appellation it bears'. Penthesilea comes in for special praise:

> In general, this queen was remarkable for her intelligence and ability as a general, and she founded a great city named Themiscyra at the mouth of the Thermodon river and built there a famous palace; furthermore, in her campaigns she devoted much attention to military discipline and at

> the outset subdued all her neighbours as far as the Tanaïs river ... and fighting brilliantly in a certain battle she ended her life heroically.[28]

Her daughter, Artemis, carried on the fine Amazon tradition: training – 'she exercised in the chase the maidens from their earliest girlhood and drilled them daily in the arts of war' – and subduing as far as Thrace 'and subdued a large part of Asia and extended her power as far as Syria'.

Some 200 years later Justinus more or less endorses this:

> Marrying their neighbours would for them be tantamount to slavery; they governed without men, and they put to death all the men who had remained at home; in order to continue their race they nevertheless consorted with the men of the adjacent nations but if any male children were born, they put them to death. The girls they kept busy not by consigning them to working in wool, but training them in combat skills, the management of horses, and hunting.[29]

Chapter 5

Women and War in Greek Tragedy

Homer's *Iliad* tells us all about the final days of the ten-year-long Trojan War. A number of the extant Greek tragedies and some of the non-extant works, incuding the earlier Theban Cycle, continue the story by revealing some of the terrible things that happened to the Trojan women captives and to the Greeks when they got home. Women, as in the *Iliad*, played a leading, and tragic, role in a number of these sequels, none of which, by definition, ended well. For our purposes, the seven tragedies which are of interest, because they demonstrate best the role of women in a war context, be it in the aftermath of Troy or in Thebes, are Aeschylus' *Seven Against Thebes*, Euripides' *Hecuba*, *The Suppliant Women*, *The Trojan Women*, *Iphigenia in Aulis*, *Phoenissae* and Sophocles' *Antigone*.

It is interesting to note just how much the issues raised by these plays still resonate today. In her perceptive article regarding Euripides in the modern world, Lucy Jackson points out that between the years 2000 and 2013 there have been 447 productions of classical plays staged by schools and amateur and professional companies in England. Every year in the last century saw a decade on decade increase in the productions of classical drama: given the prominence of women and particularly of women in war in these Greek and Roman plays, it is reasonable to assume that the issues the Greeks and Romans faced and tried to resolve on their stages still have relevance today. There is little difference between Euripides asking why go to war in *Iphigenia in Aulis* to the UK Parliament's debate over going to war against Iraq in 2003; both were agonising over the same question, both involved women, albeit indirectly. Just as interesting is the fact that the Greek tragedians, unlike the extant Greek historians, tackle the question of women as victims of war. The consequences of defeat for the Trojans are played out and agonised over time and time again: displacement, destitution, enslavement, homeland and household destruction, humiliation, ostracisation and rape are constant and common themes throughout the plays discussed here.

Aeschylus: *Seven Against Thebes*

Women play a major role in this play, in the shape of a chorus made up of Theban women, a city under siege. Much of the play comprises a troubled dialogue between the chorus and the city's king, Eteocles, brother of

Polynices. Polynices is about to attack the city to claim his rightful turn as King of Thebes, denied to him by Eteocles who is overstaying his reign and time as king. The chorus is influential in the play: it offers up prayers to the gods of Thebes in defiance of Eteocles who wants to silence the chorus.[1] The chorus pictures itself as a victim of war, fearful and frightened by the din of the siege as Eteocles formulates his plan for the defence of the city; it desperately tries to persuade Eteocles not to fight Polynices; the chorus clings to the statues of the local gods in a bid to win their protection for the city.[2] So frustrated does it all make Eteocles that he exclaims somewhat misogynistically: 'Zeus, what a race you've given us for company, these women!' to which the leader of the chorus quips in reply: 'A wretched one, just like men when their city is taken!'[3] The only benefit Eteocles can see in the womanly chorus is their support for the sacrifices which will mark his victory, the sacred lamentation and the cry of victory, the Paean.[4] The chorus vividly describes the fate that awaits them 'when a city is taken': their city is destroyed, they, young and old alike, are dragged off by their hair like horses, their clothes riped off; virgin girls are systematically raped – a fate worse than death itself. They in turn wish death on the bragging, conquering soldier: may he be struck down if ever he bursts into their homes to rape a virgin (452–6).

The women in the chorus are also influential in the shields scene in which Eteocles systematically allocates a Theban hero to face the seven Argive warriors who are detailed to assault the seven gates of Thebes. For the first six of the seven, the chorus, with increasing belligerence, asks the gods for their support, which they get. But when it comes to the seventh, everything changes. Eteocles has decided that he will face Polynices; the chorus tries to persuade him not to: for the women, the stain, the *miasma*, of brother killing brother will live for ever.[5] The last act of these powerful women immersed in a fratricidal war is to lament the two brothers united in death, fulfilling the curse inflicted on them by Oedipus and exterminating the house of Laius.

Euripides: *Hecuba*

The action of the play (first performed in 424 BCE) takes place about the same time as events described in Euripides' *Trojan Women*: that is, after the fall of Troy and before the women's dispersal to various parts of Greece as slaves. The chorus is made up of Trojan women already consigned to a life of endless servitude. Hecuba's double tragedy is the sacrifice of her daughter, Polyxena, on the altar of the ghost of Achilles, and the treacherous slaying of her son, Polydorus, by the Thracian king Polymestor. The chorus of women, Polyxena and Hecuba all pay the heaviest price for their gender in the aftermath of the war. They graphically describe the fate that awaits them, dumb with grief, husband dead and city sacked. Prayers to Artemis, goddess of virginity, are prayed but go unanswered. Polyxena faces her fate bravely, proudly asserting that she would rather die than live a life of servitude and serial rape. Her fall from her regal status is great as she finds herself reduced to slavery: 'it is that name of slave, so ugly, so foreign, that makes me want to die'. Her alternative is a life of abject misery and drudgery as described in lines 351–66, reminiscent of the fate Hector foresees for Andromache:

> Or should I live to be knocked down to the highest bidder, sold to a master man for cash? Sister of Hector ... doing the work of a skivvy, kneading the bread and washing the floors, forced to drag out endless weary days. Me, the bride of kings compelled by some gutter slave to share his fetid bed.

Hecuba dutifully attends to the funeral rites due to her daughter and vows revenge for the slaughter of Polydorus; she enlists the support of Agamemnon, taking full advantage of the fact that Agamemnon is obsessed with Cassandra – his concubine and her other daughter. Hecuba lures Polymestor into a tent where she claims her treasures are hidden: Hecuba and women from the chorus slay Polymestor's two sons and put his eyes out; blinded and furious, he casts about for the women who have committed this 'vile act'. However, justice is deemed to have been done but Polymestor, still raging, prophesies the deaths of Hecuba by drowning and Agamemnon at the hands of his wife, Clytemnestra. The Greeks set sail with the distressed women of the Chorus to their new lives as slaves.

Andromache is also a queen and may, on the face of it, seem comparatively fortunate given that she was handed over as the wife of Achilles' son, Neoptolemus. Not a bit of it: she too laments the loathsome fate that awaits her: 'A bed which from the word go I never wanted and now reject for good. God knows that that was a bed I never crept into willingly' (Euripides, *Andromache*, 36–8).

Euripides: *The Suppliant Women*

The action of this play, first performed in 423 BCE, revolves around the dire predicament the bereaved mothers of slain Trojan warriors, the Seven Against Thebes, find themselves in when they are prohibited by Creon from performing funeral rites and burying their sons, all in contravention of divine law and ancient custom. Their men had died in the battle between Polynices and Eteocles, squabbling over the right to rule Thebes. The execution of funeral rites and the organisation of proper burial was one of the few responsibilities allowed to women in a society which precluded them from public activity in most other areas of civic life. Creon's prohibition here would, therefore, have resonated badly with the grieving women. Eventually they prevail on Theseus, King of Athens, through Aethra his mother, to intercede. He attacks Thebes, although the women are fearful of yet more loss of life; Theseus himself washes the corpses of the dead sons and prepares them for burial. The women are naturally grateful that the bodies have been recovered, but distressed at the prospect of seeing them as corpses, agreeing that it would have been better had they never married. Theseus cremates and obtains the ashes of the fallen Trojan heroes – shielding the women from the horrific sight of the mutilated and rotting corpses.

As the women continue their lamentations they see Capaneus' wife Evadne in her wedding dress climbing the rocks above her husband's tomb, declaring her plan to join her husband in the flames of the pyre. Her father Iphis tries to talk her down, but she leaps to her death. As Iphis dies, the orphaned youths hand over the ashes of their fathers to their grandmothers. The boys' lamentations are mingled with promises of revenge and more war.

War is undoubtedly a high price to pay, but such is the strength of duty and feeling experienced by the bereaved suppliant women that it is considered a price worth paying to see their sons accorded the appropriate funeral rites and a proper burial.

Euripides: *Iphigenia in Aulis*

Iphigenia, eldest daughter of Agamemnon is pivotal, like Helen, to the Trojan War. She is an essential pawn in the spat between Artemis and Agamemnon; he eventually decides to sacrifice his daughter on the altar of appeasement to the goddess and to sanction the sailing of his troops to Troy to do battle, thus preserving their honour and satisfying their lust for battle. Mysteriously becalmed at Aulis, the Greek troops become restless and hover on the brink of revolt: thus threatened, Agamemnon makes the invidious and terrible decision to sacrifice Iphigenia. He sends a deceptive message to his wife, Clytemnestra, telling her to send Iphigenia to Aulis because she is to marry Achilles. Agamemnon vacillates to such an extent that he has a change of heart and sends a second message to his wife, cancelling out the first. This second missive is intercepted by Menelaus, Agamemnon's brother, who is furious over the change of plan because the whole objective of the Trojan War was to regain *his* wife Helen. Menelaus eventually comes round, only to have Agamemnon change his mind again and prepare for the sacrifice of his daughter, anxious lest the army storms his palace at Argos and massacres all his family. Clytemnestra is by now on her way to Aulis with Iphigenia.

Iphigenia, Clytemnestra and Achilles soon discover Agamemnon's duplicity; Achilles is incandescent and vows to defend Iphigenia. His attempts to rally the Greeks, however, only result in the discovery that 'the whole of Greece' – including his own Myrmidon forces – demand that the fleet sails for Troy to do battle; Achilles narrowly escapes a stoning. Clytemnestra and Iphigenia fail to persuade Agamemnon to change his mind; Iphigenia implores Achilles not to waste his life over what she sees as a lost cause and bravely and heroically agrees to her sacrifice. She proclaims that she would rather die a hero, celebrated as the saviour of Greece, than be dragged kicking and screaming to the altar. She goes to her death, leaving her mother Clytemnestra utterly distraught and presaging her murder of Agamemnon and Orestes' matricide.

Like Helen, Iphigenia has a leading role to play in the prosecution of the Trojan War. It is her life that is sacrificed to secure the war; without her death there would have been no war with Troy.

Euripides: *The Trojan Women*

Women also suffer horribly in *The Trojan Women*, first produced in 415 BCE; it is a tragic commentary on the capture of the Aegean island of Melos and the subsequent slaughter and subjugation of its inhabitants by the Athenians. The play opens with Athena and Poseidon discussing ways to punish appropriately the Greek armies for condoning the rape by Ajax the Lesser of Cassandra, the eldest daughter of King Priam and Queen Hecuba; he had sacrilegiously dragged her from a statue of Athena. Euripides shows how much the now destitute Trojan women suffer: more grief is piled on when the Greeks share out the women between them. The deposed Queen Hecuba learns that she will be appropriated by Odysseus, while Cassandra is acquired by Agamemnon. Cassandra, though, had seen all this coming: she knows how Clytemnestra will slay both her and Agamemnon. Sadly, no one believes or listens to Cassandra when she offers sound counsel: 'Would ye be wise, ye Cities, fly from war! (l. 308ff.). Yet if war come, there is a crown in death for her that striveth well and perisheth Unstained: to die in evil were the stain!' Cassandra has already bewailed the lot of the Greek wives, denied the privilege of burying their dead husbands, languishing in widowhood while their men lie rotting in a foreign field.

Things get much worse when Andromache, widow of Hector, tells Hecuba that her youngest daughter, Polyxena, has been killed as a sacrifice at the tomb of Achilles. Not only is Andromache destined to be the concubine of Achilles' son Neoptolemus, but she is informed that her baby son, Astyanax, has been condemned to die because the Greeks are afraid lest the boy grows up to avenge his father, Hector; the plan is to hurl Astyanax from the battlements of Troy to his death.

Helen of Troy does not escape the misery: cuckolded Menelaus arrives to take her back to Greece to face a death sentence; Helen begs for mercy and tries to seduce her husband into sparing her life but Menelaus is adamant – for now: he later relents and takes her back. The body of Astyanax is carried in on Hector's shield: Andromache naturally wished to bury her child herself with the proper rituals, but too late: her ship had already set sail. Hecuba it is who prepares the body of her grandson for burial, before she is taken off in Odysseus' baggage train.

The play graphically highlights the fate of these women after a devastating ten-year-long war, and the impact that war has on their later lives. They, along with countless others, are bereaved, homeless, deprived of their dignity and freedom and displaced; they are left with nothing apart from the prospect of lifelong slavery and the rape that comes with it. Hecuba, for whom Troy had been her lifelong home, has just seen Troy go up in flames, she has witnessed the deaths of her husband, children and grandchildren; the former queen and everything she ever had that was left was now the property of Odysseus. More generally, it reminds us that the horrors of war 3,000 years ago are just as bad today and women's suffering as victims is just as terrible. What Euripides showed has an eerie resemblance to the scenes we see daily of refugee camps and urban destruction in and around Syria. Euripides warns us that the more jubilant and triumphant war is, the more wretched and miserable is the fallout – for vanquished and victors like: victors beware.

The women in the chorus sum up their desperate plight: these women can only foresee a life of dreaded nightly rape; they echo the words of Polyxena in the *Hecuba*: 'Shall I be a skivvy, or forced into the bed of Greek masters? Night is a queen, but I curse her!' (202–4). Profound thoughts and words that must have been repeated by women victims of war through the ages down to this very day.

Euripides: *The Phoenissae*

The play gets its name from the chorus which comprises women from Phoenicia – war booty sympathetic to their Theban ancestors and on a one-way ticket to Delphi where their fate is to be sacrificial fodder to Apollo. Phoenicia is significant because it is the home of Cadmus and of that 'horned ancestor' Io, raped by Zeus and right there at the beginning of the tortuous sequence of events that led to the Trojan War and culminating in the curse of Oedipus bestowed on his two fratricidal sons. The Phoenician women, though, are not the only powerful women in the play: Jocasta, until her suicide after the deaths of her two sons, and her daughter, Antigone, also loom large in the action. Indeed, importantly, they open the play: Jocasta provides the back story by relating the history of the House of Cadmus and the origins of the curse of Oedipus and his descendants; Antigone has her teichoscopy in which she describes the army besieging her city from the terrace of her palace.

Jocasta soon takes control by taking command; only too aware of the fate that awaits her sons, she does everything she can to forestall the inevitable by trying to fix a last-ditch meeting between Eteocles and Polynices, the latter an unwilling exile in his own land. Assuming the role of a military commander, she, desparate for reconciliation, orders a truce between the opposing armies but Eteocles is intractable and refuses to compromise, adamant as he is that he will retain power in the land, Polynices' power that is and Polynices' land. Jocasta's sagacious appeal to fairness and equality falls on deaf ears.[6] Instead, she is forced to witness an unpleasant exchange in which the brothers verbally mutilate each other, both parting with the promise of killing the other in an impious and sacrilegious prayer which leaves Jocasta with nothing but the prospect of two dead sons and a ruined city. When she learns of the duel she is determined, with the help of Antigone, to battle her way between her sons but arrives too late. Seizing a 'bronze-hammered' sword found lying between the bodies of Eteocles and Polynices, she stabs herself and dies a soldier's death as she falls on the bodies of her boys. This battlefield death in no-man's-land between two armies, effected with a soldier's sword, is the death of a militaristic woman falling in battle.

Antigone gets her first taste, and view, of battle and war from the vantage point of the palace terrace from which she describes the military activity going on below in a teichoscopy. Like Medea in Valerius Flaccus' *Argonautica*, Antigone is initially innocent of the ways of war and is frightened by the horror of it all, although, at the same time, she is awed by the scale and magnificence of the weaponry worn by the warriors.[7] She curses them, predicting the servitude their martial behaviour will impose on the Theban women. This all changes, however, when she reluctantly descends to the battlefield with Jocasta amid the carnage engendered by the duel between her two brothers.[8] Antigone is now confronted by the horrors of war. Things go from bad to worse when her mother takes her own life and Antigone is left to perform the crucial funeral rites for all three relatives, and to break the terrible news to a pathetic Oedipus.

Both Jocasta and Antigone show immense courage and tenacity in their, ultimately unsuccessful, attempts to deflect or mitigate the sins and mistakes of their kin. But they are not just up against the fickleness and vainglory of mortal men: the gods are in control here and nothing the women can do can thwart the will of the gods. Jocasta and Antigone boldly intrude into the male world of war, but to no avail; so, all that was left to Antigone was to report the bad news and perform the traditionally female role of lamenting and organising the funerals.

Sophocles: *Antigone*

Antigone was written in 441 BCE, the third of the Theban trilogy plays and follows on where Aeschylus' *Seven Against Thebes* ends. It is another dramatic expression of the tragic predicament women find themselves in when prohibited from carrying out the funeral rites and burial of their kin: Antigone is forbidden by Creon to bury her brother, Polynices. The corpse is to remain on the field of battle to be picked over by carrion: a fate worse than death for all directly concerned. The defiant and brave Antigone goes to bury her brother herself, despite her sister Ismene's refusal to help.

When Creon learns that Polynices has received funeral rites and a symbolic burial effected with a thin covering of earth, he is incandescent with rage and orders that the culprit be found and brought to book; Antigone is duly discovered and admits her guilt, comparing the immorality of Creon's edict with the morality of her own actions. A still angry Creon assumes Ismene to be implicated; she is summoned and confesses falsely to the crime, wishing to die with her sister; Creon orders that the two women be imprisoned.

Haemon, Creon's son and Antigone's fiancé, pledges allegiance to his father and tries tactfully to persuade him to spare Antigone, claiming that 'under cover of darkness the city mourns for the girl' (l. 689). Things go badly and Haemon leaves, vowing never to see Creon again. Creon's cruel reaction is to spare Ismene and to entomb Antigone alive in a cave. She is taken away to her living death, midst lamentations from the Chorus over her impending doom.

Tiresias, the blind prophet, warns Creon that Polynices should at last be buried: the gods are displeased to such an extent that they are refusing all sacrifices or prayers from Thebes. Creon accuses Tiresias of corruption to which the seer responds that Creon will lose 'a son of [his] own loins' (l. 1066). All of Greece will despise him; a terrified Chorus asks Creon to take their advice: he agrees and frees Antigone and buries Polynices. However, in the meantime, Antigone has hanged herself; Haemon tries, and fails, to stab Creon then turns the knife on himself. Creon's wife, Eurydice, commits suicide and, with her dying breath, curses her husband.[9]

In Euripides' *Medea* the heroine turns into the antithesis of what the Greeks liked to believe was characteristic of their women: 'Usually, a woman is frightened, too weak to defend herself or to bear the sight of steel. But if she happens to be wronged in love, hers is the bloodthirstiest heart of all.'[10] The usual fearfulness and abhorrence at the weapons of war here gives way to a facility for atrocity when a woman is 'wronged in love'.

Chapter 6

Women and War in Greek Comedy

The fate and actions of women as a consequence of war is also explored in Aristophanic Greek comedy. The famous example is, of course, the *Lysistrata*, one of a number of plays based on fantasy situations in which the norms of Athenian society are absurdly contravened and contraverted.

Lysistrata

Lysistrata, the name translates as 'Army Disbander', was performed in 411 BCE, and is the story of one woman's extraordinary but mad mission to end the Peloponnesian War. Lysistrata persuades the women of Greece to withhold sex from their husbands and lovers until they negotiate a peace; however, all it does in the end is excite a battle between the sexes. Nevertheless, it does show how inept some men can be and the ability of women to intrude onto the traditional territory of men.

In the beginning, a solemn oath is struck over a wine bowl in which the women renounce all sexual pleasure, including the 'Lioness on the Cheese Grater', an obviously satisfying sexual position. Early success comes in the capture of the Acropolis, home to the treasury; the women are now able to freeze the funds that are needed to finance the ongoing war. The magistrate turns up looking for a means to pay for oars and then proceeds to trot out the usual pejorative, misogynistic stereotypes – deploring the hysterical nature of women, their bibulous behaviour, their predilection for promiscuous sex and attraction to strange and exotic cults; he also censures his fellow men for not controlling their women. Lysistrata responds by delineating the frustration felt by women during war when the men make stupid decisions that affect the whole population while their wives' opinions are ignored. She feminises him, covering his head with her headdress, and gives him a basket of wool – traditional accoutrements of women – and tells him that from now on war – man's work – is now woman's business. She explains the plight of young, childless women who sit at home, growing old while the men are away fighting endless campaigns, alluding to the double standard where women are obliged to marry very young and men can marry when they choose, having played the field to their heart's content. To Lysistrata, the magistrate, and everything he stands for, might as well be dead. Nevertheless, reconciliation eventually ensues and everyone is happy. The war continues apace but

Aristophanes has at least allowed the women a voice to adumbrate the inequalities they face in Greek society, to voice an opinion about the war and the suffering it causes the women left at home, and to criticise the handling of the war by the men of Athens.

Lysistrata was not the only comedy in which women were on top: we know of two plays called *Gynaikokratia*, one by Alexis (*PCG* 2, frags 42–3) and one by Amphis (*PCG* 2, frag. 8); *Tyrannis* by Pherecrates (*PCG* 7, frags 150–4) and Theopompus' *Stratiotides* (*PCG* 7, frags 55–9).

In Aristophanes' *Birds* (829–31) Euelpides complains about a similar gender reversal when Athena, a female deity no less, protects the city in a suit of armour while so-called warrior Cleisthenes is portrayed as an effete man armed with his shuttle; bellicosity apart, Athena is also patron of wool working and other women's crafts. The line here is a parody of Euripides' *Meleager*, frag. 522: 'if work with shuttles were to be the concern of men, while women were seized with the delights of weaponry'.

The *Miles Gloriosus* or Bragging Soldier

Most of Greek New Comedy is lost to us. We know of about 100 comic playwrights who were active between 404 and 323 BCE with a combined output of 700 plays. Sadly, what precisely happened during the eighty-one-year transition between Aristophanes and Menander, from old to new comedy – via middle comedy – is largely unknowable. However, we can say that the lost plays are populated by stock characters which are each cast in numerous different plays. These characters epitomise Greek life and real Greek people so, given the prevalence of war and the ubiquity of garrisons in the Greek world manned often by mercenaries, it is not surprising that the 'swaggering soldier' is one such type. He originates in a lost play called *Alazon – Braggart* – and is, of course, best known to us in the Roman play, Plautus' *Miles Glorioususus*, which, like the rest of his extant output, is based on Greek New Comedy originals. Plautus (*c*.254–184 BCE) set his play in the Greek city of Ephesus; Pyrgopolynices is the braggart's name and he, typically of Plautine plots, has kidnapped the former girlfriend, Philocomasium of Pleusicles. In so doing, he fulfils one of the standard roles in New Comedy as the man who temporarily blocks the relationship between two young lovers. In Plautus, sexually active mercenaries find their way into the plots of the *Bacchides*, *Curculio*, *Epidicus*, *Pseudolus* and *Truculenius*; Terence casts one in his *Eunuchus*.

Chapter 7

Women and War in Greek History and Philosophy

One of Herodotus' objectives when he wrote his history was to illuminate the differences between the foreign and the Greek. The behaviour of 'barbarians' (foreigners) in 'barbarian' (foreign) countries and civilisations is one of the prominent features in Herodotus' ethnographical excursions around the Mediterranean. Unlike Thucydides, who virtually ignores women because he believes they are not relevant either to war or history, Herodotus mentions 375 women as individuals or in groups, 76 in an ethnographical context. Additionally, there are queens and regents, princesses and royal mistresses. Four of these foreign queens, Nyssia, Tomyris, Artemisia I and Pheretime, exhibit impeccable political and strategic military skills that are as good as any man's – much to the indignation of many a Greek man, no doubt. Herodotus presumably believed that women had a valuable role to play in aspects of war and were, therefore, an essential element of his history. He is also anxious to show how their actions were in just reaction to wrongs done to them by their men, how a breach by these men of the *nomos*, the law, was avenged.

Nyssia of Lydia

Candaules, also known as Myrsilos, was a king of Lydia from 735–718 BCE. Herodotus (1.7.2–13) says his name meant 'dog throttler'. In telling this story he gives us a tale of caution in which a vengeful woman takes centre stage: Candaules is murdered by his wife, Nyssia, due to his arrogance. For Herodotus, she was responsible for allowing the Lydian throne to pass out of the hands of the Heraclid dynasty – after 505 years and twenty-two generations – to the Mermnadae dynasty in the guise of Gyges. How? Candaules was in the habit of boasting about his wife's prodigious beauty to his bodyguard, Gyges; she took exception to such inappropriate and disrespectful behaviour when he exposed her nudity: 'If you don't believe me when I tell you how lovely my wife is, a man always believes his eyes more than his ears; so do as I tell you – get a look at her taking her clothes off.'[1]

At first Gyges was outraged and refused, only too aware of the taboos surrounding nudity in Persian society and concerned how unpredictably Candaules might actually react if and when the seedy deed was done (1.8.2;

1.10.3). Candaules eventually persuaded him, by asserting that Nyssia had no shame when naked. He revealed a plan in which Gyges would lurk behind a door in the royal bedroom to watch Nyssia undressing; Gyges would then steal away unnoticed while the queen's back was turned. So, that night, Gyges took up his position and leered as planned, but the queen caught a glimpse of him and realised in an instant that she had been betrayed, shamed and humiliated by her own husband. Nyssia swore revenge, and formulated her own plan. Next day, she summoned Gyges and confronted him: 'One of you must die. Either my husband, or you, who have outraged convention by seeing me naked.' Eventually, Gyges, quite understandably, chose to betray the king and save his own skin. Nyssia's scheme involved an element of *déjà vu*, literally: Gyges hid behind the door of the bedroom again, this time armed with a knife provided by the queen, and slew Candaules in his sleep. Gyges married the queen and became king.

Candaules did not pay just with his life. He has the dubious privilege of lending his name to candaulism, a deviant sexual practice in which a man exposes his (usually) female partner, or images of her, to other people for their voyeuristic pleasure. The term is also applied to the practice of undressing or exposing a female partner's body to others, or forcing her into having sex with a third person, or into prostitution or pornography. Today the term is increasingly applied to the posting of revealing images of a female partner on the Internet, or forcing her to wear sexually suggestive clothes for depraved and prurient public consumption.

Nyssia was obviously dignified and shrewd, a fact that, for the average Greek, makes Herodotus' story all the more disturbing – all the more he would be careful to keep his woman under wraps and shielded from the prying eyes of strangers. Nyssia defends the conventions that her husband spurned and proved him wrong when he insinuated that women are shameless when naked. She punishes the violation and restores convention: in the end, only her husband had still ever seen her naked, even if that necessitated a radical change of husbands.

Queen Tomyris of the Massagetae

Cruel and unusually militaristic are two ways to describe Queen Tomyris of the Scythian Massagetae; at the same time, though, her actions can be seen as just desserts for Cyrus' duplicity. Strabo, Polyaenus, Cassiodorus and Jordanes, as well as Herodotus, all make reference to her.[2] She is famous for slaying King Cyrus the Great (*c.*576–530 BCE), the founder of the Achaemenid Empire, when he invaded her country. Herodotus shows his readers how Tomyris easily saw through the king's proposal of marriage as an ill-disguised desire to topple her regime and enslave the Massagetae; she wisely rejected him.

Tomyris showed more wisdom and foresight when Cyrus began building bridges in order to cross the frontier river dividing their territories; she warned him off three times to end his campaign against her country (1.206.1).

Cyrus then made the mistake of duping Tomyris' army, which was under the command of her son, Spargapises, into drinking copious amounts of wine while at battle readiness; Cyrus had slyly left a stash behind on the battlefield when his army withdrew. Scythians, however, were not used to drinking wine, being much more partial to hashish and fermented mare's milk; accordingly, they drank themselves to inebriation – and, while under the influence, were victoriously subdued by Cyrus' Persians. Spargapises was captured; Tomyris warned that her patience was running out, and that she would give Cyrus his fill of blood if he did not release her son (1.212.3). But Spargapises persuaded Cyrus to remove his bonds and promptly committed suicide. A vengeful Tomyris challenged Cyrus to a second battle, promising him that fill of blood. Herodotus melodramatically describes it as the 'fiercest of all the battles waged between the barbarians'. Tomyris was victorious: Cyrus was killed and Tomyris had his head cut off and his corpse crucified; she then shoved his head into a wineskin full of human blood: 'I live and have conquered you in fight, and yet by you am I ruined, for you took my son with guile; but thus I make good my threat, and give you your fill of blood.'[3]

To the Greeks, women and war simply did not mix. Herodotus' sensational telling of this escapade was designed not just to illuminate what he saw as the perverse and blood-thirsty behaviour of barbarian women compared with the civilised women of Greece, but to demonstrate just how far removed from Greek women the barbarians were. At the same time, though, he does explore Tomyris' wisdom and good political and military sense, and exhibits some sympathy when he shows that, despite her victory, Tomyris has ultimately lost because she had lost her son. Herodotus' general description of the Massagetae, including their partiality for routine human sacrifice, and the prurient descriptions of the allegedly permissive sexual mores of barbarian women are manifestations of how the Greeks attacked the soft, vulnerable underbelly of their enemy by denigrating and slurring the sexual behaviour of their women. This insidious, prejudicial and demoralising tactic continued to be deployed to good effect by later Greek historians and by the Romans.

Pheretima

The story of Pheretima (d. 515 BCE), wife of the Greek Cyrenaean King Battus III and the last queen of the Battiad dynasty, provides another cautionary lesson for Herodotus' audience. He tells us that when Battus died in 530 BCE, her son Arcesilaus III became king but was defeated in a civil war some time after 518 BCE and exiled to Samos while Pheretima went to the court of King Euelthon in Salamis, Cyprus. Arcesilaus, however, recruited an army in Samos, returned with it to Cyrenaica, and won back his regal position murdering and exiling his political opponents. In this he was encouraged, no doubt, by Pheretima. When Arcesilaus left Cyrene for Barca, Pheretima ruled the city as regent but Arcesilaus was murdered by exiled Cyrenaeans intent on revenge. Pheretima sped to Arysandes, the Persian governor of Egypt, seeking help in avenging the death of her son; Arysandes loaned her Egypt's army and navy, at the head of which she marched to Barca and demanded the surrender of those Barcaeans responsible for the murder of Arcesilaus. When the Barcaeans refused to own up, Pheretima laid siege to their city for nine months and took the city when Amasis, her Persian commander, tricked the Barcaeans: he ordered his soldiers to dig a large trench in front of the city, camouflaged with wooden planks and earth; he then lured the Barcaeans out of the city with a promise of a well-rewarded armistice. They literally fell into the trap: Pheretima, still crazed by vengeance, ordered the Barcaean wives' breasts to be cut off, and enslaved the rest of the Barcaeans to the Persians.

Such an atrocity was appalling enough, but for it to be committed by a woman, however strong her maternal love for a lost son, would have shocked many Greeks. Herodotus is at pains to show that such war crimes committed by a foreigner, and a woman foreigner at that, would not go unpunished. Pheretima returned to Egypt, and gave the army and navy back to the governor. However, while there, she contracted a contagious parasitic disease, and died in late 515 BCE. Herodotus tells us that she was eaten alive by the worms spawned by her disease – punishment by the gods for her butchery of the women of Barca.[4] She lives on in the name of the worm which infested her: *pheretima* is a genus of earthworm found in New Guinea and other parts of South East Asia; the worms are still used as a medicine in China and carry biological agents efficacious in the treatment of epilepsy. The *Pheretima aspergillum* worm contains hypoxanthine, a herb used as an antipyretic, sedative and anticonvulsant. It lowers blood pressure and contains a platelet-activating factor.

Artemisia I of Caria

Artemisia (fl. 480 BCE) was born in Halicarnassus (modern Bodrum in Turkey and birthplace of Herodotus). She was Queen of Halicarnassus and commander of the Carian navy; Herodotus reports that she acquitted herself in exemplary fashion – to him her deeds were a *thauma*, a 'marvel'; Herodotus marvels at her because she is actually a woman fighting against the Greeks (ἐπὶ τὴν Ἑλλάδα στρατευσαμένης γυναικός); she gives the best advice to Xerxes (γνώμας ἀπίστας βασιλέϊ ἀπεδέξατο, 7.99). He admires her manliness (ἀνδρηίης, 7.99) and her eagerness to step up and lead the Halicarnassians after her husband's death. The use of the epithet *andreia* is significant: it, of course, means 'bravery' but its etymology is from *andros* meaning a man –

bravery for the Greeks was inextricably bound up with manhood, it was the preserve of men so its use to describe a woman is exceptional. The Romans did similarly: *virtus* – virtue and bravery – has as its root in the word *vir*, Latin for man. Herodotus uses *andreia* only this once to qualify a woman's actions and it is to describe Artemisia's bravery when she and her navy supported the Persians against Herodotus' compatriots, the Greeks. Indeed, Artemisia had already won a prestigious seat on Xerxes' councils of war dispensing the best of advice and winning the epithet *androboulos* – advising like a man.[5] In contrast it is difficult to envisage a woman sitting on a Greek council of war. This is how Greek Herodotus introduces Persian Artemisia:

> I must speak of a certain leader named Artemisia, whose participation in the attack on Greece, notwithstanding that she was a woman, moves me to special wonder. She had obtained the sovereign power after the death of her husband ... She ruled over the Halicarnassians [despite having a grown-up son], the men of Cos, of Nisyrus, and of Calydna; and the five triremes which she furnished to the Persians were, next to the Sidonian, the most famous ships in the fleet. She likewise gave to Xerxes sounder counsel than any of his other allies.

Before the battle of Salamis, Xerxes called all his naval commanders together and sent Mardonios, his second-in-command, to ask whether or not they thought he should fight the battle. All advised him to fight – except Artemisia. Herodotus reports her wise counsel and in so doing demonstrates that she can clearly see 'the bigger picture':

> Tell the King to spare his ships and not to engage in a naval battle because our enemies are much stronger than us by sea, as indeed men are to women. And why does he need to risk a naval battle anyway? Athens, the reason for undertaking this expedition is already his and the rest of Greece too ... If Xerxes chooses not to rush into a naval encounter, but instead keeps his ships close to the shore and either keeps them there or moves them towards the Peloponnese, victory would be his. The Greeks can't hold out against him for very long. They will leave for their cities, because they don't have food stored up on this island ... But if he hurries to engage I am afraid that his Persian navy will be defeated and the land-forces will be weakened as well. In addition, he should also consider that he has certain untrustworthy allies, like the Egyptians, the Cyprians, the Kilikians and the Pamphylians, who are completely useless.

Xerxes, though impressed, fought the battle anyway. Xerxes watched Artemisia demonstrating tactical brilliance when she was pursued by the ship of Ameinias of Pallene.[6] To shake him off she attacked and rammed an allied Persian vessel fighting on her own side and so hoodwinked Ameinias into assuming that she was an ally of his; not unreasonably, Ameinias then gave up the chase. But her ingenuity came at a price: the friendly victim, sadly, was manned by people of Persian allies Calyndos and went down with all hands in what is one of the earliest recorded instances of tactical, intentional 'friendly fire', or 'blue on blue'. A confused Xerxes, looking on, assumed that she had successfully attacked an enemy Greek ship, and seeing the indifferent performance of his other commanders, famously commented 'My men have become women, and my women men'.[7] The baffled Xerxes summed it all up with: 'O Zeus, surely you have formed women out of man's materials, and men out of woman's.'

Xerxes was oviously not the only man that day who was impressed by the military actions of a woman. Herodotus continues:

> Now if [Ameinias] had known that Artemisia was sailing in this ship, he would not have given up until either he had captured her or had been taken himself; for orders had been given to the Athenian captains, and moreover a reward was offered of 10,000 drachmas for the man who took her alive; since they thought it intolerable that a woman should fight Athens.[8]

Artemisia made inspired use of the Persian *and* Greek flags she had cleverly stowed on board:

> Artemisia always ... carried on board with her Greek, as well as barbarian, colours. When she chased a Greek ship, she hoisted the barbarian colours; but when she was chased by a Greek ship, she hoisted the Greek colours so that the enemy might mistake her for a Greek, and give up the pursuit.[9]

More tactical wizardry, and that friendly fire, followed in the words of tactical expert Polyaenus:

> Artemisia ... found that the Persians were defeated, and she herself was near to falling into the hands of the Greeks. She ordered the Persian colours to be taken down, and the master of the ship to bear down upon, and attack a Persian vessel, that was passing by her. The Greeks, seeing this, supposed her to be one of their allies; they drew off and left her alone, directing their forces against other parts of the Persian fleet. Artemisia in the meantime sheered off, and escaped safely to Caria.[10]

After the battle, Xerxes rewarded her sterling performance as the best of all his commanders with a complete suit of Greek armour; rubbing salt into male wounds and exemplifying hurt pride he simultaneousely awarded the captain of her ship a distaff and spindle! Persian Xerxes and Artemisia, like

Herodotus' Egyptians whose menfolk he witnessed working the wool instead of the women, were turning the world on its head.

More sound Artemisian advice followed when Xerxes asked her whether he should now lead his troops to the Peloponnese himself, or withdraw from Greece and leave his general Mardonius to the task. Artemisia replied that he should retreat back to Asia Minor and advocated the plan suggested by Mardonius, who requested 300,000 Persian soldiers with which he would defeat the Greeks.

According to Herodotus her considered response was:

> I think that you should retire and leave Mardonius behind with those whom he wishes to have. If he succeeds, the honour will be yours because your subordinate carried it out. If on the other hand, he fails, it would be no great matter as you would be safe ... In addition, if Mardonius were to suffer a disaster who would care? He is just your servant and the Greeks will have but a poor triumph. As for yourself, you will be going home with the object for your campaign accomplished, for you have razed Athens.[11]

Xerxes took her advice this time, leaving Mardonius to prosecute the war in Greece. Xerxes sent her to Ephesus to take care of his illegitimate sons, a task considerably less challenging (and somewhat demeaning) than fighting valiantly and spectacularly in the battle of Salamis, and being de facto chief military advisor to Xerxes.

Artemisia obviously impressed both Xerxes and Herodotus, the latter despite the unpalatable fact that she was a foreign enemy of Athens. She was not without her detractors though, par for the course for a successful woman, particularly one who excelled in war: Thessalus, a son of Hippocrates, describes her as:

> a cowardly pirate in a speech in which speech Thessalus alleged that in 493 BCE the King of Persia demanded earth and water from the Coans but when they refused he handed the island over to Artemisia to lay waste. Artemisia led a fleet to Cos to hunt down and slaughter the Coans, but the gods intervened: Artemisia's ships were destroyed by lightning and she hallucinated visions of great heroes and fled Cos, but afterwards conquered the island.

Photius, referring to a work called *New History* (now lost) by Ptolemaeus Chennus cannot resist pedalling the following unedifying story:

> And many others, men and women, suffering from the evil of love, were delivered from their passion in jumping from the top of the rock, such as Artemesia, daughter of Lygdamis, who made war with Persia;

enamoured of Dardarnus of Abydos and scorned, she scratched out his eyes while he slept but as her love increased under the influence of divine anger, she came to Leucade at the instruction of an oracle, threw herself from the top of the rock, killed herself and was buried.[12]

Herodotus repeatedly shows Artemisia stepping outside the traditional boundaries ordained for women and intruding into a male world where, in Greece, she would have no place (7.99, 8.68, 8.88, 8.93). She does this because the Persians at the time were deemed to be lacking *andreia* and a suitable leader. Her demonstrable skills in warfare reduce, by implication, Xerxes to little more than a a weak leader. Artemisia is mentioned by the Old Men in Aristophanes' *Lysistrata* with awe, as something of an Amazon (675).

Artemisia II of Caria

Artemesia II was wife of King Mausolos (r. 387–353 BCE) with whom she ruled in consort. Herodotus described her as 'wondrous'. Decrees and laws were issued in joint names and honours were heaped on them, as an egalitarian regal couple. When Mausolos died, Artemisia ruled on her own for some years maintaining the reputation of the name Artemisia for prowess in battle.

Her expressions of grief for her husband were legendary: she is even reputed to have concocted and drunk a potion comprising her husband's bones and ashes. She organised prestigious poetry and oratory competitions to honour Mausolos and completed the building of his mausoleum which became one of the Seven Wonders of the World, known as the Mausoleum of Halicarnassus. She embarrassed the people of Rhodes when she beat off their attack; the Rhodians found it hard to accept that they had been repelled by a barbarian, and a woman barbarian at that (Demosthenes, 15.23).

According to Vitruvius, the conflict came about when the Rhodians tried to free themselves from the Carians after the death of Mausolos: when they sent a naval expedition to Halicarnassus, Artemisia was one step ahead: she stationed her troops and ships in a discreet harbour and positioned the town's inhabitants on the walls, whence they 'welcomed' the approaching Rhodian ships. The Rhodian soldiers disembarked, Artemesia's ships attacked the Rhodian vessels now empty of troops and executed the crews. Artemisia then commandeered the Rhodian ships and sailed them back to Rhodes, crewed by Carians in disguise; here they were welcomed by the Rhodians who thought that their navy was returning victorious. Atremisia's forces entered the town and executed the leading citizens.

Artemesia II is also an example of the pen being mightier than the sword: in the words of Polyaenus:

Artemisia planted soldiers in ambush near Latmus [in modern western Turkey]; and herself, with a large train of women, eunuchs and musicians, celebrated a sacrifice at the grove of the Mother of the Gods, which was about seven stades distant from the city. When the inhabitants of Latmus came out to see the magnificent procession, the soldiers entered the city and took possession of it. Thus did Artemisia, by flutes and cymbals, possess herself of what she had unsuccessfully tried to obtain by arms.[13]

Telesilla of Argos

The poet Telesilla (fl. 510 BCE) was from Argos; she is just as famous for her military prowess and soldierly activities as she is for her poetry. Apparently, as a young woman, Telesilla was so sickly she went to the Pythia for some medical advice: Pythia told her: 'τὰς Μούσας θερατεύειν', 'serve the Muses', and so began Telesilla's career in poetry.[14]

Pausanias describes the brave military action of this soldier-poet at Argos.[15] After invading Argive lands in 510 BCE, the Battle of Sepeia saw Cleomenes, King of Sparta, defeat and wipe out the hoplites of Argos and massacre all the survivors. So, when Cleomenes attacked Argos, there were no male warriors left to defend it. Telesilla took the initiative when she posted on the city walls all the slaves and all the males normally exempt from military service because they were too young or too old. She collected weapons from sanctuaries and homes, armed the women and drew them up in battle positions. The Spartans tried to terrify the Argive women with their battle cry but the Argives remained unperturbed and fought bravely, standing their ground. The Spartans realised that destroying the women of Argos would be a cheap success, while defeat would mean a shameful and ignominious disaster, so they left the city.

Plutarch (*Mulierum Virtutes*, 4) adds some fascinating detail relating to cross-dressing and bearded ladies. He suggests that the battle took place 'on the anniversary of which they celebrate even to this day the "Festival of Impudence", at which they clothe the women in men's shirts and cloaks, and the men in women's robes and veils'. As for ethnic integration, he tells us about some mischievious legislation:

> To rectify the scarcity of men they did not unite the women with slaves, as Herodotus records, but with the cream of their neighbouring subjects, whom they made Argive citizens. It was reputed that in their marriages the women showed little respect and obvious indifference to their husbands because they felt inferior. The Argives then enacted a law stipulating that married women sporting a beard must occupy the same bed with their husbands!

In another instance of the pen working with the sword, Pausanias concludes with a description of a statue at Argos in front of the Temple of

Aphrodite that is dedicated to Telesilla. It depicts a woman holding a helmet, which she contemplates and is about to place on her head; tellingly, there are books lying at her feet. The festival Hybristica or Endymatia, in which men and women wore each other's clothes, also celebrates the heroism of Telesilla and her female compatriots.

Herodotus only alludes to the incident when he tells us about an oracle, told by a Pythian priestess, which predicted that female should conquer male: 'But when the time shall come that the female conquers in battle, driving away the male, and wins great glory in Argos, then many wives of the Argives shall tear both cheeks in their mourning.'[16] According to Tatian, Telesilla was commemorated by a statue in the Theatre of Pompey in Rome, made by Niceratus.[17]

Hydna of Scione

Also known as Cyana (fl. 480 BCE), Hydna of Scione sounds like an early representation of a Royal Marine commando or a member of what was the Special Boat Squadron. Hydna was an accomplished swimmer and diver and was responsible for the almost single-handed destruction of the Persian fleet.

According to Pausanias (10.19.1), Hydna and her father volunteered to help in the war with the Persians during a critical battle. Hydna was schooled by her father, Scyllis of Scione, an expert diver and an expert swimmer. When Xerxes' fleet was assailed by a violent storm off Mount Pelion, Hydna and Scyllis completed its destruction when they swam 10 miles in stormy waters to where the Persian navy was moored for the night. Knives in hand, they silently swam among the boats, cutting their moorings: rudderless in the wind and waves, the ships crashed together; some sank; most were badly damaged. The Amphictyons dedicated statues to them at Delphi, the most sacred site in the Greek world. This fascinating story is discussed more fully by the United States Naval Institute (*Proceedings*, 68 (1942), p. 662).

Herodotus is more explicit in his telling of the decisive action taken by Athenian women against fellow Athenians; the first instance is in response to the treason advocated by Lycidas when he was disposed, possibly by bribery, to accept a proposal made by the Persian Mardonius to defect to the Persians and renounce Greece. This would be in return for autonomy, land and money for rebuilding wrecked temples. The Athenians were so angered by this preposterous suggestion that they surrounded Lycidas and stoned him to death. The women of the city then took it upon themselves to attack Lycidas' house and then stone his wife and children to death. Protection of the homeland was obviously the motivating factor; the incident demonstrates quite clearly how willing, and able, the patriotic women of Athens were to perpetrate terrible retribution in order to preserve their freedoms and honour in the face of a shameless enemy.[18]

Herodotus gives another instance of Athenian female bellicosity after the doomed expedition to Aegina when the Argives came to help the Aeginetans. Only one Athenian survived and returned safely to Attica. When the wives of the Athenian troops learned this they were outraged: they mobbed the man and stabbed him to death with the brooches from their clothing, each demanding from him to know where her husband was. The Athenian authorities were appalled and perplexed by this and could think of no other way to punish the women than change their dress to the Ionian style. Until then Athenian women had worn Dorian dress but this was now changed to a style that would obviate the need for lethal brooch-pins.

Women Defending and Suffering in Sieges

Diodorus has a similar example of bellicose female fortitude, this time in the face of a siege imposed by the Carthaginian general Himilcon on the people of Gela in Sicily in 405 BCE. The plan was to evacuate the women and children to Syracuse in expectation of the danger faced by the city and its inhabitants. The women, however, had other ideas and insisted they remain and endure the same fate as their menfolk. The valiant women remained in Gela and only went to Syracuse with the men when the city was surrendered. In the meantime, though, they performed a valuable role coordinating the activities of snatch parties preying on the Carthaginians.[19] They were just the sort of women a city needed when under seige; Diodorus continues:

> the Geloans defended themselves boldly; for the bits of the walls which fell during the day they rebuilt at night, with the women and children helping. Those women who were physically the strongest were armed and always in battle, while the rest of the Geloans stood by to attend to the defences and everything else with all eagerness.[20]

Another siege provides yet more evidence of women providing material logistic support. Usually reticent when it comes to reporting women in war, Thucydides tells us how in 429 BCE women were life-savers at the seige of Plataea as executed by the Spartans. The Plataeans had prudently evacuated most of their non-combatant citizens to Athens – 'their wives and children and oldest men and most of the noncombatants' – leaving a small rearguard force of 400 Plataean and 80 Athenian troops – and 110 women 'to bake their bread for them': 'γυναῖκες δὲ δέκα καὶ ἑκατὸω σιτοποιοί'.[21] What this clearly tells us is that female bakers, or cooks generally, were the norm in a garrison and, indeed, in a city under siege even when all the other civilians had left. Moreover, it was not just a case of token staffing: there was one woman catering for every four soldiers and the women accounted for one in five of those left to resist the siege. Two years, and many loaves, later the Plataeans surrendered – the men were executed and the bread-making women enslaved. Such is the importance of bread, and women.

In another rare piece of reportage on women, Thucydides tells us about the brave and resourceful womenfolk of Corcyra, modern Corfu, who made a perfect nuisance of themselves around 427 BCE hurling down roof tiles onto the heads of the besieging enemy. Contemporary Greek men would have it that such behaviour could not be more unseemly for a woman, and for a Greek woman at that. But it demonstrates again that, given the need and opportunity, women were quite able and prepared to fight valiantly for their city.[22] Slaves too joined in, further emphasising the absurdity of the situation and its distance from the social norm. Thucydides does not mince his words: to him it was simply and emphatically *para phusin* – unnatural. Be that as it may, the action of the women was intrumental in the Corcyran victory and Thucydides' blasé description, despite his reservations, suggests that this sort of thing was not exceptional.

Support of a very different kind was provided by the prostitutes who went with the Athenians on the nine-month seige of Samos – there is no reason to believe that, like the breadmaking, this was not usual practice and women formed an integral part of logistic support in all its manifestations.[23] Aptly, the prostitutes set up a temple to Aphrodite there.

Recent excavations have revealed that in the city of Eretria attempts to protect the women and children from sieges were made by building a series of forts – fortified farmhouses essentially and other forms of stronghold – which acted as emergency refuges outside the city. Besieged cities often fell eventually, so it was natural for the authorities to make what provision they could to protect non-combatants; for example Eretria submitted to sieges four times in 400 years: to the Persians in 490 BCE, to the Macedonians in 268 and to the Romans in 198 and in 86 BCE. In Greece generally there were an estimated 100 sieges between 700 BCE and the death of Alexander in 323 BCE: twenty-five ended with the massacre of the defenders with the population enslaved; thirty-four with enslavement only and forty-one with surrender. With hindsight, it is obvious that surrender was the best policy but the decision to renounce one's home, possessions, women and children would have been hideously difficult with an enemy baying for blood at the city gates.

As we have seen, the Geloans made plans to evacuate the women and children when the Persians approched; the Carthaginians planned likewise until thwarted by Archidameia. In 480 BCE the women and children of Delphi were shipped to Achaia; the Athenians evacuated slaves and children, and presumably women, to the safety of Troezen, Aegina and Salamis; in 431 BCE the Plataeans and Athenians, up against Thebes, evacuated non-combatants to Athens; Brasadas moved women and children out to Olynthus fearing the reaction of Athens when Mende and Scione defected; the Agrigentes stayed one step ahead of the Carthaginians by moving women and children first to Gela, then Syracuse and then to the Italian mainland. When Scione fell, the fighting men were executed and the remaining women and children were enslaved; the Athenians took Mende and declared it open for pillaging with no apparent slaughter; they also siezed Torone and reduced the women and children there to slavery. In a surprising reversal of practice, and a rare show of compassion, the defenders were sent to Athens and freed; things reverted to type, however, at Melos where the defenders were executed and the women and children enslaved. In 220 BCE, Lyttos in Crete was taken while its defenders were away on an expedition and its remaining population transported; the defenders returned only to find a ghost town. Worse happened at Abydos when Philip V of Macedon came calling: the defenders, fearing the worst, cut the throats of the women and children and then committed suicide themselves.

Where a surrender was negotiated, women naturally featured in the terms: there are about fifteen known instances of this in the fifth and fourth centuries BCE. Thucydides describes one such case at Potidaea (2.70.3) in an example eerily familiar today, when safe passage was allowed for the defenders, women and children and auxiliaries; the refugees could take one extra item of clothing – except for women who could take two – and a fixed amount of money for the journey. Cities helped other cities: an advantage of cities sending their vulnerable people to places of safety was the welcoming reception of those destinations, as in the case of the Troezenians when they took in the Athenians and the Athenians when they took in the Plataean women and children between 429 and 427 BCE. Individuals too showed generosity: Demosthenes records how Satyros, a comic actor, negotiated with Philip II to secure the release of the imprisoned daughters of Apollophanes of Pydna who were enslaved by the king at Olynthus; he compares this unfavourably with the Athenian ambassadors who sexually assaulted another female prisoner at Olynthus during a banquet. The Elataeans showed their gratitude to the citizens of Stymphalos in a decree for taking them in when the Romans captured their town in 198 BCE. The inhabitants of Aegiale on Amorgos did likewise when Hegesippos and Antipappos rescued girls, women and slaves kidnapped by Cretan pirates (*Sylloge Inscriptionum Graecarum*, 521):

> during the night pirates landed in our territory, and girls and married women and other people, both free and slaves, were captured – a total of more than thirty; the pirates destroyed the ships in the harbour and captured the ship of Dorieus, with which they sailed off carrying away both the hostages and the rest of the booty. Hegesippos and Antipappos, the sons of Hegesistratos, who were among the captives, jointly persuaded the leader of the pirates, Sokleidas, who was sailing with them, to

> release the free people, also some of the freedmen and the slaves, while they offered themselves as hostages, and showed great concern that none of the female or male citizens should be carried off as booty and be sold, nor suffer torture or hardship and that no free person should die. Thanks to these men the captives were saved and returned home without harm.

Timessa, from Arkesine on the same island, also helped save a number of citizens; this is according to an inscription, and is noteworthy because Timessa, unusually for a Greek woman, paid the ransom. Although it was not uncommon for wealthy citizens to pay ransoms on behalf of others, and about a hundred surviving decrees record public gratitude for such actions, Timessa is the only woman known to pay such a ransom (*Hellenistic Inscriptions*, 50).

In 305 BCE when the Rhodians were making hectic preparations for a seige by Demetrius, one of the ways in which the loyalty of the citizens was fostered included the social support for the elderly and the granting of dowries to unmarried daughters at public cost.

Defensive work undertaken by women was not just carried out in a siege situation. In 417 BCE the Argives began a massive construction project, building walls down to the sea in anticipation of a Spartan attack. The entire able-bodied citizenry was mobilised, including women and slaves.[24]

Pausanias leaves us an account of Marpessa (or Choira, Sow), who was 'braver than the rest', in the defence of Tegea against the Spartans under King Charillus in the seventh century. The battle was inconclusive until Marpessa rallied her fellow women to take up arms. Marpessa's contingent made all the difference and the Tegeans won the day, taking many prisoners including the prize catch that was the King of Sparta.[25] The jubilant women marked the occasion with a celebration in honour of Ares in which Marpessa's weapon rather than an effigy of Marpessa was displayed; at the same time the women denied the men any share of the meat from the sacrifices. These women were not just being mean to their men or greedy with their meat: they were reinforcing the role reversal implicit in taking up arms – traditionally men's business – with a further reversal of roles when they sacrificed and ate sacrificial meat – an action which was, in this context, normally the exclusive preserve of men.

Plutarch describes the Phocian women who offered to commit suicide if their army was beaten by the Thessalians. This may have been partly motivated by the very real fear of rape – thousands of German women committed suicide at the end of the Second World War to avoid rape by the advancing Russian armies – it nevertheless also demonstrates extreme loyalty and bravery. Luckily, the Phocian army won the day (Plutarch, *Moralia* 244b–d).

We have already seen how Thucydides, somewhat grudgingly perhaps, describes an episode of targeted tile slinging in his male-focused *History of the*

Peloponnesian War. Thucydides' belligerent Corcyran women were not the only women to pick up and aggressively hurl a roof tile in anger. Four years before this episode in 431 BCE, roof tiles were also the weapon of choice for the Plataeans when attacked by the Thebans. The Plataean women and slaves repaired to the roof tops and orchestrated a fusillade of tiles, setting the scene for an eventual rout of the attackers.[26] Indeed, Diodorus insists that it was the intervention of the women that tipped the balance in favour of the defenders. When the Thebans panicked, they left a number of their men behind who were taken prisoner.

Bombardment by tile, though, was no guarantee of military success for the women slingers: Polyaenus recalls the story of the female inhabitants of an unspecified Acarnanian town who made a rooftop assault on the Aitoleans attacking them. The defence failed and men and Acarnanian men and women died clinging together as one; significantly, the women gave up their resistance once their menfolk had perished or were captured.[27]

The biggest defeat in this situation, though, was suffered by the Selinuntians in 409 BCE at the hands of a 100,000-strong Carthaginian force, equipped with siege equipment, rampaging through Sicily. The whole town of Selinus was mobilised including elderly men, children and women, the latter providing missiles and provisions. But the wall was breached and the subsequent hand-to-hand fighting involved not only the Selinuntian soldiers but the women too; according to Diodorus: 'many gathered to the aid of the defenders'. The Carthaginians retreated and messages were sent out from Selinus to Acragas, Gela and Syracuse requesting relief. The following day saw a repeat with the inhabitants fighting to the last roof tile until they ran out of ammunition. The Carthaginians razed the town killing 16,000 citizens and capturing 5,000; only 2,600 escaped. A shocked Diodorus describes euphemistically how the women suffered 'terrible indignities, and some had to watch their daughters of marriageable age suffering treatment inappropriate for their years'. This may have included emulation of the old Assyrian atrocity of impalement where hands and heads were cut off, the hands tied around the necks of the victims, their heads stuck on swords and spears. That same year, Hannibal Mago took Himera: all the surviving soldiers were executed as a sacrifice to the memory of his grandfather Hamilcar and the women and children shared out among his troops. The city itself was totally destroyed, its buildings, including the temples, were razed to the ground.[28]

Diodorus is important because first and foremost he describes the atrocities endured by women and girls: a life of slavery awaited the women who survived 'under masters who spoke an unintelligible language and were nothing short of bestial'. He also clearly tells how women became embroiled in the close combat when the walls were breached, and adumbrates how essential it was for women to participate in the conflict for the good of the city and its survival: in the chaos, 'the magnitude of the crisis was so great that it called for even the aid of their women'.[29]

A report by Pausanias would suggest that roof-top tile-slinging by women defenders was typical behaviour and roof tiles unleashed by women were raining down on besiegers all the time.[30] Apparently, at the beginning of the Second Messenian War at the start of the seventh century BCE a bad storm prevented the women of Eira from taking to the rooftops for a tile barrage on the attacking Spartans below.

Plutarch reports on the unusual, almost comic, death of Pyrrhus in 272 BCE after he was wounded by an Argive – not a hero of any kind, simply 'the son of a poor old woman' who was viewing the action in a teichoscopy:

> His mother, like the rest of the women, was at this moment watching the battle from the roof, and when she saw that her son was fighting with Pyrrhus she was distressed by the danger he was in, so, picking up a tile with both hands hurled it at Pyrrhus. It hit his head just below his helmet and crushed the vertebrae at the base of his neck, making his sight blurred and his hands drop the reins. Then he sank down from his horse …

Plutarch continues to describe how the old mother's well-aimed tile was followed by a very messy decapitation of the dazed and injured Pyrrhus. Her missile has been called 'the most historically significant roof tile'.[31]

Other instances include the Messenian women who repulsed an army of Macedonians in 214 BC.[32] Women took up arms in 278 BCE when Aetolians fought 40,000 invading Galatians. Pausanias describes how the women were lined up alongside the men and shot missiles at the Galatians with some considerable accuracy; he notes that the women showed greater bravery than men in the action, adding that the enormity of the situation made it necessary for the women to be enlisted to help: earlier the Galatians had captured the Aetolian city of Callion, massacring its inhabitants: 'the fate of the Callions is the most wicked ever heard of, and is without a parallel in the crimes of men'.[33]

Plutarch gives us the pleasing story of the plucky and gutsy women of Chios who were appalled when their menfolk signed a treaty with the hostile and agressive Erythraeans. One of the injurious terms imposed on the Chians was that they were to surrender on oath all of their clothes, apart from one cloak and a tunic – *himation* and *chiton*. To extricate their men from such a humiliation the women, keen to protract hostilities, told them to explain to the Erythraeans that in their dialect cloak and tunic translated as shield and spear. The Chians thus retained their weapons and were able to make a safe escape.[34]

That was not the only act of belligerence demonstrated by the women of Chios. When Philip V was laying siege to Chios in 201 BCE his plan to take the city backfired badly. He promised the slaves two things if they defected to him: first, their liberty (nothing surprising there); second, and much more controversially, the chance to marry their masters' wives – a 'barbarous and insolent' proposition according to Plutarch. The Chian women obviously were of the same mind because they, being 'dreadfully and outrageously incensed', 'rushed forth furiously and climbed the wall, bringing stones and darts, encouraging and chivvying the soldiers; so that in the end these women unnerved and repulsed the enemy, and caused Philip to raise his siege, while not so much as one slave went over to him'.[35]

Polyaenus has the story of the women of Cyrene in North Africa assisting their men against Ptolemy VIII in about 163 BCE by building fortifications, making munitions, preparing food and tending the injured.

In about 379 BCE the women of Sinope, a city on the Black Sea, were particularly inventive: when attacked by the Persian Datames, they dressed up in imitation armour – actually bronze pots and pans – to give the impression that the defending forces were more numerous than they really were. Aeneas Tacticus adds the footnote that the women were not permitted to throw anything because you can spot a woman by the way she throws a mile off! He also tells that Pisistratus had women board his Athenian ships to fool the Megarians into believing that they were up against a much stronger force.[36]

A reference in Plutarch's *Philopoemen* to women engaged in the skilled work of fitting crests to helmets might suggest that women were involved in the armaments industry, even in times of peace.[37] This may well be supported by an Athenian epigraph which tells of a certain Artemis who was married to Dianysios, a helmet maker – it is quite conceivable that she helped him in this, making plumes, for example.[38]

Women as Camp Followers

We have already noted the activities of women as camp followers, an essential component of the army baggage train. Perhaps the most famous example of this is given by Xenophon in his *Anabasis*: the 'March of the 10,000' when in 401 BCE the baggage train supported by Greek mercenaries hired by Cyrus the Younger in his bid to take the Persian throne certainly included women.[39] After the Battle of Cunaxa when the Persians reached the Black Sea the baggage train was deemed too cumbersome – literally too much 'baggage' – so they put everyone age 40 or over, the sick, injured, children and women and actual baggage onto boats while the younger people marched along the coast. However, we do know that some women remained with the troops: there was at least one slave dancer still in the company of a (lucky) Arcadian when the army reached Paphlagonia.[40] There were attempts to conceal women and/or boys which were obviously successful because women provided life-saving encouragement to floundering soldiers during a perilous river crossing. They were also busied preparing food for the troops.

Baggage trains were, of course, valued highly by the troops, for many different reasons: so important were they that victorious soldiers were known to defect to the defeated army if the baggage train was captured by them and herded away behind enemy lines. This confirms that, particularly among mercenaries, when their wives, concubines or mistresses were in those trains, the bond of affection between them was greater than loyalty to one army or another.

Baggage trains, however, were often just that, and the women therein were sometimes considered little more than excess baggage. Philip II lightened the load and increased mobility by banning women; Alexander followed suite – initially, but relented as his army advanced farther from Greece and he allowed his troops to take captured women along with them. An unfortunate incident occurred when Alexander encamped in a wadi in the Gedrosian desert: the soldiers were on higher ground and survived a flash flood: the women, children and livestock situated on the river bank were swept to their deaths. Some historians suggest that this was a deliberate ploy on the part of Alexander to unencumber his army.

It was inevitable, perhaps, that camp followers and baggage trains assumed strategic importance. Populated as they were with women, attacking a baggage train was an easy and effective way of attacking its army; it was a soft target. An example of this involved Eumenos, former treasurer of Alexander the Great, who came up against Antigonos Monophthalmos in the maelstrom following Alexander's death. At Gabien, Antigonos saw the strategic potential of the women and their families in the baggage train and attacked it. Although he lost something like 5,000 men in the battle (and Eumenos only 300), Antigonos won the day when Eumenos' men defected on hearing news of the baggage train attack: their commander turned Eumenos over to Antigonos who had him executed.[41]

Diodorus Siculus tells the remarkable story of how Alexander, during his Indian campaign, attacked a contingent of mercenaries in mid-mutiny. The mutineers formed a defensive circle, enclosing their wives and children in the centre. When the soldiers in the perimeter rank fell, the baggage train women and children bravely stepped up to take their places, picking up their weapons and aggressively retaliating against the continuing brutal attack (17.84.4–6):

> They fought hand to hand ... every type of death and injury could be seen. The Macedonians thrust with their long spears through the light shields of the mercenaries and pressed the iron tips right into their lungs, while they in turn flung their javelins into the close ranks of their enemies – unable to miss the mark, so near was the target. As many were wounded and not a few killed, the women picked up the weapons of the fallen and fought beside their men, since the acuteness of the danger and

the ferocity of the action forced them to be unnaturally brave. Some of them, clad in armour, sheltered behind the same shields as their husbands, while others rushed in without armour, grasped the enemies' shields, and hindered their use by the enemy. Finally, fighting women and all, they were overcome by sheer weight of numbers and cut down, winning a glorious death in preference to shamefully saving their lives at any cost. Alexander moved the infirm and unarmed together with the surviving women to another place, and put the cavalry in charge of them.

Women Against Alexander the Great

At Multan (in modern day Punjab in Pakistan), an injury sustained by Alexander so angered the Macedonians that they slaughtered all the inhabitants including women and children (Arrian 6.9.5–6.11.1). During the seige of Massaga (in Pakistan's Swat Valley) the Indians reneged on a ceasefire agreement with Alexander who promptly took the undefended city and captured the mother and daughter of the King of Massaga (Arrian 4.26.4–4.27.4). Diodorus (17.84) recounts how the Indians fought ferociously – not just the men but the women too.

The outcome at the fall of Persepolis in 330 BCE was particularly repellent. The Macedonian troops were crazed with greed and lust to the extent that Macedonian soldier cut off the hands of Macedonian soldier in the frenzied scramble for loot. Women were stripped of their finery and jewels and paraded naked towards a future life of slavery; such was the devastating effect on the Persians that entire families committed suicide by leaping from walls or torching their own houses to avoid such a repugnant indignity. Diodorus (17.70) and Quintus Curtius (5.6.1–8) agree that Alexander eventually called an end to the rapine.

The siege of Thebes (292–291 BCE) was led by King Demetrius I of Macedon after it had revolted against Macedonians. The population, including women and children, was butchered after Alexander cruelly decided to allow their fate to be decided by his allies who had suffered at the hands of the Thebans up to 150 years earlier. Ancient vengeance won the day when 6,000 Thebans died and 30,000 were sold into slavery (Diodorus 17.14.2–3; Arrian 1.9.9).

What the historians are unequivocally telling us, then, is that women often suffered incalculably. Their lives and the lives of their families were utterly wrecked. At the same time, perhaps because they may have been mindful of this shocking eventuality, women were able to provide, and were presumably sometimes required to provide, a range of logistical support services to the men fighting in and for their cities. They executed this either as camp followers or as an integral part of a mobilised citizenry. Moreover, women on occasions, if the situation demanded, were not averse to taking the initiative and taking things into their own hands by going on the offensive themselves.

There was one form of support that Alexander could have done without during his frontier wars with the Sogdians – the longest and most bloody campaign Alexander ever fought. All seven of the old Persian cities on the Jaxartes border were laid waste and the surviving populations enslaved to the Greek and Macedonian settlers of Alexandria-Eschate. Thousands of Alexander's men died; tens of thousands of the natives were slaughtered or captured as slaves. In 328 BCE, the chief of the Sogdians, Spitamenes, was double-crossed and beheaded by his own wife – intervening in the progress of the war in no small way. When she brought the head of her husband to Alexander's tent, the king sent her packing lest she offend his 'mild-mannered' troops with her aggressive barbarism.

The Hellenistic era saw a boom in the establishment of the garrison as a convenient and all-round economical alternative to direct, expensive occupation – expensive both in terms of manpower and resources. It is reasonable to expect that, as in the Roman Empire later, mixed marriages took place in these garrisons, particularly where the length of service and the postings were extensive. A soldier could leave his wife or partner at home, but he could not leave behind his natural sexual needs and desires. Inscriptional evidence supports this in Egypt; in Crete, a fertile source of mercenaries, the Cretans took their women to war with them: wives and daughters as well as sons can be attested in Attica, Euboia and Thessaly. Women from Aspendos, Euboia, Crete, Arabia and Byzantion were to be found in the Pergamon garrisons in Cypriote cities: all of the above, with the exception of Arabia, were popular recruiting grounds. A late third-century BCE funeral inscription from Palestine tells us about Charmadas, who hailed from Anopolis on Crete: he served in the Ptolemaic army in Syria where his daughter married Machaios, a comrade from Aitolia (*SEG*, 8.269). Somewhere between 146 and 116 BCE, in another Ptolemaic garrison – in Kition in Cyprus, Aristo, daughter of high-ranking Dion from Crete, married equally high-ranking Melankomos, also from Aitolia.

Philosophy, Women and War

In the late 360s BCE Xenophon set the philosophic agenda when he wrote in his dialogue, *Oeconomicus*, that women's place was in the home dealing with 'indoor tasks', while men were more hardy and could endure outdoor activities such as journeys and military campaigns.[42] This is, of course, a reiteration, in the words of Ischomachus, of the social norm in which women were largely excluded from public life and, therefore, ineligible for military activity. This is the same Ischomachus who is looking for a clean slate in a wife, a woman with nothing on her mind or in her brain. She is the type of woman who crops up later in the jurist Gaius' *Institutiones* (fl. 130–80 CE) who confirmed that, on reaching puberty, Roman boys relinquished their guardians, but not girls – the reason being *propter animi levitatem* – girls were considered to be what today some would disparagingly call 'airheads'. Ischomachus is happy for his wife to be completely clueless so that he can mould and shape her as he sees fit, impressing on her only the characteristics and knowledge he wants. This is one of Ischomachus' many rules for training and controlling the good wife – the secret of which is having her run the household effectively and efficiently – to his specific requirements – because that is what the gods, and he, decree. Ischomachus' view is no more than an early expression of male attitudes to the role of women in Greek society – a role that cannot accommodate any direct position for women in military life, or indeed the overt military activity we have seen.

Plato is ambivalent on the subject on women and war. In *Laws* he asserts that women take no part in wars, adding that they are unlikely ever to cause the enemy any trouble in the event that they are ever called up to fight.[43]

On the other hand, from a theoretical and utopian standpoint, women seem to come out well as potential warriors. In the *Republic*, written in about 380 BCE, Plato had stated quite categorically that women, in keeping with their eligibility to be guardians of the ideal state, should be taught music and gymnastics and also the art of war, which they must practise like men – wearing armour and riding on horseback.[44] For Plato, women had a part to play in the defence of the ideal state. Moreover, women are to have an indirect but nevertheless very important military role: the brave man is to have more wives than other men in order that he may have as many children as possible. Women are to help in the production of more brave warriors; they are required to provide childcare for soldiers' children and accompany them on battlefield visits to learn about war from an early age.

Again in the *Laws*, Plato modifies this when he advocates that women play a defensive, reservist role defending the state during a siege with sorties outside the walls while male soldiers are away fighting.[45] Plato has Socrates recommend military and athletic training for women with exercises in drill and the bearing of arms. Deprived of this valuable military contribution, the state would be all the weaker:

> Even if it should happen to be necessary for them to fight in defence of their city and their children, they will be unable to handle with skill either a bow [like the Amazons] or any other missile, nor could they take spear and shield, after the fashion of the Goddess [Athena], so as to be able nobly to resist the devastation of their native land, and to strike terror – if nothing else – into the enemy at the sight of them marshalled in battle-array?

For Plato, women had a valuable role to play in the military organistion of the ideal state: their contribution to the defence of the state, while limited according to their physical strength, was vital.

Aristotle believed that courage had value during wartime, citing how useless the Spartan women were during the Theban invasion when they caused more havoc than the enemy with their behaviour. Aristotle says that women in other cities did exhibit bravery; was this just another way of discrediting the Spartans for whom he had no time?[46]

Writing some 100 years later, Philon of Byzantium (*c.*280–*c.*220 BCE), also known as Philo Mechanicus, the Greek engineer and writer on mechanics, was, as his name suggests, an authority on mechanical engineering, including its application to warfare. His *Mechanike Syntaxis* or *Compendium of Mechanics*, comprises nine parts, three of which deal specifically with military matters. These are *Belopoeica* (βελοποιικά) – on artillery; *Parasceuastica* (παρασκευαστικά) – preparation for sieges; and *Poliorcetica* (πολιορκητικά) – on siegecraft. In among the information relating to deployment of war machines – such as catapults and other war engines, starving the inhabitants of the besieged town, bribing suitable people to assist, using poisonous recipes to kill the inhabitants and employing cryptography to pass secret messages – there is interesting comment on the role of women in war, particularly their usefulness in defending sieges.

To Philon, the deployment of women was essential: 'children, female slaves, women and girls'.[47] Women had officially become part of the military, especially in their role as defenders of cities; their participation in combat and warfare was now enshrined in influential works of military strategy and tactics.

We have seen how war in ancient Greece was highly gendered: the Greek man's ability to wage war for his *polis* was a requirement for citizenship, and citizenship was everything. Fighting for the *polis* was a man's badge of engagement with the local community. Although women had little or no public role or profile in Greek society, they did play an active part in war and military life from feeding their troops to fighting the enemy, from hurling missiles to manufacturing armour. When and where the occasion was required, women were ready and able to (wo)man the parapets, and help defend the *polis* for which their husbands, brother and fathers were also risking their lives.

Chapter 8

Women Warriors Catalogued

It is a short step from Philon's military engineering textbooks to stand-alone reference books or lists embedded in literary works: this is the catalogue. Catalogues and handbooks of one kind or another were common in the Graeco-Roman world, perhaps reflecting a desire for orderliness and categorisation. 'Types' of men and women feature prominently and include various war-related catalogues of armies in the *Iliad* and the *Aeneid*; chthonic catalogues of women in the *Odyssey* and the *Aeneid*; and Neanthes of Cyzicus' *About Illustrious Men* in the third century BCE. Charon of Carthage compiled two collections four books' long, listing illustrious men and women in short biographies and anecdotes. Photius (codex 161) tells us about the fourth-century CE sophist Sopatros and his twelve books, one of which extracts brave exploits of women by Artemon of Magnesia and another describes 'women who achieved a distinguished name and great glory'. Theophrastus (frags 625–7) gives us a list of women who caused wars or destroyed houses. Of less interest here are the various compilations of courtesans: Athenaeus (583d–e) notes lists by Aristophanes of Byzantium, Apollodorus and Gorgias of Athens while Suetonius is supposed to have catalogued prostitutes; those mentioned by Photius include lists of female dancers and actresses, flute players and prophetesses. Athenaeus has his own list of courtesans in his *Deipnosophistae* (571e ff.). Nevertheless, they all probably do show that cataloguing, including the cataloguing of warlike women, was good publishing business and many more were compiled than have survived.

By the end of the fifth century CE there was also no shortage of advice on how to prosecute a war or win a battle. Of the extant works of military strategy and warfare there was Aeneas Tacticus (fl. fourth century BCE), the *Poliokretika* or *How to Survive under Siege*; Polybius (*c.*200–*c.*118 BCE), 19–42 of his *Histories* on military matters, especially camps, and author of the lost *Tactica*; Julius Caesar's *Bellum Gallicum* and *Bellum Civile* from the first century BCE; Onasander (fl. first century CE), a Greek philosopher and author of *Strategikos*, one of the most important treatises on ancient military matters with information not available in other works on Greek military tactics, especially concerning the use of light infantry in battle; Frontinus (late first century CE): *Stratagemata* – a compendium of over 500 examples of military devices and ploys, intended by Frontinus as a sort of *vade mecum* (practical handbook) for the military commander – the work is an appendix to his the *Art of War*, which has not survived; Vegetius: *De Re Militari* (late fourth century CE) which covers training of soldiers, strategy, maintenance of supply lines and logistics, leadership and tactics including deception; Zosimus' *Historia Nova* (fl. 490s–510s); Polyaenus' *Stratagemata*, divided into eight books and written in the second half of the second century CE: the first six contain the stratagems of the most celebrated Greek generals, the seventh of those of foreign militarists and the eighth of the Romans and illustrious women. It is this last section which interests us along with Plutarch's *Mulierum Virtutes*, *On the Bravery of Women*, compiled some 150 years earlier, and the anonymous, undatable *Tractatus de Mulieribus*, an obscure Greek work describing fourteen mainly valiant Greek and barbarian women.

Aeneas Tacticus is, of course, famous for telling us about the deception that involved arming women with pots and pans to make it seem to the enemy that they were additional defending troops. He also tells how Dionysius got over the problem of garrisoning a city by leaving behind a few men and marrying off some of the slaves to the daughters, wives and sisters of their masters so as to make them hostile to their masters and boost their loyalty to himself.[1]

Frontinus describes another deception involving women in 179 BCE, or at least the illusion of women:

> When the Voccaei were hard pressed by Sempronius Gracchus in a pitched battle, they surrounded their entire force with a ring of carts, which they had filled with their bravest warriors dressed in women's clothes. Sempronius rose up with greater daring to assault the enemy, because he thought he was fighting against women, when those [men] in the carts attacked him and put him to flight.[2]

Plutarch's *De Mulierum Virtutibus*

In Plutarch's *De Mulierum Virtutibus*, a section of his *Moralia*, there are fifteen ethnic groups of women from various parts of Greece, and twelve individuals.[3] Plutarch starts his treatise by distancing himself from Thucydides when he asserts unequivocally:

> Regarding the virtues of women, Clea, I do not hold the same opinion as Thucydides. He says that the best woman is she about whom there is the least talk among persons outside be it in censure or commendation; he feels that the name of a good woman, like her person, ought to be shut up indoors and never let out. But I believe that Gorgias appears to display better taste when he says that not the form but the fame of a woman should be known to many. Best of all seems the Roman custom, which publicly renders to women, as to men, a fitting commemoration after the end of their life.

He also tells of his intention to spare us any repetition of well-known stories that have been frequently trotted out down the years (*Moralia*, 243d).[4] He wants to keep his work fresh and original while at the same time reminding us that there were other, similar works already out there in the market.

The following are the women, groups and individuals from Plutarch not already discussed.

The Trojan Women

From *De Mulierum Virtutibus*, 1:

> Most of those that escaped from Troy ... because of their inexperience in navigation and ignorance of the sea, were driven onto the shores of Italy ... While the men were wandering about the country gathering local intelligence, it suddenly occurred to the women that for a happy and successful people any sort of a settled habitation on land is better than all that wandering and voyaging, and that the Trojans must create a homeland, since they could not recover what they had lost. So, altogether they burned the ships, with one woman, Roma, taking the lead. Having done this, they went to meet the men who were racing down to the sea to save the ships; worried that the men may be angry, some embraced their husbands and some their relatives, and kissed them coaxingly, and softened them up. This is the origin of the custom, which still persists among Roman women, of greeting their kinsfolk with a kiss. The Trojans, realizing that this was inevitable, and after having met some of the local inhabitants, who received them kindly and generously, came to be happy with what had been done by the women, and took up their home there with the Latins.

So, in one fell sweep, Trojan women asserted a dominant role in the establishment of Rome, their new homeland, and introduced kissing as a social greeting.[5]

The Persian Women

These plucky women lifted up their clothes and made heroes out of their formerly cowardly warriors (*De Mulierum Virtutibus*, 5):[6]

> When Cyrus incited the Persians to revolt from king Astyages and the Medes he was defeated in battle. The Persians fled to the city, with the enemy not far from forcing their way in with them, so the women ran out to meet them, and, lifting up their tunics, said, 'Where are you rushing to so fast, you who are biggest cowards in the whole world? Surely you cannot, in your retreat, slink in here from out there.' The Persians, mortified at the sight and the words, chided themselves for being cowards, rallied and, engaging the enemy once more, put them to rout.

> As a result of this, it became an established custom that, whenever the king rode into the city, each woman would receive a gold coin; the author of the law was Cyrus but Ochus they say, being a mean man and the greediest of the kings, would always make a detour round the city and not go in, so as to deprive the women of their gift.[7] Alexander, however, entered the city twice, and gave all the women who were pregnant a double amount.[8]

The Celtic Women

Celtic women exhibit marvellous skills of diplomacy, thus preventing internecine bloodshed (*De Mulierum Virtutibus*, 6):[9]

> Before the Celts crossed over the Alps and settled in that part of Italy where they now live, a dire and persistent factional discord broke out among them which went on and on to the point of civil war. The women, however, put themselves between the armies, and, taking up the issues, arbitrated and decided them with such irreproachable fairness that a wonderful friendship of all towards all came about between both states and families. As a result of this the Celts continued to consult with the women regarding war and peace, and to decide through them any disputed matters with regard to their allies. At all events, in their treaty with Hannibal they wrote the provision that, if the Celts complained against the Carthaginians, the governors and generals of the Carthaginians in Spain should be the judges; and if the Carthaginians complained against the Celts, the judges should be the Celtic women.

The Etruscan Women

More inspired cross-dressing, by brave and inventive Etruscan women (*De Mulierum Virtutibus*, 8):

> When the Etruscans took Lemnos and Imbros, they forcibly abducted Athenian women, to whom children were born. The Athenians expelled these children from the islands on the grounds that they were half-barbarian; nevertheless, the refugees put in at Taenarum and made themselves useful to the Spartans in the war with the Helots. For this they received citizenship and the right of intermarriage, but were not deemed worthy to hold office or to be members of the Senate, lending credence to the idea that they had some radical plan, and that they intended to undermine the established institutions. Accordingly, the Spartans locked them up in prison, placed a strong guard over them, with a view to convict them on good evidence. The wives of the prisoners came to the jail, and after many prayers and entreaties, were permitted by the guards to go just close enough to greet and to speak to their

> husbands. When they had gone inside the women made their husbands quickly change their clothes, leaving theirs for their wives; then, putting on their wives' clothes they walked out with their faces covered. This done, the women waited there, ready to face all terrors, but the guards were deceived and allowed the men to pass, believing, of course, that they were women.

They then seized the strongholds on Mount Taÿgetus, incited the body of Helots to revolt, and gladly received them as an addition to their forces. The Spartans were terrified and, sending heralds, made peace with them, the conditions being that they should get back their wives, receive money and ships, sail away and, having found land and a city elsewhere, be considered as colonists and relatives of the Spartans.

The Women of Salmantica

Another instance of brave female ingenuity and initiative (*De Mulierum Virtutibus*, 10):

> When Hannibal, the son of Barca, before making his campaign against the Romans about 220 BCE, attacked the great city in Spain that is Salmantica, the besieged were at first terrified, and agreed to do what was ordered by giving him 6,000 pounds and 300 hostages. But when Hannibal raised the siege, they changed their minds and did nothing of what they had agreed. So he returned and ordered his soldiers, with the promise of plunder, to attack the city. At this the barbarians were panic-stricken, and came to terms, agreeing that the free inhabitants should depart clad in one civilian garment, and should leave behind weapons, property, slaves, and their city. The women, thinking that the enemy would search each man as he came out, but would not touch the women, took up swords, and, hiding them, hurried out with the men. When they had all come out, Hannibal set a guard of Masaesylian soldiers over them ... the rest of the soldiers rushed into the city in disorder and set about pillaging. When they saw how much booty was being carried off, the Masaesylians could not bear to be mere spectators, nor did they concentrate on the guard, but became annoyed and started to move to take their share of the spoils. At this point the women, calling upon the men, handed them the swords; some of the women themselves attacked their guards ... the men struck down some, routed the rest, and forced a way out together accompanied by the women. Hannibal, learning of this, sent men to pursue them, and caught those who could not keep up. The others reached the mountains, and, for the time being, escaped. Afterwards, however, they sent a petition to him: they were allowed back to their city, and received immunity and humane treatment.[10]

Chiomara
From *De Mulierum Virtutibus*, 22:

> Chiomara, the wife of Ortiagon, was taken prisoner-of-war along with the rest of the women at the time when the Romans under Gnaeus Manlius Vulso in 189 BCE subdued the Galatians. The officer who took possession of her used his luck, as soldiers do, and raped her. He was, naturally, an ignorant man with no self-control when it came to either pleasure or money and fell victim to his love of money. When a very large sum in gold had been mutually agreed upon as the price for the woman, he brought her, to exchange for the ransom, to a place where a river formed a boundary. When the Galatians had crossed and given him the money and received Chiomara, she, by a nod, indicated to one Galatian soldier that he should strike the Roman as he was affectionately taking leave of her. When the man obediently sliced off the Roman's head, she picked it up and, wrapping it in the folds of her tunic, departed. When she reached her husband she threw the head down in front of him, at which he said in amazement, 'Fidelity is a noble thing, dear wife.' 'Yes,' She said, 'but it is a nobler thing that only one man can be alive who has been intimate with me.[11]

* * *

Plutarch's women exhibit *virtus* in every sense of the word: bravery, virtue, manliness, strength of body and mind, excellence, goodness, power, ingenuity and military talent. They are indeed well named by Plutarch: their military talent extends to tactical excellence, clever deception and other elements of psychological warfare, leading from the front, setting a good example, ruthlessness, compassion and modesty. We can identify the same laudable traits in the women described by Polyaenus.

Polyaenus' *Stratagemata*[12]

As with Plutarch, a number of Polyaenus' forty-five women are dealt with in other parts of this book where more appropriate. Polyaenus covers much of the ground and characters we have met in Plutarch with the addition of more women, and some extra detail. Five of Polyaenus' women feature in the *Tractatus de Muliebris*: Semiramis, Rhodogune, Tomyris, Pheretime and Artemisia, but, interestingly, only Polyaenus' Rhodogune and Artemisia bear any similarity to the account given in the *Tractatus*. Some of the entries are biographies while others describe a particular stratagem. Here are the women who are not in Plutarch: what they demonstrate is a wide range of involvement by women in military affairs from right across the classical period: these women between them exhibit bravery, self-sacrifice, patriotism and diplomacy.

Semiramis

We have already met the resourceful Semiramis (*c.*811–806 BCE) but this inscriptional information is of additional interest; it reads like an extract from her curriculum vitae – a complete military all-rounder (*Stratagemata*, 8.26):

> Semiramis received intelligence of the revolt of the Siraces while she was in her bath; and without waiting to have her sandals put on or her hair done, she got out immediately and took to the field. Her exploits are recorded on pillars, in these words: 'Nature made me a woman, but I have raised myself to rival the greatest of men. I swayed the sceptre of Ninus; and extended my dominions to the river Hinamames on the east; on the south, to the country which is fragrant with the production of frankincense and myrrh; and northward to the Saccae and Sogdians. No Assyrian before me ever saw the sea; but distant as the seas are from here, I have seen four. And who can set bounds to their proud waves? I have directed the course of rivers at my will; and my will has directed them where they might prove useful. I have made a barren land produce plenty, and fertilised it with my rivers. I have built walls which are impregnable; and with iron forced a way through inaccessible rocks. At great expense I have made roads in places which before not even the wild beasts could cross. And great and various as my exploits have been, I have always found leisure time in which to indulge myself and my friends.

Rhodugune of Parthia

Rhodogune was the daughter of Mithridates I (171–138 BCE), and sister of Phraates II (r. 138–127 BCE). Like Semiramis, she was not going to miss a fight by something so mundane as a bath (*Stratagemata*, 8.27):

> Rhodugune was just getting out of her bath, with her hair still undone when she received news of the revolt of a subject nation. Without waiting to do her hair, she got on her horse, and positioned herself at the head of her army. At the same time, she vowed not to have her hair done until she had subdued the rebels; this she eventually achieved after a tedious war. She then bathed, and had her hair dressed. From this incident, the seal of the kings of Persia bears on it Rhodogune with dishevelled hair.

Nitetis

Nitetis wins back Egypt with some determined action (*Stratagemata*, 8.29):

> Cyrus king of Persia asked Amasis king of Egypt (r. 570–526 BCE) for his daughter in marriage. But instead of sending his own daughter, Amasis sent him Nitetis, the daughter of king Apries, whom he had murdered,

and whose throne he had taken. Nitetis passed for the daughter of Amasis for many years while she lived with Cyrus. But after giving him children, and making herself mistress of his affections, she informed him who she really was: that her father was Apries, the king and master of Amasis. 'And now,' she said, 'since Amasis is dead, it will be a generous act to revenge the injury of my family on Psammetichus his son.' Cyrus consented but he died before the attack could take place. However his son Cambyses was urged by his mother to do the deed which he accomplished successfully, and transferred the crown of Egypt once more into the hands of the family of Apries.

Philotis

Philotis proves to be an adept military schemer to beat the Fidenites (*Stratagemata*, 8.30):

> The Fidenites, in the fourth century BCE, under the command of Postumius, began a war against the Romans who were weakened by the recent sack of Rome by the Gauls; at the same time they offered to form an alliance with them, a condition of which was that the Romans give them their daughters in marriage; this would cement relations between the two nations, as they had themselves done with the Sabines. The Romans were then in no position to wage a war, and yet were unwilling to part with their daughters. Philotis, or Tutela, a young and attractive slave, proposed that they dress her up, and other good-looking slaves, and send them to the Latins in place of their daughters; she would let the Romans know by lighting a torch when in the night the Latins went to bed. So, as soon as they had retired with their new brides, Philotis lit the torch and the Romans surprised the Latins in bed, and killed them. Plutarch adds that the women were rewarded with freedom and a dowry at public expense.[13]

Cheilonis

Cheilonis also uses cross-dressing to effect the release of her husband (*Stratagemata*, 8.34):

> When Cheilonis, the daughter of Cleadas, and wife of Theopompus, learnt that her husband was taken as prisoner-of-war by the Arcadians, she went to Arcadia to visit him. The Arcadians, seeing the affection she had shown, allowed her to visit him in his cell; once there she changed clothes with him, and so he made his escape, while she remained in prison. Before long Theopompus had the chance to seize a priestess of Artemis while she was celebrating in a procession at Pheneus; the inhabitants of Tegea released Cheilonis in exchange for the priestess.

Pieria

A clever Pieria uses some effective guile and wins a lasting peace for her country (*Stratagemata*, 8.35; see also Plutarch, *Moralia*, 253):

> After they had revolted against the house of Neleus, a large number of the Ionians who inhabited Miletus separated and established themselves at Myus where they lived in a state of hostility with their former countrymen, though not quite in an actual state of war. Occasionally they used to meet each other at festivals, and on other public occasions. At a solemn festival called Neleis, Pieria, the daughter of Pythus a man of distinction, went to Miletus. Phrygius, a descendant of Neleus, met her there; he was smitten and asked her how he could most agreeably serve her. 'By giving me a chance,' replied Pieria, 'of coming here often, and with as many friends as I please.' Phrygius got her meaning; he effected a permanent peace, and a re-establishment of the union of the two states. The love of Phrygius and Pieria became famous ever after in the annals of Milesian history or, in the words of Plutarch: 'insomuch that the Milesian women do to this day make use of this benediction to new married wives, that their husbands may love them so as Phrygius loved Pieria'.

Polycrete

Polycrete uses shrewd subterfuge to rid her county of the Milesians (*Stratagemata*, 8.36):

> The Milesians, with the help of the Erythraeans, made war on the Naxians; Diognetus, the Milesian general, ravaged their country, and brought away considerable booty, including a number of women, among whom was Polycrete. Diognetus fell in love with Polycrete who lived with him not as a slave, but as his wife. In the Milesian camp a local festival was celebrated, at which the Milesians give themselves up to drinking and pleasure. Polycrete requested Diognetus' permission to send a small present of the sumptuous fare that was prepared to her brothers back home; so, she moulded up a piece of lead in a cake, and ordered the messenger to tell her brothers that it was intended only for their use. On the lead she inscribed a message to the effect that if they attacked the Milesian camp, they might surprise the enemy while they were drunk and sleeping. The Naxian generals accordingly attacked, and were victorious. Polycrete was highly honoured by her citizens for her service; and at her insistence they allowed Diognetus to continue his reign and keep his possessions.

However, Plutarch has a tragic postscript to Polycrete's story (*Moralia*, 254):

> She was met by her countrymen at the gate, who received her with shouts of joy and garlands, and applauded her action. But she could not bear the

extreme joy, but died, falling down at the gate of the citadel, where she was buried. It is now called the Sepulchre of Envy, as though some envious fortune had begrudged Polycrete the chance to enjoy so great an honour.

Lampsace

Lampsace too proves to be an effective spy in delivering her country from the barbarians (*Stratagemata*, 8.37):

> In 654 BCE the Phocaeans under the command of Phoxus marched to the assistance of Mandron, king of the Bebryces, who had been attacked by the neighbouring barbarians. As a reward for their service, Mandron granted to the Phocaeans a part of the country, and city, and invited them to settle there. By their courage and actions they had won many victories, and had enriched themselves with great spoils; this attracted the envy of the barbarians, so that in the absence of Mandron, the barbarians conspired to massacre them. But Lampsace, the daughter of Mandron, got wind of the plot, and though she could not prevent it, she secretly revealed it to the Greeks. The Phocaeans prepared a magnificent sacrifice in the suburbs, and invited the barbarians to come to it. They then divided themselves into two groups, one of which secured the walls while the other slew the banqueters, and took control of the city. They afterwards honoured Lampsace, and named the city Lampsacus after her. Lampsace died soon after her heroic act.

Plutarch adds (*Moralia*, 255):

> The citizens gave her a magnificent burial and went on to worship her as a heroine, renaming Pityussa to Lampsacus in her honour; later on a vote was held to promote her to the status of a goddess.

Aretaphila of Cyrene

Aretaphila proves a wise politician and skilfull negotiator in the removal of the tyrant Nicocrates (*Stratagemata*, 8.38):

> Nicocrates, tyrant of Cyrene around 50 BCE, among a number of other oppressive and atrocious acts, slew Melanippus, priest of Apollo, with his own hands and married Aretaphila, Melanippus' wife, a woman of exquisite beauty. She tried by poison, and various other methods, to take revenge on the tyrant on behalf of her distressed country, and her husband's death of which she was accused, and brought to trial. But, despite the tortures which she endured, she confessed nothing, except that she had administered to him a love potion, in order to win his affections. She was finally acquitted by the tyrant's order; and deciding that she had suffered for no reason, he afterwards treated her with great attention

and affection. Aretaphila had a daughter, who was extremely beautiful, and she introduced her to Leander, the tyrant's brother. He fell in love with her, and with the consent of Nicocrates married her. Leander was to be won over by the frequent remonstrations of his mother-in-law, and resolved to free his country by killing the tyrant, his brother, which he managed to achieve after much difficulty, and with the assistance of Daphnis, the servant of his chamber.

Plutarch adds that (*Mulierum Virtutes*, 19):

> 'Aretaphila was not deficient in political wisdom' and that 'the piteous and undeserved suffering of her country distressed her the more; for one citizen after another was slaughtered, and there was no hope of vengeance from any quarter'. Unfortunately, Leander removed his brother but not the tyranny because he ruled in much the same way. In the background Aretaphila secretly stirred up a war with an African potentate Anabus to overrun the country and lead his army against the city; then she accused Leander and his generals of not being zealous in their prosecution of the war preferring peace and quiet to secure his power over the citizens. She offered to broker a peace and would get Anabus to come to meet him. She had a chat with the African beforehand, in which she asked him, on the promise of many gifts and much money, to seize Leander when he came to meet him. Finally she bravely and determinedly dragged him over to the African. Instantly he was seized and made prisoner, and, after being clapped in irons, was guarded by the Africans until Aretaphila's friends arrived with the money. Calbia, Nicocrates' mother, they burned alive, and Leander they sewed up in a leather sack and dumped in the sea. They asked that Aretaphila should share in the new government, but she, seeing the city free, withdrew from public life and spent the rest of her days quietly at the loom in the company of her friends and family.

Camma

Camma cleverly avenges her husband's murder and takes her own life (*Stratagemata*, 8.38; see also Plutarch, *Moralia*, 257):

> Sinorix and Sinatus possessed tetrarchies in Galatia. Camma, the wife of Sinatus, was esteemed as virtuous, and beautiful; she was priestess of Artemis, the highest rank of office that a woman can hold in Galatia. Sinorix, however, fell in love with her, despairing that he would never be gratified either by force or entreaties, while Sinatus was still alive. He, therefore, had Sinatus assassinated and soon afterwards approached Camma, who repeatedly rejected his advances. At last however, yielding to the persistence of her friends and acquaintances, she pretended

> to consent, but on these terms: 'Let Sinorix come to the temple of Artemis, and there we will make our marriage vows in the presence of the goddess'. On the day, Sinorix, attended by many Gauls, both men and women, waited for her; she accompanied him to the altar with fond words and tenderness. There she drank to him from a golden cup, and bade him share with her in the drink. He received it with pleasure, as a token of bridal love, and drank it down. But the bridal cup was a potion of strong poison. As soon as she saw that he had drunk it, she fell down on her knees, and shouted: 'I thank you, venerable Artemis, for granting me in this your temple a glorious revenge for my murdered husband.' After saying this, she dropped down dead herself while Sinorix died with her, at the altar of the goddess.

Eryxo

Eryxo plans the successful assassination of the tyrant Laarchus (*Stratagemata*, 8.41; see also Plutarch, *Moralia*, 260):

> Around 560 BCE Laarchus was declared regent of Cyrene, during the minority of Battus son of Arcesilaus; but, intoxicated by power, he soon became not only a king, but a tyrant, inflicting on the citizens the most atrocious acts of cruelty and injustice. Battus' mother was Eryxo, a woman of great modesty and exemplary virtue. Laarchus was passionately attracted to her, and proposed marriage. This she discussed with her brothers. When they, as agreed between them and their sister, delayed, she sent a servant to Laarchus, informing him that her brothers seemed to disapprove of the marriage; but, if he would agree to meet them at her house, she expected that the discussion might remove their present objections. This seemed to promise a favourable outcome, so Laarchus visited Eryxo's house by night without a guard. There he found Polyarchus, her eldest brother, together with two youths, armed and waiting to receive him; they immediately fell upon him, and killed him. Then they proclaimed Battus king; and restored to the inhabitants of Cyrene their old form of government.

Chrysame

Chrysame takes control and wins the day with some judicious use of magic (*Stratagemata*, 8.43):

> When the Ionian colonists came to Asia, Cnopus, who was descended from the family of the Codridae, made war on the inhabitants of Erythrae. He was directed by the oracle to send an expedition to a Thessalian priestess of Hecate Enodia which returned with the priestess Chrysame. Possessing great skill in the occult qualities of herbs, she chose out of their herd a large and beautiful bull, gilded his horns, and

> decorated him with garlands, and purple ribbons embroidered with gold. She mixed in his fodder a medicinal herb that would drive him mad; the efficacy of this medicine was so great, that not only the beast was seized with madness but also anyone who later ate its meat. When the enemy camped opposite, she raised an altar in their line of sight: the bull was brought out, let loose and ran wild into the plain, roaring, and tilting at everything he met. The Erythraeans saw the bull, intended for the enemy's sacrifice, running towards their camp, and considered it a good omen. They seized the beast, and offered him up in sacrifice to their gods; everyone, as part of the sacrifice, ate a piece of the flesh. The whole army was soon afterwards seized with madness, and exhibited the same wildness and frenzy as the bull. When Chrysame saw this, she directed Cnopus immediately to unleash his forces, and charge the enemy. Incapable of making any defence, the Erythraeans were cut to pieces; Cnopus took control of Erythrae.

Polycleia

Shrewd and ambitious Polycleia inherits her kingdom (*Stratagemata*, 8.44):

> Aeatus the son of Pheidippus had an only sister called Polycleia … the oracle had declared that whichever of their family first crossed the river Achelous, should possess the city, and occupy the throne. While Aeatus was engaged in a war with the Boeotians … and his army was preparing to cross the Achelous, Polycleia bandaged up her foot, pretending to have hurt it, and requested her brother to carry her across the river. He, not suspecting any tricks, readily complied with her request; he gave his shield to his armour-bearer, and took his sister on his shoulders. But as he approached the opposite bank, she leapt from him onto the bank. Turning to Aeatus, she said: 'Remember the oracle, by whose declaration the kingdom must be mine; for I was the first to reach the shore.' Aeatus was pleased with the trick, and captivated by the girl's actions, he married her, and shared the kingdom with her. Their marriage produced a son, whose name was Thessalus from whom the city was afterwards called Thessalia.

Leaena

Courageous Leaena renders herself speechless under torture (*Stratagemata*, 8.45; see also Pausanias, 1.23):

> How Aristogeiton and Harmodius delivered Athens from the tyrant's yoke is known to every Greek. Aristogeiton had a mistress, whose name was Leaena. Hippias ordered her to be interrogated by torture, so as to determine what she knew of the conspiracy; after she had long bravely

endured the various tortures that were visited on her, she cut out her tongue with her own hand, lest any more pain should extort from her any information. The Athenians in memory of her, erected in the Propylaea of the Acropolis a statue of a lioness in brass, without a tongue.

Axiothea

Proud and noble Axiothea prefers death to slavery (*Stratagemata*, 8.48):

> When Ptolemy, king of Egypt, sent a powerful force to dispossess Nicocles (fl. *c.*374 BCE) of the kingdom of Cyprus, both Nicocles and his brothers, rather than submit to slavery, committed suicide. Axiothea the wife of Nicocles, wishing to emulate them ... assembled their sisters, mothers, and wives and barred the doors of the women's quarters. While the citizens were crowding into the palace, with their children in their arms they set fire to the house. Some killed themselves with a sword, and others bravely jumped into the flames. Axiothea ... first stabbed herself, and then threw herself into the fire to save even her dead body from falling into the hands of the enemy.

Laodice

From *Stratagemata* 8.50:

> Antiochus married Laodice (died before 236 BCE), his sister on the father's side, and had a son by her, Seleucus. He afterwards also married Berenice, daughter of king Ptolemy, by whom he likewise had a son; but he died while this son was in his infancy, and left his kingdom to Seleucus. Laodice did not think her son [Seleucus] was secure on the throne, while the son of Berenice was alive, and sought ways to bring about his death. ... The assassins however showed the people a child very like him whom they had murdered; they declared him to be the royal infant, whom they had spared, and a guard was appointed to protect him. Berenice also had a guard of Gallic mercenaries, and a fortified citadel appointed for her residence. The people swore allegiance to her ... but she was secretly assassinated. Several of the women in attendance also died while attempting to save her. However Panariste, Mania, and Gethosyne buried the body of Berenice, and placed another woman in the bed where she had been murdered pretending that she was still alive and likely to recover from the wound she had received. They persuaded her subjects of this, until Ptolemy, the father of Berenice arrived. He dispatched letters to the countries around in the names of his daughter and her son, as if they were still alive; and by this stratagem devised by Panariste he secured for himself the whole country from Taurus to India, without a fight.

Deidameia

A tough negotiator whose tenacity was her downfall (*Stratagemata*, 8.52):

> Deidameia, the daughter of Pyrrhus, attacked and took Ambracia, to avenge the death of Ptolemy. And when the Epirots sued for peace as suppliants, she granted it only on condition that they acknowledged her hereditary rights, and the honours of her ancestors. This they agreed to do, without any intention of observing their agreement. For some of them immediately formed a plot against her life, and bribed Nestor, one of Alexander's guards, to murder her; but he, struck by her regal dignity, fixed his eyes on the ground as if in meditation, and returned without finishing the job. She then retired to the temple of Artemis Hegemone, where Milon, who had been guilty of murdering his own mother Philotera, pursued her with a drawn sword. She had just time to call out to him, 'Slaughter, you matricide, on slaughter raise' [Euripides, *Orestes*, l. 1587], before Milon aimed a blow, and slew her in the temple.

Mania

Mania was a woman-warrior of the first order (*Stratagemata*, 8.52; see also Xenophon, *Hellenica*, 3.1):

> Mania, the wife of Zenis prince of Dardanus, governed the kingdom after the death of her husband, with the assistance of Pharnabazus. She always went to battle drawn in a chariot; she gave out orders while in action, formed her lines, and rewarded every man who fought well, as she saw he deserved. And – what has scarcely happened to any general, except herself – she never suffered a defeat. But Meidias, who had married her daughter, and might from that close relationship have been thought to be loyal to her, secretly entered her apartments, and murdered her.

Tirgatao

Tirgatao – guerrilla fighter and tough negotiator (*Stratagemata*, 8.55):

> Tirgatao of Maeotis married Hecataeus, king of the Sindi, a people who live just north of the Bosphorus. Hecataeus was expelled from his kingdom, but was reinstated by Satyrus, tyrant of Bosphorus. Satyrus gave him his daughter in marriage, and urged him to kill his former wife. As Hecataeus passionately loved Tirgatao, he could not bear to think of killing her, but confined her to a fortified castle from which she escaped. Fearing lest she should incite the Maeotians to war, Hecataeus and Satyrus made a thorough search for her, which she skilfully eluded, travelling through remote and deserted ways, hiding herself in the woods in the day, and continuing her journey by night. At last she reached the country of the Ixomatae, where her own family was on the throne. Her

father was dead, and she afterwards married his successor. Then she roused the Ixomatae to war, and encouraged many warlike nations around the Maeotis to join her alliance. The confederates first invaded the country of Hecataeus, and afterwards ravaged the dominions of Satyrus. Harassed by a war in which they found themselves inferior to the enemy, they sent an embassy to sue for peace, accompanied by Metrodorus the son of Satyrus, who was offered as a hostage. She granted them peace, on stipulated terms, which they bound themselves by oath to observe. But no sooner had they made the oath, than they made plans to break it. Satyrus prevailed on two of his friends to defect to her, and put themselves under her protection so as to find an opportunity to assassinate her more easily. On their defection, Satyrus wrote a letter asking for them to be handed back; she answered alleging that the law of nations justified her in protecting those who had placed themselves under her protection. The two men one day asked to meet her. While one distracted her with a pretended matter of importance, the other levelled a blow at her with a drawn sword, which hit her belt; the guards immediately seized and imprisoned them. They were afterwards interrogated by torture, and confessed the whole plot upon which Tirgatao ordered the hostages to be executed, and devastated the territories of Satyrus with fire and sword. Stung with remorse for the calamities he had brought upon himself and his country, Satyrus died in the middle of an unsuccessful war, leaving his son Gorgippus to succeed him on the throne. He renounced his father's actions and sued for peace, which she granted on payment of a tribute, and put an end to the war.

Amage

A determined leader of men and ruthless warrior who successfully defeated the Scythians (*Stratagemata*, 8.56):

Amage, wife of Medosaccus king of the Sarmatians, who inhabit the coast of the Euxine sea, saw that her husband was totally given up to luxury, and took the reins of government into her own hands. She judged pleas, stationed garrisons, repulsed the invasions of enemies, and directed everything with so great ability, that her fame extended through all Scythia. The inhabitants of the Tauric Chersonesus, who had been greatly harassed by a king of the neighbouring Scythians, had heard of Amage's fame, and they requested an alliance with her. In consequence of a treaty formed between the two nations, she wrote to the Scythian prince, requesting him not to repeat his devastation in the Chersonesus. When he treated her prohibition with contempt, she marched against him with a hundred and twenty men of tried and tested courage, and extraordinary strength, each of them with three horses. In one night and

> day she covered a distance of 1,200 stades, and arrived unexpectedly at the palace where she slew all the guards. And while the Scythian, bewildered by this moment of sudden danger, conceived her force to be much greater than it really was, Amage rushed into the palace, where she had made her first attack, and slew the Scythian, along with his friends and relations. She put the inhabitants of Chersonesus back in free possession of their country, and gave his hereditary dominions to the son of the Scythian prince; warning him to take heed from his father's death, and not to invade the territories of the neighbouring Greeks and barbarians.

Arsinoe

Guile allows Arsinoe to make a narrow escape (*Stratagemata*, 8.57):

> After the death of Arsinoe's husband Lysimachus, the city of Ephesus was in chaos with riots during which Seleucus knocked down the walls, and threw open the gates. Arsinoe placed a slave in the royal litter, whom she dressed in her own robes, and posted a strong guard around her. Then, dressing herself in ragged clothes and disguising her face, she slipped through a private door, ran to her ships, and getting on board, immediately weighed anchor and made her escape. In the meantime Menecrates, one of her opponents' generals, attacked the litter and slew the servant she had left in it, mistaking her for Arsinoe.

Cratesipolis

Deception and cunning wins the day for Cratesipolis (*Stratagemata*, 8.58):

> Cratesipolis, who had for a long time fought in vain for an opportunity to betray Acrocorinth to Ptolemy, applauded the loyalty and bravery of her guard of mercenaries when they repeatedly assured her that the place could be defended; however, she said, it may be wise to send for reinforcements from Sicyon. For this purpose, she openly sent a letter of request to the Sicyonians; and privately an invitation to Ptolemy. Ptolemy's troops were dispatched in the night, admitted as the Sicyonian allies, and put in possession of Acrocorinth without the agreement or knowledge of the guards.

The Priestess of Athene

The priestess fools the Aetolians into believing she was Athena and brings about their defeat (*Stratagemata*, 8.59):

> During the siege of Pellene, which was conducted by the Aetolians, the priestess of Athene, on the occasion of the festival of the goddess, led the procession of the day from a high hill, opposite to the tower where the men of Pellene used to take on arms. She was the tallest and most

attractive maiden who could be picked out, dressed in a full suit of elegant armour and a three-plumed helmet. The Aetolians, seeing a maiden come out in arms from the temple of Athene, and advance at the head of the armed citizens, supposed that she was the goddess herself, who had come to the protection of the city. They immediately raised the siege, and the men of Pellene pursued them in their retreat, and killed many of them.

Mysta

A cunning plan saves and restores Mysta (*Stratagemata*, 8.61):

> When Seleucus Callinicus was defeated by the Gauls at Ancyra, and fell into the hands of the enemy, his wife Mysta discarded her royal robe, put on the ragged dress of a lowly slave, and in such a guise was sold among the prisoners. After being shipped with the rest of the slaves to Rhodes, she revealed her true identity. The Rhodians immediately re-purchased her from the buyer, dressed her in a manner appropriate to her station, and escorted her to Antioch.

Epicharis

Epicharis is tortured by Nero but bravely outwits him by committing suicide (*Stratagemata*, 8.62):

> Piso and Seneca were accused of a conspiracy against Nero; Mela, Seneca's brother, had a mistress, Epicharis (d. 65 CE). Nero cross-examined her by torture in a bid to discover what she might know of the plot but she resolutely bore the torture without revealing anything. She was therefore dismissed for the time being; but three days afterwards she was ordered to be brought back in a litter. While she was being conveyed, she pulled off her belt, and strangled herself with it. As soon as the men who were in charge of the litter had brought it to the place of torture, they set it down, and told Epicharis to come out; but on looking inside the litter, they found only a corpse. This event irritated Nero, who found that he had been outwitted by a prostitute.

Tacitus describes the scenario more poignantly (*Annals*, 15.57):

> She was, however, kept in custody. Subsequently, when the conspiracy was discovered, Nero ordered her to be tortured on the rack because she refused to name any of the accomplices; but neither blows, nor fire, nor the increased fury of her tormentors, could extort any confession from her. When on the second or third day after she was carried in a sedan chair – for her limbs were now broken – to be tortured a second time, she throttled herself on the way with her belt, which she fastened to the

> chair. She thus acted more nobly than many a noble eques or senator, who without even being tortured betrayed even their nearest relatives.

Caphene and the Melian Women

Caphene's military intelligence and bravery proves successful (*Stratagemata*, 8.64):

> After the Melians under Nymphaeus had established themselves in Caria, the Carians, who were settled at Cryassus, grew jealous of their power, and were anxious to get rid of them. With that in mind, they held a public festival, and invited the Melians to come. But a Carian maiden [Caphene], who had fallen in love with Nymphaeus, revealed their plot to him. He then answered the invitation of the Carians, that it was the custom of the Greeks never to attend such an entertainment without their wives. They were, therefore, invited to bring their wives with them. The Melians accordingly went in their tunics, and unarmed; but each of their wives carried a sword under her clothing, and sat next to her husband. In the middle of the show, observing a signal, the women instantly opened their clothing, and gave each man his sword. The men fell upon the barbarians, and cut them to pieces; then they took possession of their city and lands.

Plutarch adds a happy conclusion, appreciating the achievement of the women (*Moralia*, 246):

> Then, taking possession of the land and razing that city, they built another, to which they gave the name New Cryassus. Caphene married Nymphaeus and received the honour and gratitude merited by her valuable services. It is right and proper to admire both the silence and the courage of the women, and that not a single one of them among so many was led by timidity to turn coward, even involuntarily.

The Phocian Women

We have already noted how the Phocian women's pride and fear of inevitable rape in defeat proved a morale booster for their warriors (*Stratagemata*, 8.65; see also Plutarch, *Moralia*, 244):

> The Phocians and Thessalians fought a war with such animosity, that the Thessalians made a resolution to give no quarter to any Phocian who bore arms, and to reduce their wives and children to slavery. Before the battle, Phocian women collected a great quantity of wood, which they piled up, and climbed on it with their children; they vowed that, as soon as they saw their husbands defeated, they would set fire to the pile, and die in the flames. This resolution of the women produced corresponding bravery in the men; they fought obstinately, and obtained the victory.[14]

(The Anonymous) *Tractatus De Mulieribus*

The title is problematic. (The Anonymous) *Women Intelligent & Brave in War* is often given as the title of this somewhat obscure catalogue of women: we do not know the author, when it was written, what genre it was intended to be in or what the real title was. It sometimes goes under the title *Tractatus De Mulieribus Claris in Bello* but *Gunaikes en Polemikois Sunetai kai Andreia* may be nearer the mark, given that this what is in the manuscript. The first edition was published in 1789 (Heeren); in 1839 it appeared in Westermann's *Scriptores Rerum Mirabilium Graeci* – a motly collection of works; the last publication was an edition by Landi in 1895. Whatever the intended title, it is probably not quite accurate because, of the featured fourteen women included, two are not warrior women at all and some, for example, Argeia and Lyde, do not exhibit any military qualities.

Deborah Gera has published the seminal work on the tract: she suggests a publication date some time in the late second or early first century BCE; as for putative authorship, she contradicts this when she nominates Pamphile of Epidaurus as a possibility (fl. first century CE).[15] A prolific historian in the reign of Nero, Pamphile's works include the thirty-three book *Historical Commentaries*, *Epitome of Ctesias* in three books, numerous epitomes of histories and other books including *On Disputes* and *On Sex*.[16] The emperor Julian (r. 361–363 CE) may have known the work as he mentions in a list of warring women Semiramis, Nitocris, Rhodogune and Tomyris in exactly the same order as they appear in the *Tractatus*; concidence?[17] Despite the difficulties, the *Tractatus* remains a valuable adjunct to Plutarch, Polyaenus and to the other primary sources of women war warriors.

The fourteen women are Semiramis, Zarinaea, Nitocris the Egyptian, Nitocris the Babylonian, Argeia, Dido, Atossa, Rhodogune of Parthia, Lyde, Pheretime, Thargelia, Tomyris, Artemisia I of Caria and Onomaris. They are all described in short, pithy thumbnail sketches.

Zarinaea, Nitocris of Egypt, Argeia, Theiosso (Dido), Atossa, Lyde and Thargelia are of particular interest because they do not feature in Plutarch or Polyaenus, although we do know them from other sources; Onomaris is more interesting still as the *Tractatus* is the only surviving source for her.

The women who do not feature in the catalogues already described are detailed below.

Zarinaea

The *Tractatus* entry tells us how, when her husband and brother Cydraeus King of the Sacians died she married Mermerus, ruler of Parthia. Zarinaea fought in a battle against the Persians and was wounded; she was pursued and caught by a Stryangaeus who spared her life. Mermerus later captured and killed him despite Zarinaea's plea that he be spared. An indignant Zarinaea then released some prisoners with whom she conspired to kill Mermerus; she then allied with the Persians. The author's source is Ctesias (*FGrH*, 688, F7).

Nitocris of Egypt

Nitocris, Queen of Egypt, did not exhibit military skills: the *Tractatus* tells us that she exacted revenge on her brother's murderers by inviting them to an entertainment in a large hall and drowned them by diverting the river through the hall. She then 'flung herself into a room full of ashes'. The source is Herodotus, 2.100. Nitocris is the first known woman ruler of Egypt.

Nitocris of Babylon

This Nitocris, however, was militarily adept and cunningly deceptive. She was, apparently, cleverer even than Semiramis, diverting the river running through her city in order to hamper the progress of any enemy incursions. She also built her tomb over the city gate to trick Darius who would expect to find treasures inside. All he got was an inscription berating him for his greed. The source is Herodotus, 6.52.

Argeia

Argeia demonstrates no military skill. See Herodotus, 6.52.

Theiosso (Dido)

The *Tractatus*, after Timaeus, tells how Dido founded Carthage and later committed suicide. The source is Timaeus (*FGrH*, 566, F82). We know from other sources, not least Virgil in Book 4 of the *Aeneid*, how she was a strong and able leader of her rich and prosperous country.

Atossa

Atossa, according to Hellanicus, 'was most warlike and brave in every deed'. More than that, though, she was brought up by her father, Ariaspes, as a man and inherited his kingdom; she was the first queen to sport a tiara, and the first to wear trousers; she could write and she introduced eunuchs to the world. Hallanicus is the source (*FGrH*, 4, F178a).

Lyde

The *Tractatus* reveals no military activity, just an example of exemplary parenting of a very difficult child. See Xenophilus (*FGrH*, 767, F1).

Thargelia

Thargelia of Milesia married Antiochus, King of the Thessalians; when he died she ruled Thessaly for thirty years, repelling a Persian invasion through diplomacy. The source is Aeschines, frag 21 Dittmar; Hippias (*FGrH*, 6, F3).

Onomaris

Onomaris was a distinguished Galatian, a Gaulish-Celtic tribe. She showed great leadership and military prowess. When her country was beset by 'scarcity' she took control of events because no man was willing to lead the Galatians to a new, more rewarding, life elsewhere. In this respect she is reminiscent of Artemisia I, who also came forward to take up power when no man seemed capable of doing so. Onomaris pooled all the resources owned by her tribe, in order presumably to deter envy and superiority and to foster communal ownership, and led her people over the Ister in a mass emigration; she then defeated the locals there and ruled the new land. These events probably took place in the fourth or third centuries BCE. Onomaris typifies the not unusual high social status of Celtic women, some of whom rose to prominence as leaders of men: Boudica and Cartimandua are famous examples. Four out of Plutarch's twenty-six women are Celts.

* * *

Ten of the fourteen *Tractatus* women are non-Greek, nine of the ten are from different countries while the four Greeks are each from different *poleis*; they are all queens. The geographical diversity and regal status may suggest a deliberate decision to demonstrate the ubiquity of warrior women in the Mediterranean world and the relatively high number of queens who exerted independence and power. Most got to be where they were by dint of their being wives, mothers or widows of reigning or former reigning kings; none of the widows show a need or desire to remarry. They are their own women, women powerful now in their own right; some go on to be more famous than their husbands, as in the case of Tomyris and Artemisia. They all hold on to their power tenaciously. Physical appearance is irrelevant to the author of the *Tractatus*: we know from other sources that some of the fourteen were beautiful, but our author focuses, by and large, on their military or political qualities. Guile and ingenuity are key weapons and stratagems, part of the 'intelligence' alluded to in the work's apparent title which some of our women have in spades: Semiramis, Artemisia, the two Nitocrises, Dido and Atossa all use deception to good effect.

Chapter 9

Spartan Women: Vital Cogs in a Well-Oiled War Machine

Women played an active and vital role in keeping the renowned war machine at Sparta well-oiled and efficient. Since Spartan men were preoccupied with military training, bonding with comrades in mess life and constantly doing battle, it fell to women to run the farms back home and keep the *polis* going in their absence. Working the wool was never as important a part of the Spartan woman's life as it was in the rest of Greece: she had many more important things with which to fill her day. No history written by a Spartan survives; we have, therefore, to rely on the prejudiced, xenophobic and hostile writings of other Greeks for our picture of Sparta and Spartan society. Polemic and propaganda no doubt stain what has come down to us.

Sparta was a military society.[1] The army and the wars it fought were everything. Lycurgus in the eighth century BCE summed it up nicely when he said that Sparta's walls were built of men, not bricks. Women of childbearing age were essential in keeping the war machine in good working order. Indeed, the zeal with which they applied themselves to this work for the state, this war-related industrial baby production, may in fact have diminished the Spartan woman's natural maternal instincts if the following encounter is true; it occurred when a mother is told of the death of her five sons in battle and she retorts to the messenger: 'don't tell me about that, you fool; tell me whether Sparta has won!'. And when he declared that Sparta was victorious, 'Then,' she said, 'I accept gladly also the death of my sons' (Plutarch, *Lacaenarum Apophthegmata*, 6.7). Just as 'sensitive' was the wife and mother who told her son or husband departing for yet another war to come home carrying his shield, or, if not, carried on it (*Lacaenarum Apophthegmata*, 6.16). The bereaved mothers of the fallen at the Battle of Leuctra in 371 BCE are said to have had smiles beaming on their faces out of sheer pride.

Women's production of male children was just as important to the Spartan *polis* as a man fighting in the Spartan army. Women who died in childbirth and men who died in battle were honoured in equal measure – with their names inscribed on their gravestones. By the same token, producing a son who turned out to be a coward was a cause for great shame and sorrow. One traitor, Pausanias, met a rerrible end when he took refuge in a sanctuary to

Athena. His mother, Theano, instead of pleading for his life, picked up a brick and placed it in the doorway: very soon, others followed her lead and completely bricked up the temple door. Pausanias eventually died a slow, suffocating and starving death inside. Plutarch, in his *Lacaenarum Apophthegmata*, *Sayings of Spartan Women*, cites three Spartan mothers who killed their cowardly sons with their own hands, as quoted below.

At the siege of Sparta in 272 BCE, King Pyrrhus of Ephesus hesitated with his crack mercenary troops before attacking the walls of Sparta because they were defended by women, children and old men.[2] However, he was less concerned about the very real possibility of killing women and children than he was about the fearsome female opposition he faced.

As noted, Spartan men were preoccupied, obsessed even, with their military careers and, though usually marrying from their mid-twenties, did not see very much of domestic or family life before the age of 30. Their wives, therefore, played a vital and active economic and domestic role in raising their children and managing the household. It was they who were wholly responsible for raising sons until they were aged 7 when they left to join the junior army (*agoge*) – the start of their extensive and intensive training. It was, therefore, crucial that women of the citizen class be in tip-top condition physically and mentally to prepare them for quality conception and the very best in motherhood. Like women in the rest of Greece, the wife stayed at home but, unlike her sisters elsewhere, she was educated in the arts and took training in athletics, dancing and chariot racing: a strong, fit and educated mother delivered strong babies for a strong army for a strong Sparta.

In the early fourth century BCE, we have an example of the ideal active, competitive Spartan woman in Cynisca.[3] She (born *c.*440 BCE) was the wealthy daughter of the King of Sparta, Archidamus II.[4] Xenophon tells us that she was urged by her brother, Agesilaus II, to compete in the Olympic Games in the prestigious four-horse chariot race as an owner and trainer of horses. Agesilaus was ever keen to instill bravery and belligerence in the Spartans, and to raise the profile of women. However, an alternative explanation has it that he wanted to discredit the sport by having a woman win it.[5] Cynisca duly won the four-horse race (*tethrippon*: τέθριππον) in 396 BCE and in 392 BCE. She was honoured with a bronze statue of a chariot and horses in the Temple of Zeus in Olympia. The inscription read that she was the only woman to win in the chariot events at the Olympic Games. Pausanias reminds us that usually only Spartan kings were honoured in this way.[6] This is what the statue tells us: 'My fathers and brothers were Spartan kings. I won with a team of fleet footed horses and put up this monument. I am Cynisca: I declare that I am the only woman in Greece to have won such a wreath.'

Other female chariot race winners include Zeuxo of Argos, Euryleonis, Bilistiche, Timareta, Theodota, Arstocleia and Cassia. Spartan Euryleonis was victorious in the two-horse chariot races in the 368 BCE games; she was only the second female *stephanite* (crown-bearer) in Olympic history. A statue of Euryleonis was erected in Sparta in about 368 BCE and is one of few bronze statues that survives anywhere in the Greek world. Belistiche was a courtesan, a *hetaira* – she won the four-horse and two-horse (*synoris*) races in the 264 BCE Olympic Games. Ptolemy II Philadelphus was so impressed that he took her as his mistress and deified her as Aphrodite Belistiche.[7] According to Clement of Alexandria, she had the further honour to be buried under the shrine of Sarapis in Alexandria. Aristocleia won a two-horse chariot race in Larisa.[8] The Panathenaic victor lists tell us that Zeuxo was victorious in the four-horse chariot race in about 194 BCE.[9]

The military achievements and aspirations of Sparta and the Spartans held such a fascination to other Greeks that Plutarch compiled a book on the famous sayings of Spartan women, the *Lacaenarum Apophthegmata* in the *Moralia*; some of these refer to the roles played by women in the Spartan military world. Two are attributed to Gorgo (d. between 518 and 508 BCE), the wife of Leonidas I: Gorgo was famed for her military and political judgement and wisdom. She was the daughter of a king of Sparta, the wife of another and the mother of a third. When she was about 18 (Herodotus says 8 or 9) Gorgo precociously but astutely told her father, the vacillating King Cleomenes, to dismiss the tyrant Aristagoras of Miletus, who requested military aid for cash from Sparta for his rebellion against Persia. Gorgo also rose to the occasion when a mysterious blank wax tablet was sent to Sparta from the exiled King Demaratus, then residing at the Persian court, regarding a Persian attack on Greece:

> when it had arrived at Lacedemon, the Lacedemonians were not able to make head or tail of it until at last, as I am informed, Gorgo, the daughter of Cleomenes and wife of Leonidas, suggested a plan of which she had herself thought up, bidding them scrape the wax and they would find writing on the wood; and doing as she said they found the writing and read it, and after that they informed the other Greeks.[10]

Gorgo also had the good sense to make prudent preparations for her later life as a war widow. Before the Battle of Thermopylae, knowing that her husband's death was inevitable, Gorgo asked him how she should spend her widowhood. Leonidas replied 'marry a good man who will treat you well, bear children for him, and live a good life'.

Here are more excerpts from Plutarch's *Lacaenarum Apophthegmata*, all illustrating the bellicose mindset of Spartan women.

The high regard with which women held war heroes:

> When Brasidas, Argileonis' son, was killed [at the Battle of Amphipolis in 422 BCE] some citizens of Amphipolis arrived at Sparta and visited

> her; they asked if Brasidas had met his death with honour and in a way worthy of Sparta. And when they proceeded to tell of his greatness, and declared that he was the best of all the Spartans, she said, 'My son was a good and honourable man, but Sparta has many a man even better than him.'[11]

To Spartan women, cowardice is anathema:

> When a messenger came from Crete bringing the news of the death of Acrotatus [son of Areus I, King of Sparta who fell at the Battle of Megalopolis in 265 BCE], she said, 'When he met the enemy, was he not bound either to be killed by them or to kill them? It is more pleasing to hear that he died in a manner worthy of myself, his country, and his ancestors than if he had lived forever a coward.'[12]

Cowardice carries a heavy price: 'Damatria heard that her son had been a coward and unworthy of her, and when he arrived home, she murdered him. This is the epigram referring to her: Sinner against our laws, Damatrius, slain by his mother, was of the Spartan youth; she was of Sparta too'.[13]

Another Spartan woman murdered her son, who had deserted his post, on the grounds that he was unworthy of his country, saying, 'Not mine this son.' This is the epigram referring to her:

> Off to your fate through the darkness, vile son, who causes so much hatred that the Eurotas flows not even for the timid deer. Worthless whelp that you are, obnoxious remnant, be off now to Hades; Off! for never did I bear Sparta's unworthy son.[14]

And: 'Another, hearing that her son had been saved and had run away from the enemy, wrote to him, "Bad news is being spread about you; either clear your name of this or die".'[15]

There is no hiding place or escape for cowards: 'Another, when her sons had run away from battle and come to her, said, "Where have you come now in your cowardly flight, vile sinners? Do you intend to slink in here where you came out of?" And so saying she pulled up her dress and showed them where'.[16] And: 'One woman, seeing her son coming towards her, asked, "How is our country?" And when he said, "everyone's' dead," she picked up a tile and, hurling it at him, killed him, saying, "And so they sent you to bring the bad news to us!"'[17]

On the other hand, there is pride in a brave hero: 'Another was burying her son, when an ordinary old woman came up to her and said, "Ah bad luck, you poor woman." "No, by Heaven," she said, "it's good luck; for I gave birth to him so that he might die for Sparta, and this is exactly what has happened."'[18]

Disability is no bar to potential bravery; no help for heroes here: 'Another, as she accompanied a lame son on his way to the field of battle, said, "At every step, my child, remember to be brave."'[19]

War wounds bring no sympathy: 'Another, when her son came back to her from the field of battle wounded in the foot, and in great pain, said, "If you remember to be brave, my child, you will feel no pain, and be quite happy"'.[20]

The Spartans and their ways were viewed with fascination by the rest of Greece. One example of the involvement of their women in things military typifies the role women played out in the Spartan military sphere generally. In 272 BCE Pyrrhus was on the verge of attacking Sparta, and the Spartans were about to evacuate their women and children to Crete. But like the Geloan women some 130 years before, the Spartan women had decided that they were going nowhere. The daughter of King Cleonymus II, Archidameia, entered the *gerousia*, the Spartan Council of Elders, and declared in no uncertain terms that it was not right that they should live after Sparta had died.[21] So, they stayed and helped in the defence. Traps, like modern tank traps, comprising waggons half sunk in mud were contructed to block off Pyrrhus' elephants; significantly, though, 'when they began to carry out this project, the women and girls turned up, some of them in their robes, with tunics belted, and others in their tunics only, to help the elderly men in the work'. The women were able to relieve the younger, fighting men and allow them to rest, 'assuming their share of the task they completed with their own hands a third of the trench'. Next day, 'these women handed the young men their armour, handed over the trench to them, and told them to guard and defend it, in the sure knowledge that it was sweet to conquer before the eyes of their fatherland, and glorious to die in the arms of their mothers and wives, dying in a manner worthy of Sparta.' Polyaenus adds that, apart from providing invaluable logistical support, the women boosted the morale of the Spartans no end: 'some fetched the tools, other dug in the ditches, some again were employed in sharpening the weapons, and others assisted in dressing the wounded. The spirit of the women gave new resolution to the Spartans, who again took the field; they engaged Pyrrhus, and defeated him.'[22]

Next day the Spartans led by Arcotatus, Chilonis' lover, were encouraged by their women defenders and by the sight of Chilonis, daughter of Leotychidas, who had ostentatiously placed a halter around her neck, vowing she would commit suicide rather than go back to her abusive husband, Cleonymus, if and when Pyrrhus captured the city. Cleonymus, a bit of a mercenary and a pretender to the Spartan throne, had been denied that throne because of his violent behaviour; to make matters worse he was gravely humiliated when it became known that Chilonis was being seduced by arch-enemy Arcotatus. This drove Cleonymus to leave Sparta and enlist the help of Pyrrhus. Pyrrhus subsequently led the attack on the city but was repulsed. The Spartans, overjoyed by Arcotatus' superb leadership, jubilantly told him to withdraw from the battle and return to Chilonis in order to father more children of warrior calibre for Sparta. The following day saw a new attack by Pyrrhus but the Spartans held out: the Spartan women dutifully acted as ammunition loaders, taking away the wounded and providing food and drink for the sick and the fighters. Chilonis and Acrotatus had a child, who later ruled as Areus II, Agiad King of Sparta.

About 300 years earlier, in the first half of the seventh century, King Polydorus of Sparta defeated the Argive army but then, when he attacked the city of Argos, found himself up against the women of Argos, the only survivors there.

Spartan women rose to the occasion against the Messenians when the Spartan forces were busy fighting outside the city: a squadron of Messenians stationed itself menacingly outside the apparently defenceless city so the women took up arms and repelled the attackers. When the Spartans returned they believed their women, clad in armour, to be Messenians and would have attacked them had not their wives and daughters removed their armour to reveal who they really were. The Christian author Lactantius (*c.*250–*c.*325 CE) primly reports that the men and women soldiers lost no time in reacquainting themselves with each other; his suggestion is that they just had sex with the first woman they came across, regardless of whose wife she was: 'But the men, recognising their wives, and excited to passion by the sight, rushed to promiscuous intercourse, for there was no time to discriminate.'[23]

The war monument at Messene bears the names of twenty-four fallen, ten of whom are women.

Timycha of Sparta (early fourth century BCE), and her husband Myllias of Croton, were part of a group of Pythagorean pilgrims, who were attacked by Syracusan soldiers on their way to Metapontum, in defiance of the tyrant Dionysius the Elder. Beans were taboo to Pythagoreans so when they had the option of running through a field of beans to escape, they declined. Instead, they fought and died, with the exception of the pregnant Timycha and her husband, who were taken prisoner. Dionysius interrogated her about the taboo, but, mindful of her duty to preserve Pythagorean mysteries, she refused to answer. Instead, she bit off her tongue and spat it out at his feet in a supreme gesture of defiance, and to ensure her silence.

These were not the only Pythagoreans who died due to a phobia of a field of beans: beans were the death of Pythagoras himself when people hostile to the Pythagoreans set fire to Pythagoras' house, sending him running out towards a bean field; when he realised where he was, he stopped, declaring that he would rather die than go through the field – the mob promptly slit his throat. According to Diogenes Laertius, the fava bean was sacred to the Pythagoreans because it has a hollow stem, and it was believed that souls of the dead would travel through the ground, up the hollow stem and lodge in the bean (Stobaeus, 1.49.27).

But it was not always unalloyed bravery in Sparta. When the Theban army invaded Sparta after the Battle of Leuctra in 371 BCE the Spartan women were beside themselves, hysterical with fear – a reaction that was probably much, much more common than the bravery and defiance described. The terror and extreme distress exhibited by a civilian population – non-combatant men, women and children – on the approach of a hostile and foreign army must have been, and always has been, unbearable. In this case, the Spartan women had never seen an enemy before or experienced the atrocities and indignities which were no doubt rumoured to unfold. Xenophon describes the ineffably tense situation: 'keeping the River Eurotas to their right they moved on, burning and plundering houses full of many valuable things … the women could not even endure the sight of the smoke, since they had never seen an enemy before' (Xenophon, *Hellenica*, 6.5.28).

Plutarch adds the detail: 'Agesilaüs was even more harassed by the commotion and shrieks and the running about throughout the city, where the elder men were enraged at the state of affairs, and the women could not keep quiet, but were totally beside themselves when they heard the shouts and saw the fires of the enemy.' A boast Agesilaus had often made, and was now regretting, was that 'no Spartan woman had ever seen the smoke of an enemy's fires' (Plutarch, *Agesilaus*, 31.4–5).

According to Plutarch, an anonymous woman was personally responsible for ending the occupation of Thebes by the Spartans and driving the Spartans out of the city. The Thebans in exile in Athens had planned a coup but those in occupied Thebes had second thoughts at the last minute and called off the attack. A messenger was sent to Athens to apprise them of this change of plan but when said messenger went home to prepare his horse, he found, to his consternation, that his wife had loaned his only bridle to a friend. Borrowing another, or demanding his own back, would have aroused suspicion, so he did nothing and stayed at home in Thebes. Consequently, the exiles in Athens, none the wiser, launched the attack as planned and regained their city from the Spartans. The subsequent rise in power enjoyed by the Thebans was all due to the neighbourliness of one exiled Theban woman in Athens.

Chapter 10

Macedonian Women at War: Pawns and Power-Players

By the time of the Hellenstic age (usually defined as running from the death of Alexander in 323 BCE to the Roman occupation of Egypt in 30 BCE) it is probably fair to say that women in Greek society, law and politics were enjoying greater freedoms and opportunities; education was more extensive and women 'got out more'. The elite women of Macedonia in particular, like their Spartan counterparts in some respects, seem to have enjoyed greater liberty, greater respect from their men and greater social and political responsibility than women in many of the Greek *poleis*, not least Athens. Nowhere is this more evident than in the elevation of women to the highest levels of state as queens or regents in the various Macedonian dynasties. Some elite Macedonian women were more cosmopolitan, better travelled, more politically astute and considerably more dangerous and volatile than their sisters elsewhere in ancient Greece. Some displayed a breathtaking knowledge of military matters and impressive skills in tactics and strategy. The ability of elite women to rise to the higher echelons of Macedonian society and power may have something to do with the hereditary basis of power there which permitted women to be honoured and pandered to in much the same way as their menfolk were. Elite women, as elite mothers, worked ceaselessly and sometimes ruthlessly to secure or sieze power for their sons, not always with justification.

Polygamist Philip II's wives are typical of elite Macedonian women of the highest rank and, in a number of cases, are politically and militarily capable. Audata (r. *c*.359–336 BCE) was an Illyrian princess, daughter of Bardylis, the Illyrian king of the Dardanian state when she married Philip II of Macedon and became queen – she was a key player in a political stand-off between Macedonia and Dardania. Philip's marriage to Audata deterred a Dardanian invasion of Macedonia which enabled him to consolidate his power and then defeat Bardylis in a major battle in 358 BCE. Soon after the wedding, Audata changed her name to Eurydice I for political reasons, and to make her sound more Greek. But she maintained her Illyrian identity, passing it on to her daughter, Cynane, and to her granddaughter. It was not that unusual for Illyrian women to be leaders of men in battle and it was these military skills that she passed down to Cynane, half-sister to Alexander the Great; they included schooling in riding, hunting and armed combat.

Polyaenus tells us, 'Cynane, the daughter of Philip was famous for her military knowledge: she conducted armies, and in the field charged at the head of them. In an engagement with the Illyrians, she with her own hand slew Caeria their queen; and with great slaughter defeated the Illyrian army.'[1]

Philip II married Cynane off to her cousin Amyntas, but he died and she was left a widow in 336 BCE. Alexander promised her hand to Langarus, King of the Agrianians as a reward for his services, but Langarus also died following an illness.

Cynane went on to train her own daughter, Eurydice II of Macedon, in all things military. It was Cynane's ambition to have Eurydice married to her half-brother, Philip Arrhidaeus, successor to the throne of Macedon, and so she travelled to Asia to bring this about. Her move greatly worried both Perdiccas, the regent, and Antipater his general, so Perdiccas sent his brother Alcetas to murder her – despite the sympathy his troops felt for Cynane and Eurydice, as surviving members of the royal house. Perdiccas had second thoughts, spared Eurydice and permitted the marriage to Philip Arrhidaeus to go ahead; the king struggled with 'learning difficulties' brought on, according to Plutarch, by Olympias (another of Philip's wives) who attempted to kill him with *pharmaka* (drugs and spells), so as to eliminate him as a possible rival to her own son, Alexander the Great. The wedding took place, but both Philip and Eurydice were eventually murdered by Olympias.

Before that, though, Eurydice had been at the forefront of Macedonian politics and military policy: in 321 BCE she made a bid for power, demanding that the new regents of Macedon, Peithon and Arrhidaeus include her in the regency. Eurydice's close relationship with the Macedonian army, and her status as a king's wife, contributed much to her growing influence to the extent that she did succeed briefly in becoming a de facto regent. She took an active role in the proceedings at the Treaty of Triparadisus in 321 BCE – a power-sharing agreement between the generals (Diadochi) of the late Alexander the Great, in which they appointed a new regent and allocated the satrapies of Alexander's empire among themselves. It superseded the Partition of Babylon struck in 323 BCE on Alexander's death.

But Eurydice had not reckoned with Antipater, Alexander the Great's general whose star was in the ascendant and who staked his claim to the vacant regency. Eurydice tried to block this and to foster the support of a Macedonian army that was decidedly unhappy because Antipater could not pay them. Eurydice's rousing speech to the troops fell flat, though, and the Macedonian army went over to Antipater anyway; he was appointed regent and guardian of the king.[2] Eurydice saw her chance when Antipater died in 319 BCE, and was succeeded as regent by the much weaker Polyperchon. She formed an alliance with Cassander, and herself recruited an army, taking to the field in person to engage Polyperchon who advanced against her from Epirus, accompanied by Aeacides, the Epirote king. Critically, Polyperchon was supported by Olympias, along with Roxana, Alexander's widow, and by her infant grandson. The presence of Olympias was enough to decide the matter: the Macedonian troops would never fight against the mother of Alexander the Great, and went over to Olympias. Eurydice fled from the field at the Battle of Amphipolis, but was captured and imprisoned.

Eurydice and Arrhidaeus were incarcerated in a cramped dungeon with little food. Olympias, however, was increasingly alarmed at the support Eurydice enjoyed among the Macedonians, and so resolved to dispose of her. She sent the young queen a sword, a rope and a cup of hemlock, and invited her to choose her preferred method of death. Ever resilient, Eurydice remained strong and defied Olympias, praying that she too might soon be forced to make a similar choice. Finally, she bravely and resolutely ended her life by hanging, 'without giving way to a tear or word of lamentation'.[3] In 317 BCE, Cassander, on defeating Olympias, buried Cynane with Eurydice and Arrhidaeus at Aegae, the royal burying place.[4]

So ended the life of a valiant and determined young woman who transgressed the usual boundaries imposed on women in ancient Greece by taking a prominent and obtrusive role in political and military matters – only to be thwarted by another, equally resilient but much more malevolent woman.

How did this astonishing run of powerful, influential political and militaristic females come about? Eurydice I (born 407 BCE) was a Greek queen from Macedon, wife of king Amyntas III of Macedon. She was the daughter of Sirras of Lyncestis and had four children: including Alexander II, Perdiccas III and Philip II; she was the paternal grandmother of Alexander the Great. All the evidence tells us that she played a revolutionary public role in Macedonian life and was aggressively influential in a political world hitherto dominated by men. Her political activities changed Macedonian history: Eurydice I was the first known royal woman to be active in the political arena and to successfully exert political influence.

Nearly a century later Deidamia was a princess of Epirus, a daughter of Aeacides, King of Epirus and his wife, Queen Phthia, and sister of King Pyrrhus. As a young girl she was betrothed by her father to Alexander IV, the son of Roxana and Alexander the Great; she went with her fiancé and Olympias to Macedonia where they were besieged at Pydna in 316 BCE.[5] After the death of Alexander in 323 and Roxana in 309 BCE, she was married to Demetrius Poliorcetes and became a factor in the alliance between him and Pyrrhus.[6] While Demetrius was warring in Asia with his father Antigonus, Deidamia stayed at Athens; but after his defeat at Ipsus in 301 BCE, the Athenians sent her to Megara, though she was still accorded regal privileges.

She then joined Demetrius in Cilicia; he had just given his daughter Stratonice in marriage to Seleucus. Deidamia fell ill and died in 300 BCE leaving one son by Demetrius, called Alexander, said by Plutarch to have spent his life under house arrest in Egypt.[7]

Another, later Deidamia, also known as Laodamia, was daughter of Pyrrhus II, King of Epirus. The death of her father and of her uncle Ptolemy left her as the last surviving representative of the royal Aeacid dynasty. Her sister, Nereis, married Gelo of Syracuse; during a rebellion in Epirus, Nereis sent 800 mercenaries from Gaul to support Deidamia. The Molossians (a tribe native to Epirus noted for their vicious dogs) supported her, and with the aid of the mercenaries she took Ambracia for a short time. The rebel Epirotes, however, planned to exterminate the whole royal family, including Deidamia; she took refuge in the Temple of Artemis, but was slain in the sanctuary itself by Milo, a convicted matricide, who then committed suicide.[8] This all happened early in the reign of Demetrius II in Macedonia (239–229 BCE).

Thessalonike (352 or 345–295 BCE) was the daughter of King Philip II of Macedon by his Thessalian wife-cum-concubine, Nicesipolis, from Pherae. Thessalonike was another powerful woman, connected as she was to three of Macedonia's powerful men: apart from being daughter of King Philip II, she was half-sister of Alexander the Great and wife of Cassander. Thessalonike was born on the same day that the armies of Macedon and the Thessalian League won the decisive Battle of Crocus Field in Thessaly over the Phocians; King Philip proclaimed, 'Let her be called victory in Thessaly': Thessaly and *nike*. Her mother died soon after her birth leaving her to be brought up by her stepmother Olympias as her own daughter, close friend of Nicesipolis.

Thessalonike took refuge with Olympias and the rest of the royal family in Pydna when Cassander came calling in 315 BCE.[9] Pydna fell, Olympias was executed and Thessalonike was married to Cassander, who saw this as a chance to ingratiate himself with the Argead dynasty; Thessalonike became Queen of Macedon and the mother of three sons, Philip, Antipater and Alexander; her husband honoured her by naming the city of Thessaloniki after her, which he founded on the site of ancient Therma.[10] When Cassander died, Thessalonike was still able to exert influence over her sons, but Antipater had her killed over the ill-fated power sharing arrangement in 295 BCE.

Phila (d. 287 BCE), daughter of Antipater the regent of Macedonia, is regarded by ancient authorities as one of the finest women of her age; her conspicuous acuity and political astuteness made her much in demand as an advisor in her father's political affairs. She married three times: Antonius Diogenes tells us that she was married to Balacrus, the Satrap of Cappadocia by 332 BCE.[11] Ten years later, in 322 BCE, her father married her to Craterus as a reward for his help in the Lamian War.[12] Craterus died in 321, when she married a youthful Demetrius Poliorcetes, the son of Antigonus sometime between 319 and 315 BCE.[13]

Phila played an influential part in the aftermath of the Battle of Ipsus in 301 BCE when Demetrius sent her as envoy to her brother Cassander in Macedonia, to effect a reconciliation and forge a treaty between him and Demetrius. After this she returned to Cyprus, where, in 295 BCE, she was besieged in Salamis by Ptolemy I, King of Egypt, and forced to surrender. The king, nevertheless, treated her with dignity and respect and sent her and her children back to safety in Macedonia. Unfortunately, in 287 BCE, Demetrius was deposed in a sudden coup and exiled by the popular Pyrrhus; Phila could do nothing to resolve the opposition from and lack of support by the people of Macedonia at large. Rather than face exile from Macedonia, she took her own life by drinking poison at Cassandreia.[14]

Aside from her political and military acumen, Phila was one of the first activists and workers for women's welfare and rights, a veritable crusader against oppression and injustice, providing dowries for less fortunate women from her personal wealth. The Athenians consecrated a temple to Phila, in the name of Aphrodite.[15]

Phthia was a bargaining piece in the shady machinations of her mother, Olympias. Phthia's father was Alexander II (272–260 BCE), King of Epirus. Olympias contrived to marry her off to Demetrius II (r. 239–229 BCE), King of Macedonia. When Olympias discovered that the Aetolians wanted to annex from her a part of Acarnania which the father of her boys, Pyrrhus and Ptolemy, had received as a reward for military aid, she arranged the match in a bid to win the help of Demetrius to elevate her to the throne of Epirus after the death of Alexander.[16] Justinus points out that: 'he was already married to a sister of Antiochus, king of Syria, and a marriage was accordingly solemnized, by which Demetrius gained the love of a new wife, and the hatred of his former one; who, as if divorced, went off to her brother Antiochus, and urged him to make war upon her husband'. The wronged woman in question was Stratonice of Macedonia, the daughter of Stratonice of Syria and of the Seleucid King Antiochus I Soter (281–261 BCE). While married to Demetrius II, Stratonice had a daughter called Apama II. When she left Demetrius in humiliation she repaired to Syria and also tried in vain to coax her nephew, Seleucus II Callinicus (246–225 BCE), to avenge the insult by declaring war against Demetrius; she may even have sought to marry Seleucus who was preoccupied with fighting in Babylonia. Whatever, Stratonice took advantage of his absence to raise an unsuccessful revolt against him at Antioch; she took refuge in Seleucia, where she was besieged, held prisoner and killed.[17]

Etazeta (fl. 255–254 BCE) was the second wife of Nicomedes I, king of Bithynia; she was ambitious to say the least, and persuaded Nicomedes to disinherit his sons by his former marriage, paving the way to his throne for her own children. When Nicomedes I died, she ruled on behalf of her infant sons but her ambitions for them were thwarted when Nicomedes' first-born, Ziaelas, ignored the new ruling and declared war against his stepmother to claim the throne. Despite her resistance, and a desperate marriage to Nicomedes' brother to patch things up, in about 254 BCE she was evicted by Ziaelas and forced to flee to Macedon with her sons.

Prominent among the women in Alexander the Great's life was Olympias, his mother. We have seen much of Olympias gliding menacingly in and out of what can only be called, even if anachronistically, these Byzantine, Machiavellian games of power. Olympias was the daughter of Neoptolemus I, King of the Molossians, a tribe in Epirus, and sister of Alexander I. Plutarch in his *Moralia* asserts that she was originally called Polyxena, changing her name to Myrtale before her marriage to Philip II of Macedon. Philip and Olympias fell for each other when both were initiated into the mysteries of Cabeiri on Samothrace.[18] She was given the name Olympias in 356 BCE, when Philip's horse came home in the Olympic Games. The omens were good: the night before the wedding Plutarch tells us how Olympias dreamed that a thunderbolt struck her womb and started a fire. Soon after, Philip dreamed that he put a lion seal on Olympias' womb, interpreted by Aristander that Olympias would give birth to a son who bore the characteristics of a lion: Alexander the Great.

Olympias (*c.*370–316 BCE) was nothing if not dedicated to the best interests of her son: in many ways she foreshadowed the pernicious and obsessive wives and mothers of the early Roman emperors who were determined to ensure the elevation of their sons to the very highest office: Livia, the Agrippinas and Messalina could have learnt much from Olympias. She may well have been involved in the assassination of Philip II and certainly had a hand in the murder of one of his wives and their daughter.

Olympias was by nature jealous and unpredictable; her husband's marriage in 337 BCE to Cleopatra (called Eurydice by Philip) did nothing to calm the waters so she took off with Alexander to voluntary exile in Epirus at the court of Alexander I. Things went from bad to worse when Philip offered their daughter, another Cleopatra, to Alexander in marriage. The wedding did not go well: Philip was murdered by Pausanias, one of his *somatophylakes*, his personal bodyguard; Olympias was suspected of being implicated. She then had Eurydice and her child executed to reinforce Alexander's claim as King of Macedonia, confirming that Philip was not his father, but that Zeus was.

While Alexander was away campaigning she kept in constant touch with her son and assumed de facto control of Macedonia with Cleopatra, her daughter. On Alexander's death, she did everything she could to protect and promote her infant grandson, particularly in the face of competition from Cynane.

Alexander married three times: to Roxana of Bactria, Stateira and Parysatis, daughter of Ochus. Parysatis was the youngest daughter of Artaxerxes III of Persia who was murdered in 338 BCE and eventually succeeded by her second cousin, Darius III, in 336 BCE; Parysatis and her sisters continued to live at the Persian court, accompanying the Persian army during Darius' campaign against Alexander. After the Battle of Issus in 333 BCE, Parysatis and many of her family were captured in Damascus by Alexander's general Parmenion. According to Arrian, in 324 Parysatis married Alexander at Susa, just one of the famous Susa Weddings. On the very same day, Alexander married Darius' eldest daughter, Stateira, thus strengthening his ties to both branches of the royal family of the Achaemenid Empire. The marriage celebrations went on for five days during which time ninety other Persian noblewomen were married to Macedonian and other Greek military leaders loyal to Alexander.

Drypetis was the younger daughter of Darius III. She was taken prisoner by Alexander after Issus in 333 BCE, along with her sister Stateira, her mother (another Stateira) and her grandmother Sisygambis. The Persian women then joined Alexander's baggage train for the next two years.[19] Significantly, and unlike the fate of many other women in the train, Alexander treated Darius' daughters with 'as much respect as if they were his own sisters'.[20]

Roxana's (*c.*343–*c.*310 BCE) reputation preceded her; she was by repute the most beautiful woman in all of Asia and deserving of her Persian name, Rauxsnaka, meaning 'little star'. She was the daughter of Oxyartes, a Bactrian nobleman who served Bessus, the Satrap of Bactria and Sogdia. Oxyartes had sent his wife and daughters, including 15-year-old Roxana, to take refuge in the Sogdian Rocks, the reputedly impregnable fortress which, with some foresight, had been provisioned for a long siege. When Alexander suggested that the defenders surrender, they refused, telling him that he would need 'men with wings' to capture the Sogdian Rocks. Nevertheless, Alexander found his winged men and took the rock and its defenders: among them was the family of Oxyartes, including Roxana.

Love at first sight it may have been, but there was a political element to the match – Sogdia had proved somewhat intractable and the political alliance can only have helped. Apart from a reference to a miscarriage in India there is no further mention of Roxana until Alexander's death.

When Alexander died in 323 BCE, Roxana was pregnant; quite simply, the sex of the baby was pivotal to the history of the region: a son was born six months later; she wasted no time in trying to ensure the succession of her infant son, also called Alexander. After some hostilities he was declared joint king with a mentally challenged son of Philip II, Philip III Arrhidaeus.[21] Perdiccas was made administrator.

Plutarch tells that with Perdiccas' support Roxana promptly murdered one of Alexander's other wives, Stateira, and her sister, Drypetis, to clear the way for young Alexander IV, but this just made her and her son puppets in the ongoing power struggles. As a daughter of Darius, if Stateira should also fall pregnant, then any boy born might hold a stronger claim on Alexander's throne. Stateira and Drypetis were poisoned, and their bodies thrown down a well.

When Perdiccas was assassinated in 320 BCE and Antipater died, mother and son passed into the ineffective and weak protection of Polyperchon, who was soon up against Cassander, Antipater's son. In 317 Alexander IV was stripped of his royal title, and Roxana fled with him to Olympias, who was given the responsibility of raising him as a true pedigree Macedonian: Roxana, on the other hand, was still regarded by many as a barbarian. As we have seen, the army went over to Olympias; she captured Philip and his wife, and had them tortured and killed.[22] Cassander invaded Macedonia, facing an army antagonised, and their morale weakened, by Olympias' cruelty. She, with Roxana and the young Alexander, took refuge in fortified Pydna. Polyperchon was abandoned by his army, and Pydna was soon starved into surrender. Olympias gave herslf up on promises of safety, but she was stoned to death on Cassander's orders, a fate that she faced with dignity and stoicism. Cassandra compounded the atrocity by denying her funeral rites – literally a fate worse than death. Roxana and her son were imprisoned at Amphipolis and finally killed in 310 after Cassander had duplicitously promised to return the kingship to Alexander when he came of age.[23]

Polyaenus leaves us the story of a brave and resourceful woman called Timocleia.[24] While the armies of Alexander were plundering Thebes, a Thracian called Hipparchus broke into Timocleia's house; after they had eaten he raped her and forced her to reveal the whereabouts of her treasures. Timocleia admitted that she had hidden her vases, cups and other ornaments and furniture in a dry well. She took the Thracian to the well; he climbed down only to be met with a shower of stones which Timocleia and her servants rained down on him, eventually burying him. The Macedonians heard about this and arrested Timocleia who was hauled before Alexander. She was unrepentant and proud that she had avenged the brutality shown to her by Hipparchus: Alexander was impressed and applauded her chutzpah; allowing her and her relatives to go free.

Ada I of Caria (fl. 377–326 BCE) was a member of the House of Hecatomnus and ruler of Caria, first as Persian satrap and later as queen under the control of Alexander. She married her brother Idrieus, who died in 344 BCE but as satrap was expelled by her other brother Pixodarus in 340 BCE and fled to Alinda, where she carried on her rule in exile. When Alexander invaded Caria in 334 BCE, Ada adopted Alexander as her son and gave up Alinda to him. In return, Alexander bestowed on Ada the command of the siege of Halicarnassus and, after its fall, made Ada Queen of Caria (Arrian, *Anabasis*, 1.23.7–8).

Thais was an Athenian *hetaira* and companion to Alexander the Great on his campaigns in Asia Minor. She is remembered for her presence at a *symposium*, urging the burning down of the palace of Persepolis in 330 BCE; the palace was the main residence of the defeated Achaemenid dynasty. Thais persuaded Alexander to raze the palace: Cleitarchus says that it was done on a whim; Plutarch and Diodorus believe that it was retribution for when Xerxes burnt down the old Temple of Athena on the Acropolis in 480 BCE during the Persian Wars. This is how Diodorus describes the incident and Thais' role in it:

> When the king [Alexander] was inflamed by their words, they all leaped up from their couches and passed the word along to form a victory procession in honour of Dionysus. Promptly many torches were gathered. Female musicians were present at the banquet, so the king led them all out for the revel to the sound of voices and flutes and pipes, Thais the courtesan leading the whole performance. She was the first, after the king, to hurl her blazing torch into the palace. As the others all did the same, immediately the entire palace area was consumed, so great was the conflagration. It was remarkable that the impious act of Xerxes, king of the Persians, against the acropolis at Athens should have been repaid in kind after many years by one woman, a citizen of the land which had suffered it, and in sport.[25]

Thais was also the lover of Ptolemy I Soter, one of Alexander's generals and may also have been Alexander's lover, if Athenaeus' statement that Alexander liked to 'keep Thais with him' means any more than he simply enjoyed her company. On Alexander's death, Thais married Ptolemy and bore him three children, two boys and a girl: Lagus, Leontiscus and Eirene. Thais was never Ptolemy's queen, nor were their children heirs to his throne. Ptolemy, of course, had other wives: Eurydice of Egypt and Berenice I of Egypt.

The memory of Thais endured into the Roman Republic and early Empire, and beyond into the Middle Ages. Terence has a female protagonist who is a courtesan named Thaïs in his *Eunuchus* and Cicero quotes Thais' words in *De Amicitia*. In Ovid's *Remedia Amoris*, Thais' behaviour and *mores* are contrasted with Andromache's: Andromache is the epitome of the loyal and chaste wife, while Thais is the embodiment of sex and is, says Ovid, what his art is all about.[26] Thais is down there in Hell, in Dante's *Divine Comedy*,[27] in the circle of the flatterers, immersed in an excrement-filled trench for having told her lover that she was 'marvellously' fond of him. Thais' words here derive from Cicero's quotations from Terence. She emerges with Alexander, conjured up by Faustus in Marlowe's *Doctor Faustus* and appears as Alexander's mistress in Dryden's *Alexander's Feast*, or *The Power of Music* (1697), in which Alexander is enthroned with 'the lovely Thais by his side' who sat 'like a blooming eastern bride'. The poem later became an oratorio, *Alexander's Feast*, by Handel. Robert Herrick (1591–1674) comes to the following happy conclusion in his 'What Kind of Mistress He Would Have: Let her Lucrece all day be, Thais in the night to me, Be she such as neither will, Famish me, nor overfill.'

Cleophis was Queen of Assacana, a small Indian city besieged by Alexander in 326 BCE and where he sustained an injury. With her people's interests at heart, Cleophis surrendered the city to Alexander.[28] Quintus Curtius Rufus gossips that Cleophis placed her son or grandson on Alexander's lap and that Alexander showed her mercy because of her beauty; Justinus goes a step further by alleging that Cleophis was able retain her kingdom and position by sleeping with Alexander and that she bore Alexander a son.[29]

The Kambojas

The Kambojas (or Ashvakayana or Ashvakas) were famous for their equestrian and horse breeding skills; they also enjoy a reputation for clashing with the Macedonians and deploying ferocious women to fight alongside their husbands. K.S. Dardi's *These Kamboj People* (2001) describes how:

> in the entire ancient world history, the honour of participating, fighting and then attaining supreme mass martyrdom in the active and hot battle field, and that too, against, Alexandra the Great . . . goes only and only to the heroic Kamboja queen Kripya and the brave Kamboja women of the famous Assakenian Kamboja clan. They have the supreme honour of being martyred, not in a direct and straight fighting, but only through demeaning and unashamed treachery resorted to by none other than 'Alexander the Great' himself.

Diodoros Siculus describes how:

> Assakenian Kamboja women were pouncing upon the fighting Greek soldiers with an elemental fury and grappling with them and snatching away their swords, spears and shields . . . While the Assakenian soldiers were crossing swords with the enemy, their wives were covering their fighting husbands with the shields they had snatched from the Macedonian army . . . And still other Kamboja women were picking up arms of those who had fallen or were wounded or cut . . . and were fighting side by side with their husbands in the active field. . . . The Assakenois who had fought valiantly along with their women, could not this time frustrate the well trained and numerous army of Alexandra and thus met with a glorious death which they preferred to lives of disgrace.[30]

PART TWO

WOMEN AS VICTIMS OF WAR

Chapter 11

War Rape and Other Atrocities in the Classical World

'It is a law established for all time among all men that when a city is taken in war, the people and property of the inhabitants belong to the captors.'

Xenophon, *Cyrus*, 7.5.73

Wherever and whenever there is war there are victims. Many are male combatants but many more usually are civilians – non-combatants who include women among their number. Women, as we have seen, can and did participate in classical battle, but often they are left to pick up the pieces during and after war, sometimes literally. For women, more often than not, their war is not over when the war is over: wartime sexual and gender-based violence has a real, enduring impact on women's lives long after the fighting has stopped. Women suffer abject shame and widowhood; they wait anxiously at home always expecting the worst of bad news; they are left to grieve and mourn and to struggle on with their lives, often working their farms or businesses alone and bringing up young fatherless children; where the husband-soldier is wounded they may have to spend their lives as carers, tending limbless or otherwise traumatised ex-servicemen, coping with all the physical and psychological issues disabling injury brings; if raped they are ostracised, rejected by husbands and families; they submit to body shame and loss of personal esteem; sometimes they are displaced – their cities and homes wrecked or requisitioned – forced to move on as penniless refugees, carted off to strange and inhospitable lands with foreign languages and customs where they may suffer more prejudice and sexual and gender-based violence. Just as often they are sold into slavery or become concubines, considered no better than just another bit of the war booty. Women and girls suffer unspeakable and abhorrent abuse – physical, sexual and psychological – they are raped, sometimes orally and anally; they might be gang-raped or repeatedly raped over long periods of time. They may be plagued with sexually transmitted infections, they endure ad hoc abortions with the infection these often bring; there is the possibility of unwanted pregnancies, the half-foreign offspring which are a lifetime's haunting reminder of the violence and trauma they endured. They may be tortured, horribly mutilated and murdered – sometimes in front of their husbands and children as they too await a similar fate.

History tells us unequivocally that women are constant and persistent victims of war: our survey here will demonstrate that the experience of ancient Greek and Roman women, and of some of the foreigners they subdued, was no different from what went before and what has come after, with a relentless inevitability, predictability and monotony. This hideous by-product of war is what allowed the soldier, philosopher and historian Xenophon in the fourth century BCE to make that chilling, ghastly but true, statement quoted at the beginning of this chapter, some 2,450 years ago with all its foul ramifications and consequences for women.

When dealing with reams of numbers and descriptions relating to atrocities and war crimes perpetrated many years ago, it is very easy to become blasé and blunted to the dreadful reality of these events – but *these events were real*, they were actual events which all deserve the same horror and disgust which rightly attends all too similar events presented to us on our television screens today. For us, dealing with the classical era, it all started with the Trojan War – but it predates that, and continues rampant to this day. For example, in Syria in 2016, there were forty-seven or so active sieges affecting an estimated 1,099,475 people (http://syriainstitute.org/siege-watch/).

Before the Greeks and Romans? Women were being sexually assaulted the moment man started fighting fellow man. Physical and sexual violence against women has been with us since the dawn of time: a 2,000-year-old adult female skeleton excavated in South Africa reveals that she was shot in the back with two arrows. A late Ice Age discovery from Sicily has unearthed a woman with an arrow in her pelvis.

More disturbing déjà vu comes when we learn that the Israelites possibly introduced ethnic cleansing to the world when they theologised their reprisals after Yahweh decreed that the raiding Amalekites would be expunged from memory for all time – an early *damnatio memoriae* – for their attacks on and appropriation of Israelite settlements. Again, hauntingly familiar. This hateful curse endured into the time of Samuel and Saul (*c.*1100 BCE) when the former ordered the latter to exterminate the Amalekites down to the last woman and child, and to erase their agrarian economy.[1] Yahweh's sanction of the Amalekite genocide was indicative of the worrying fact that a warring state's actions, however execrable, could be justified and mitigated by the will of God. Divine approval has been necessary before the opening of and during hostilities in many theatres of war throughout the ages; omens in the ancient world were interpreted and had to be favourable before battle, while failure to observe them could be very costly. An imprudent and impious Naram-Sin lost 250,000 men when he chose to ignore ill omens before one battle. Divine sanction has often also been a useful tool in the convenient and conscience-salving justification and mitigation of unspeakable atrocities.

The Assyrians had a shocking reputation, notorious for ripping open the stomachs of pregnant enemy women. As did some Egyptians – not least Amenophis II (1439–1413 BCE), who massacred all surviving opposition at Ugarit. The Egyptians were also infamous for systematically mutilating the defeated; for example, the detritus from the Battle of Megiddo yielded eighty-three severed hands. The walls of the temple at Medinet Habu show piles of phalluses and hands which were hacked from Libyan invaders and their allies by Ramesses III (1193–1162 BCE). Ashurbanipal, the Assyrian king who reigned from 668–627 BCE, rejoiced in his violence, boasting: 'I will hack up the flesh [of the defeated] and then carry it with me, to show off in other lands'. His ostentatious brutality is widely depicted – one picture shows him implanting a dog chain through the cheek and jaw of a vanquished Bedouin king, Yatha, and then reducing him to a life in a dog kennel, where he guards the gate of Ninevah or pulls the royal chariot.[2] Babylon was a particular threat; when Ashurbanipal destroyed the city he tore out the Babylonians' tongues before smashing them to death with their shattered statuary, and then fed their corpses – cut into little pieces – to the dogs, pigs, zibu birds, vultures and fish. Ashur-etil-ilani (627–623 BCE), Ashurbanipal's heir, had a predilection for cutting open the bellies of his opponents 'as though they were young rams'. In the Christian Holy Bible we read of the children of the defeated being dashed to death, and, again, pregnant women having their stomachs ripped open.[3] War rape was a constant; again in the Bible, the prophet Zechariah exults in the sexual violation of women; Isaiah's vision was equally apocalyptic when he states: 'Their little children will be dashed to death before their eyes. Their homes will be sacked, and their wives will be raped.' In *Lamentations* it is written that 'women have been violated in Zion, and virgins in the towns of Judah'.[4]

For the ancient Greeks, too, rape came with the sanction of the gods. The Greek gods and heroes were rape role models: Zeus raped Leda in the form of a swan, Europa in the guise of a bull. He raped Danae disguised as the rain. He raped Alkmen masquerading as her own husband. Zeus male-raped Ganymede. Antiope was raped by Zeus; Cassandra was raped by Ajax the Lesser; Chrysippus was raped by his tutor Laius; Persphone was raped by Hades; Medusa was raped by Poseidon; Philomela was raped by her brother-in-law; the daughters of Leucippus, Phoebe and Hilaeira, were abducted, raped and later married to Castor and Pollux. Homer rated a slave woman at four oxen and an iron tripod at twelve oxen (*Iliad*, 23.703–5) although it must be said in a glimpse of mitigation that both the enslaved Briseis and Chryseis are loved or admired by Achilles and Agamemnon. And so on into Rome where, in legend, the Sabine women were abducted and raped, and that epitome of wifely chastity, Lucretia, was raped and Verginia was slain by her own father to avoid the very real prospect of rape ... both for political and constitutional motives.

The Greeks did little, though, without the sanction of the gods: war and battle were no exceptions. Religion safeguarded the sacred nature of treaties; it demanded the security and inviolability of envoys and it upheld the sanctity of temples and anyone taking refuge in those temples. Armies on the move were accompanied by a flock of sheep fattened up for sacrifice: the rites of bloodletting, or *sphagia*, prevailed at every critical juncture, whether it was before crossing a river, invading a border, striking camp or even starting the battle. Often the sacrifices were preceded by a meal, washed down by generous amounts of wine – that meal could be the soldiers' last supper, the alcohol injected courage and unrestrained abandon, fuelling pillage and rape in victory when a city or town was mercilessly sacked – men, women and children raped and mutilated.

Moreover, the Greeks considered the rape of women acceptable behaviour within the 'rules' of warfare; vanquished women were just nother item of war booty, later to be redeployed as wives, concubines, slaves or recycled war trophies. The practice goes back at least to the *Iliad*, where Andromache speculates on what will happen to her when Troy loses the war. Recalling the destruction of Cilicia, she reminisces how Achilles took her 'mother, who ruled' and 'hailed her ... with his other plunder'. The fact that she was a queen did not save Andromache's mother from slavery, nor did it save Andromache – she too was a queen reduced to a spoil of war. As soon as Troy falls she will be forced from regal splendour and privilege into a life of homesick domestic servitude, working the wool at another's bidding; she will, she fears, be raped on a regular basis. Homer's Greek heroes glory in the fact that they have widowed Trojan women: Diomedes sees widows' nail-torn cheeks as a mark of victory, while Achilles considers weeping women a sure outcome of his fighting. The rape of one's wife is one of the punishments that awaits any Greek or Trojan who dares to break Priam's treaty (having your brains dashed out was the other), a sentiment echoed by Agamemnon; Hector sees the carrying off of the losing side's women an inevitable consequence of defeat. When Achilles skulks off to sulk in his tent, Phoenix, his old tutor, tells the story of Meleager who, in a similar fit of pique, withdrew his forces defending his city because he was annoyed with his mother. The day was saved by Cleopatra, 'well-girdled' wife of Meleager, who tearfully explained to him the multitude of 'sorrows that come to men when their city is taken' – slaughtering the men, firing the city and abducting the children and the 'deep girdled women' (*Iliad*, 9.590–4). Priam for whom, along with rape, the massacring of sons, pillaging and the dashing of babies to the ground was routine post-battle behaviour thinks likewise; as does Andromache, as noted above, despite the solitary efforts Hector, her now dead husband, had made to protect Trojan wives. Early on in the *Iliad*, Nestor discourages the Greek soldier from thinking about going home until he has slept with a Trojan wife, to avenge the abducted Helen of Troy (2.354–6).

At the same time, though, the defence and protection of women was one of the requirements of a warrior in war and was a significant motivating factor. Meleager was certainly fired up after his wife's vivid description of the consequences of war; Hector is berated by Ares, in the guise of Thracian Acamas, for not urging his men to fight for their women; the Greek soldiers are driven by their need to fight for their wives and children; Hector twice rallies his troops, asserting that death in battle is good because it makes wives and children – and the *oikos* – safe; Polydamus advises retreat into Troy to defend against Achilles who will be targeting the Trojans' women; Priam advises Hector likewise for Hecuba; Hector died fighting for the Trojan women and children. Trojan Agenor warns Achilles that there are many Trojans who are standing up to defend the wives, children and parents of Troy.[5]

The rape and enslavement of a woman war victim is used as a metaphor for the sacking of a city when Homer describes how a woman's veil is torn off. Archaeological evidence from the thirteenth century BCE supports this – ration lists for palace servants show the names of women from various towns and cities along the coast of Asia Minor who were, no doubt, captured in war, enslaved and dispersed. Most repugnant is Agamemnon's denial of any quarter to the Trojans when he sees Menelaus showing clemency to a prisoner: Agamemnon berates him, ordering death to all Trojans – even to the foetuses in the wombs of pregnant Trojan women (*Iliad*, 6.55–60). Elsewhere Agamemnon promises twenty captured Trojan women to Achilles and boasts that vultures will feed off dead Trojan warriors while they, the Greeks, lead their beloved wives and innocent children onto the Greek boats (9.139–40, 4.237–9).

If Herodotus had got his way then posterity would be convinced of the fact that it was women who were largely responsible for all the bad things in the history of his Greece. Greeks, Phoenicians and Persians all blame women as the causes of conflict between the three mighty Mediterranean powers. Women-snatching appears to Herodotus to have been at the root of all Greek history, as indeed it was, but masquerading as nation building; it was also at the heart of the Roman civilisation when the Sabines lost the flower of their womenfolk to their bullish, rapacious Roman neighbours. The day the mercantile Phoenicians landed on a beach in the then pre-eminent Argos to sell their goods to the native Greeks was pivotal. Those exotic-looking wares obviously caught the eye and interest of many a Greek woman who proceeded to browse and buy. Many a Greek woman also caught the eye and interest of many a Phoenician man. Herodotus takes up the story: 'On the fifth or sixth day after their arrival, when their wares were almost all sold, many women came to the shore and among them especially the daughter of the king, whose name was Io (according to Persians and Greeks alike)'.[6]

So, the Phoenicians reciprocated first by eyeing up the Greek women and then by abducting some, including Io, whom they shipped off to Egypt. The first blow had been struck. It was never that simple, though, as Herodotus reveals:

> The Phoenicians do not tell the same story about Io as the Persians. They deny that they carried her off to Egypt by force. [According to them] she had sex in Argos with the captain of the ship. Then, when she discovered she was pregnant, she was ashamed to tell her parents, and so, lest they discover her condition, she willingly sailed away with the Phoenicians.[7]

In retaliation, 'some Greeks' – Herodotus diplomatically suggests Cretans – landed at Tyre in Phoenicia and snatched *their* king's daughter, Europa, whence they sailed to Aea, a city of the Colchians; from there they also snatched the king's daughter, Medea. Despite protests, the Greeks refused to return Medea because Io had never been returned to them. A de facto legal precedent had been set and so, some fifty years later, Trojan Paris, son of Priam, was able to abduct Helen with what he thought would be impunity. The Greeks protested but were told that she was staying with Paris because they had been denied when those Greeks had refused to return Medea. The result, of course, was a serious escalation into the ten-year-long Trojan War. Retribution for stolen women was now loose in the world.

But the real tragic denouement to this tangled issue of international elite women trafficking was that, despite its sheer moral and legal wrongness, in the words of Herodotus: 'So far there had been nothing worse than "women stealing" on both sides ... to be anxious to avenge rape is foolish: wise men take no notice of such things. For clearly, the women would never have been carried away if they had not wanted to be.'[8] This sad and sorry refrain has echoed down the ages – from Herodotus here via Ovid who believed that women like a 'bit of rough' and really meant 'yes' when they said 'no' – to today, lamely justifying man's brutal treatment of women when man rapes woman.[9] It is but a short, ambivalent step from Herodotus' assertion to a defence today for rape: the woman consented. Such a sentiment would have resonated with the readers of Herodotus, giving some the all clear to treat their women as they wished, just like Xenophon's war victims.

One of the key issues in the abduction of Helen was a political and military issue: the ensuing war, and Helen's role as a catalyst, willing or otherwise, became a cause of enduring enmity between the Greeks and their Mediterranean neighbours.

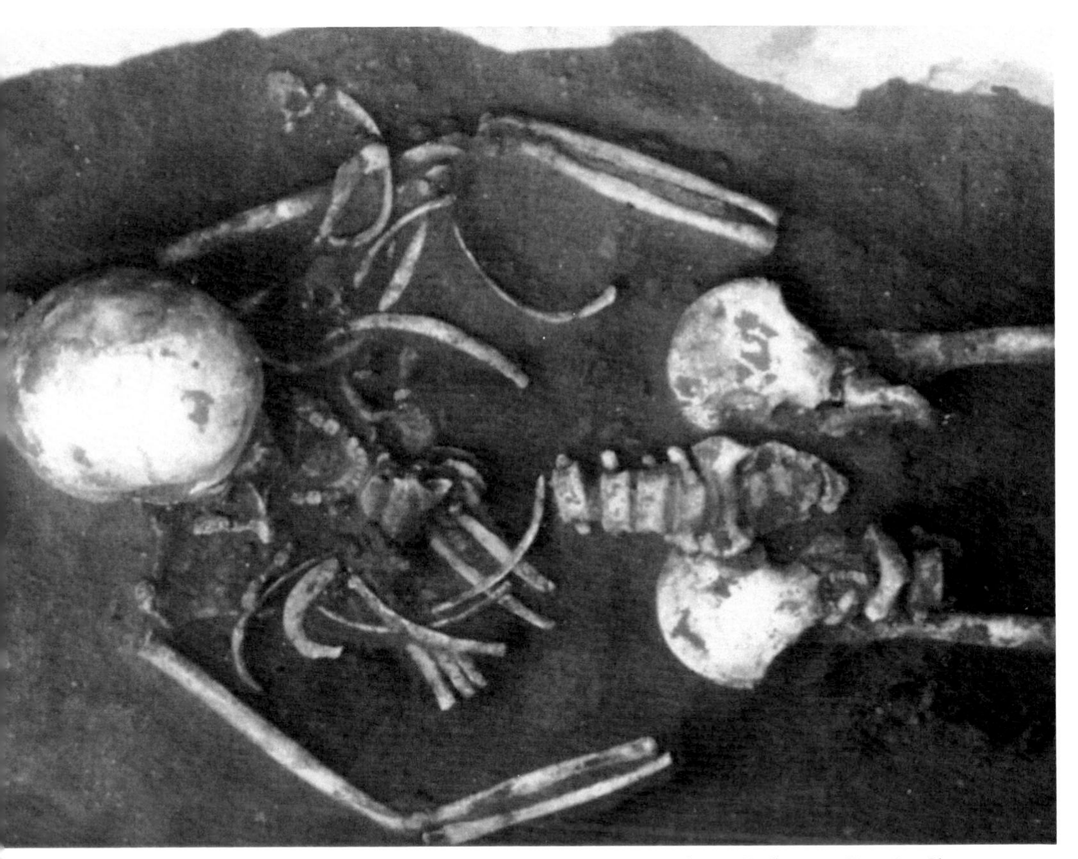

he skeleton of a young woman from Cemetery 2, Kurgan 8, Burial 4 at Pokrova, Russia. She was uried with a dagger and arrowhead, suggesting that she was a woman warrior.

.eft) Andromache, Astyanax and Hector in a touching scene from the *Iliad*. Astyanax, on ndromache's knee, reaches out to touch his father's helmet before his duel with Achilles (Apulian ·d-figure column-crater, *c*.370–360 BCE). Now in the Museo Nazionale of the Palazzo Jatta in Ruvo i Puglia (Bari).

Right) Menelaus goes to strike Helen but stunned by her beauty, he drops his sword. A flying Eros ıd Aphrodite (on the left) watch the scene. Detail of an Attic red-figure crater, *c*.450–440 BCE, found Gnathia (Egnazia, Italy). Louvre Museum Department of Greek, Etruscan and Roman Antiquities, ılly wing, Campana Gallery.

A scene showing Achilles fighting with the Amazon Penthesilea; on a sarcophagus dated *c.*250 CE now in the Vatican museum.

Papyrus fragment with a drawing of the abduction of Briseis by Talthybius and Eurybates in the *Iliad* (Bk 1, ll. 330–48), fourth century BCE. Bayerische Staatsbibliothek pap. gr. 128.

Amazonomachy marble, sarcophagus panel, *c.*160–70 CE. This served as a basin for the Tigris fountain.

Amazon mosaic from Paphos found in the Orpheus House. Note the exposed right breast, the Phrygian cap and the double-headed axe. Some argue that this is Hippolyta in the ninth labour of Hercules.

mazonomachy scene on a kythos, *c*.420 BCE. The patterned ggings and short tunics gether with crescent-shaped ields give the Amazons away being 'Persian'. Now in the etropolitan Museum of Art, ew York.

Jacques-Louis David (1748–1825), *The Intervention of the Sabine Women*, 1799. The Louvre.

Jean Jacques François Lebarbier, *A Spartan Woman Giving a Shield to Her Son*, 1805. A Spartan woman is bidding her husband, or maybe her son, farewell in the traditional manner, 'Return carrying your shield or on it'. All elements of the painting reinforce its message of civil duty. The children playing with the warrior's lance allude to Spartan military training, which began in infancy. The simplicity of the stone-walled interior underscores the austerity of Spartan existence, while the dog is both a symbol of fidelity and a reference to the famed dogs of Sparta.

Teichoskopia: how women would have viewed the battle scene below. This is the rampart walk on the East Wall in Jerusalem with the Dome of the Rock in the background. Thanks to Professor Tod Bolen.

sölchem lob vnd eer/dz er zů allen werken .

CORIOLAN

VETVRIA

der ritterschaft für ander berůmet wåre. Vnd über

Woodcut illustration of Veturia and Volumnia confronting Coriolanus, hand-coloured in red, green, yellow and black, from an incunable German translation by Heinrich Steinhöwel of Giovanni Boccaccio's *De mulieribus claris*, printed by Johannes Zainer at Ulm, *c.*1474.

detail from the frieze of the asilica Aemilia depicting the unishment of Tarpeia, *c.*14 BCE.

incenzo Camuccini, *Roman Women Offering their Jewellery in Defence of the State*, *c.*1825. © Glasgow useums.

(*Left*) Paul Jamin, *Brennus et sa part de butin – Brennus and his Share of the Spoils*, 1873.
(*Right*) Statue of a mourning barbarian woman, a victim of war, which stands in the Loggia dei Lanzi, Florence. It was originally thought to be Thusnelda, wife of the captured Arminius, but is probably a personification of Germania. A.W. Lawrence described the statue's 'intensity of grief' in *Classical Sculpture*.

The terrible scene that is the end of the Roman Empire, with women pursued in the chaos – Thoma Cole, *The Course of Empire 4: Destruction* (1836). Thanks to The Schiller Inc, Washington DC.

eft) Pavel Svedonsky (1849–1904), *Fulvia with the Head of Cicero*. Now in the Kizhi State Open Air useum of History and Architecture, Russia.

ight) Agrippina I on the Rhine fortifications. © Jasper Burns; thanks for his permission to produce this, originally published in *Great Women of Imperial Rome*.

ter Froste (b. 1935), *The Temple of Claudius* in Camelodunum, going up in flames at the hands of udica in 60 CE. © Colchester and Ipswich Museums Service.

A detail from the Column of Marcus Aurelius showing captured women and children.

(*Left*) Woodcut illustration showing Artemisia II of Caria drinking the ashes of her husband Mausolus. It is hand-coloured in red, green, yellow and black, from an incunable German translation by Heinrich Steinhöwel of Giovanni Boccaccio's *De mulieribus claris*, printed by Johanne Zainer at Ulm, *c.*1474.

(*Right*) Marble relief from Halicarnassus showing two female gladiators fighting, first or second century CE; the inscription tells us that their names were Amazon and Achillia. Now in the British Museum (GR 1847.4.24.19).

Von Arthemesia der küngin
in Caria das lv .

Rthemesia ist ain küngin gewesen in Caria
ains höhen starken gemütes. hailiger und
überseltzemer liebi zu ierem gemahel / und
so fester küschait in ierem witwen stät dz ir

Although normal conduct and laws are often suspended during times of war, it is still useful to look at what the norms were in cases of rape in Greece and Rome. Rape from a legal standpoint was complicated. The gods, as noted, were serial rapists; Zeus uses non-consensual sex not just for sexual gratification, but to demonstrate and exert his virility and omnipotence. If you were to ask him why he raped, he may well have glibly replied, 'Because I can'. Some men and many victorious soldiers would have sympathised with this. Rampant Greek and Roman soldiers would have needed little encouragement to rape and loot, but if they did then they had to look no further than their gods or legendary heroes to sanction their brutish behaviour.

The vocabulary only confused the issue: classical 'rape' was not what we today understand to be rape, revolving around the issue of consent.[10] The verb *harpazein* meant 'to carry off' or 'abduct' as *raptare* did in Latin; to convey the connotative meaning of non-consensual sexual violation, the Greeks added *bia* – 'with force'. Forced sex in Latin was usually expressed as *stuprum* (fornication) with the addition of *cum vi* or *per vim*, with violent force. *Raptus ad stuprum* was abduction with a view to committing a sex crime. The legendary 'rape of the Sabine women' was literally the 'abduction of the Sabine women', an undisguised and shameless act of nation-building; no doubt the actual raping, as we understand it, took place when the Romans got the Sabine women and girls home. To honour the (reluctant and involuntary) role played by women in this early example of empire building, the thirty individual political wards of the new Rome were triumphantly named after thirty of the women.

Moreover, in Greece and Rome, women came under the definition of property, because legally speaking in Rome at least they were under the power of a man (*patria potestas*) or guardian. The rape of a woman was a property crime, committed against the man who owned the woman. If this was the case in everyday, normal civilian life, what chance did the vanquished woman have after a siege? In wars fought by the Greeks and Romans and way before that, particularly in the often repellent postscript to sieges, the systematic rape of women, men and children on the losing side was, as with Xenophon, par for the course, with the odious assumption that it was an expectation, a duty even. It has been so, to a greater or lesser extent, ever since.[11] This may account for the absence in Herodotus of descriptive episodes of rape: rape was simply quite normal battle or siege aftermath behaviour. His mention of the rape of Phocian woman by Persian troops in 480 BCE is a rare case. 'Barbarian' Xerxes impaled Sataspes for raping the daughter of Megabyzus, but probably because it was an infringement of Megabyzus' property rights; the same is probably true in his description of Paris' abduction of Helen where the offence is committed against Menelaus and not Helen, who is merely his property. Herodotus points out that the higher the status of the victim then the heavier the penalty for committing rape; he compares the penalties for rape enshrined in the Gortyn Code which vary according to the standing of the victim, and describes how sexual voracity corresponds with political lust and tyranny.

Generally speaking, the historians are largely silent on the issue of war rape. Other than the case of the Phocians referred to above there is nothing else that has survived. Whether this is due to the fact that it was standard practice and not worthy of mention, or because of some shame or embarrassment on the part of the historian, it is impossible to tell. However, it is difficult to believe that, in reality, rape was not a terrible finale to the numerous sackings and defeats where women and children were taken into captivity and enslaved – given what we know about warfare before and after the Greeks. In 1919, in the wake of the First World War, Helen Law posed the question 'how far the silence of the historians is to be regarded as proof that other and perhaps worse atrocities were not committed. It is conceivable that other acts of violence were not recorded because they were regarded as unimportant or not unusual enough to be interesting, or were suppressed to shield the perpetrator of the deed'.

Thucydides describes (7.29) how the Thracians sacked Mycalessus and the women and children were put to the sword; but were the women raped beforehand? The pupils in a boys' school in Mycalessus were certainly massacred. Women were enslaved in 427 BCE when the Spartans established that the Plataeans had never helped them against the Thebans and up to 200 Plataean men were slaughtered and the town was razed (Thucydides, 3.68). The women and children of Mytilene had a lucky escape when the Athenians rescinded a command that they be enslaved (3.50). The women of Scione were not so lucky when they were enslaved and their men slain (4.122) after the town revolted in about 423 BCE. When Cleon captured Torone he sold the women into slavery and sent the men away to Athens (5.3) in 422 BCE. Isocrates tells us that when the Melians refused to compromise after the Athenians rejected their request to remain neutral Melos was besieged, the men were butchered, the town was destroyed and the women and children found themselves enslaved (Isocrates, 4 and 12). Mass deportation of all inhabitants including wives and children occurred at Aegina in 431 BCE and Histiaea 480 BCE (3.27 and 1.114). Philip II and Alexander the Great were brutal besiegers: at Sarnus in Illyria Philip deported more than 10,000 inhabitants; there were also mass deportations at Potidaea, Stageira and Olynthus.

Sataspes was Xerxes I's cousin and, as stated, had been condemned to death for abducting and raping Megabyzus' daughter. However Sataspes' own mother, Atossa, persuaded Xerxes to escalate the punishment – Sataspes was then commanded to circumnavigate Africa, a Herculean task that he failed to accomplish. He feebly claimed that at the farthest point he reached he met a 'dwarfish race, who wore a dress made from the palm tree', and that his ship was becalmed (the Benguela Current?). Xerxes was unimpressed and had him impaled. Atossa, it seems, was much more concerned about the seriousness of her son's crime than either Sataspes or Xerxes.

Sexual molestation was not covered by diplomatic immunity, according to Alexander the Great. When he killed the Persian ambassadors for fondling what they thought were Macedonian women at a banquet, he agreed to give his sister Gygaie in marriage to Bubares in compensation for the murders. These 'women' were in fact 'smooth-chinned young men' who had been put at the disposal of the unsuspecting and inebriated Persians and reciprocated their guests' attentions by wielding their daggers on the befuddled Persians. Servants and carriages disappeared too but the promise of Gygaie and a large sum of money kept the incident a diplomatic secret (5.20ff.).

Roman soldiers were free to seize a captured woman and effectively enslave her for their own personal use – for life; a manifestation of the mass abduction of the Sabine women on a individual level.

Further down the pecking order, things were yet worse. Raping a slave – many of whom would have become slaves as victims of war – was only considered a criminal offence when the slave's owner had not given consent – more a case of damaged goods than atrocious sexual violence against a fellow human being.

Some more enlightened generals and politicians did, nevertheless, try to curb the practice of routine war rape. In the late Republic, Cicero was at pains to separate the men from the brutes. According to him, the acquisition of plunder and booty was a legitimate reason for going to war, but fair play was important too. Fair play in the prosecution of war (*mores belli*), however, was notably absent in Verres' sacking of Syracuse, where massacre and rape went hand in hand and the law of the victors (*ius victoriae*) was taken to extremes.[12]

Atrocity was sometimes a means to an end: in the Jugurthine War, the city of Capsa was razed to the ground by Marius – not for reasons of greed or brutality, but to avoid any future resurgence of the strategic city. Moreover, the people there were stereotyped as a duplicitous and fractious lot: so, the men were butchered, the women and children were sold into slavery and the booty was divided up. To Sallust this was a breach of the *ius belli*: 'as for the rules of war, the crime against the rules of war was not committed because of the greed or criminality of the consul, but because of the strategic advantage of the place to Jugurtha'.[13]

The Romans were generally ruthless and thorough in their pursuit of victory, and by the same token, in their avoidance of defeat: they probably felt that they could justify their behaviour because, for all they knew, it might be them next on the receiving end when Roman women, *their* wives, mothers, sisters and daughters would be subjected to violence, enslavement and rape. Calgacus speaks for many in his *declamatio* at Mons Graupius (in modern

Scotland) in 83 CE when he exhorts his army before the decisive conflict with Agricola's legions. The Romans, he asserts, are rapists on a global scale, *raptores orbis*; to plunder, butchery and rape they give the misnomer 'government'; they call the social and physical desolation they create 'peace'.[14] Earlier in the British campaign Boudica's daughters had been raped and she had been flogged by rapacious and venal Romans. But we do not have to take Calcagus' word for it: Boudica herself vocalises her anger and contempt after the ignominious and cruel treatment meted out to her and to her daughters. She rallied her troops from a chariot, her violated daughters standing before her: in a rousing speech recorded by Tacitus she declared that, to the Britons, there was nothing unusual about a woman commander and invoked her lineage from mighty warriors; *solitum quidem Britannis feminarum ductu bellare testabatur*. She was fighting for her kingdom, her freedom and to avenge her battered body and her raped daughters: *contrectata filiarum pudicitia*. Crucially, she was intent on erasing the *stuprum* (depravity) inflicted on her daughters and restoring their violated *pudicitia* (chastity); the Romans slay the elderly and rape virgins: *ut non corpora, ne senectam quidem aut virginitatem impollutam relinquant*.[15]

Generally speaking, though, the earlier historians of Rome, particularly Livy who describes rape in legend but infrequently in history, are relatively reticent on war rape – possibly, as has already been suggested, because it was the norm and taken for granted that it would and did happen? But who is to say that the countless women subjugated by the Romans or subdued by their enemies did not suffer rape and its consequences on a scale relative to, say, the 2 million or so German women and girls raped by the Red Army in 1944 and 1945 on the roads to Berlin in revenge for the rapine inflicted on Russian women by the invading Wehrmacht earlier in the war?

Rape, of course, was never the exclusive preserve of the Romans; all armies did rape. Perhaps an illustration of the general ubiquity of rape comes in Livy's description of the sack of Victumulae by Hannibal in 218 BCE when all manner of atrocity was visited upon the hapless inhabitants by the Carthaginians:

> They had just completed the surrender of their arms as commanded, when orders were suddenly given to the victors to treat the city as though it had been taken by storm, and no form of mutilation, which on such occasions historians usually mention, was left undone – so terrible was the example set of every type of depravity and cruelty and brutal tyranny visited on the poor inhabitants.[16]

Indeed, Livy uses rape or the threat of rape, as a vehicle to explain some of the most momentous events in Roman history. The first, and most famous, episode was the 'rape', or abduction, of the Sabine women in 750 BCE, alluded to above.[17] This was an exercise in nation building – the Romans needed women to prolong their race so they took what they could find, married them and produced the required progeny.

The rape of Lucretia by monarchist Sextus Tarquinius, and her subsequent suicide, firmly roots brutish rape and matronly *pudicitia* and valour in the traditional foundation of Rome. Livy tells how Tarquinius blackmailed Lucretia by threatening to put it about that she had been *in flagrante delicto* with a slave – a terrible, shameful crime; he threatened to have her and a slave killed, their corpses placed next to each other in bed, if she did not yield to him. Lucretia could not live with the shame that such a calumny would bring and succumbed; her body is defiled but, she protests, her mind remains pure. Despite the absolute forgiveness of her father, Lucretius and of Collatinus, her husband, Lucretia commits suicide: brave, virtuous and inextricably linked with Rome's proud early beginnings. Rome, like Lucretia, was compromised and violated by the Tarquins; the noble reaction of Lucretia and her avengers symbolises Rome's honourable struggle against regal tyranny and led to the overthrow of monarchy in preference to a republic. Livy makes her an unimpeachable *exemplum* of feminine virtue in early Empire Rome, his Rome, then beset by increasing adultery and failing marriages.

To Valerius Maximus Lucretia is *dux Romanae pudicitiae* – a leader in Roman sexual propriety – who, by a wicked twist of fate, possesses a man's spirit in a female body. Brave Cloelia is also a *dux*, leader, of a column of girls, as we shall see. Agrippina is a 'big-hearted woman who took on the role of a leader'. *Dux* is almost always used in classical Latin to describe a man, although in two further instances it is used to signify two powerful, unnerving militaristic foreign women: Dido and Boudica.[18] Military leadership to Roman historians was exclusively a male quality – any woman who possessed it was defeminised accordingly.

It is important to note, though, that such military related masculinisation is not confined to the classical world. As recently as the 1940s, Austrian Princess Stephanie Hohenlohe (1891–1972) was interned by the US army after the battle at Pearl Harbor on suspicion of being a spy and described as 'worse than ten thousand men'. On release in 1945 she simply reverted to her old ways as a fixer, enjoying good relations with US Presidents John F. Kennedy and Lyndon B. Johnson. Her earlier successes included paving the way, for a fee, for Viscount Rothermere to meet Hitler in the 1930s, and organising the visit of the Duke and Duchess of Windsor to Nazi Germany in 1937.

The rape of Lucretia is all the more shocking because Livy prefaces it with a story highlighting Lucretia's virtue. During a drinking session, which included Tarquinius and Collatinus, at the siege of Ardea, the subject of wives, good or otherwise, came up; an alcohol-fuelled decision was then made to ride back to Rome and Collatia to establish whose wife was the most virtuous. The princes' wives were found ensconced in a sumptuous dinner party, *in convivio luxuque*. On the other hand, even though it was late at night, Lucretia was discovered sitting surrounded by her maids in the hall working away at the wool by lamplight. This, of course left, no doubt as to who had won the contest.[19] According to Ovid, Lucretia was working on a cloak for Collatinus.[20]

The death of Verginia in 449 BCE had not dissimilar causes and ramifications. It occurred at a highpoint of the Conflict of the Orders – the struggle between the patricians and the plebeians – during the *decemvirate*, an ultimately unsuccessful attempt to reorganise Roman administration and codify the law (which found voice as the enduring Twelve Tables) while all other magistracies were suspended and official decisions were not subject to appeal. In this case the actions of Verginius, father of Verginia, provoked a secession of the plebeians, and precipitated the abolition of the *decemvirate* and the restoration of the consulship and tribunate of the peoples, and with it the right of appeal. The *decemvirs* had become more and more tyrannical and refused to concede power, even though their tenure was limited to two years. This corruption and despotism was eerily reminiscent of the rule of the kings, overthrown some sixty years earlier and due, as we have seen, in no small part to Lucretia. Members of the second *decemvirate* either committed suicide or were exiled. Livy explains the constitutional significance and how the Republic was restored because of this tragic incident:

> This was followed by a second atrocity, the result of brutal lust, which occurred in Rome and led to consequences no less tragic than the outrage and death of Lucretia, which had brought about the expulsion of the royal family. Not only was the end of the decemvirs the same as that of the kings, but the cause of their losing their power was the same in each case.[21]

Appius Claudius, a *decemvir*, had his eye on Verginia and stalked her. His attempt to seduce her and win her favour with promises fell flat. Verginia was from a humble, honourable, principled family, and although she was already betrothed to Lucius Icilius, Claudius contrived a plan whereby Marcus Claudius – 'the decemvir's pimp' – claimed that Verginia was actually his slave and had been illegally adopted by Lucius Verginius. While on her way to school one day, Marcus Claudius attempted to abduct Verginia. A shocked and outraged crowd gathered round to help the girl. Marcus Claudius insisted that the matter be dealt with in court before a judge, namely Appius Claudius; in the meantime Verginia was to be held with Claudius while her father rushed back from active service for the hearing. An enraged Lucius Icilius also arrived, threatening to die rather than betray his fiancée; Appius relented and allowed Verginia to go home. Despite attempts by Appius to have Verginius arrested before he could reach Rome, the distraught father arrived and next morning led his daughter into the Forum; he is in mourning, she is dressed in rags, attended by *matronae*. Appius, to everyone's astonishment, found in favour of Claudius. Livy tells us that the supportive women in the crowd were 'more effective with their silent tears than words could be'. Nevertheless, the incident ended tragically when Verginius exercised his right as *paterfamilias*, grabbed a butcher's knife and stabbed his daughter in the heart to liberate her from the shame (*stuprum*) a marriage to Appius Claudius would inevitably bring: *in libertatem vindico*. Verginius cursed Appius; the women angrily and indignantly demanded to know if this be the reward they get for raising children and remaining chaste: *pudicitiae praemia*. According to Livy, *muliebris dolor*, 'the grief of women', is so much more powerful, their grief so much deeper due to their lack of restraint. The upshot was that the *decemvirate* was disbanded and the Republic was restored.

Two women, Lucretia and Verginia, who paid for their *pudicitia* with their lives; two women who, as symbols of virtue, helped to change the course of Roman history; two women who, like Rome itself, were vindicated and defended by men against the tyrannical abuse of power. 'The rape of women became the history of the state', as Joshel puts it.[22]

Not that rape was endorsed by all. The Capuan Vibius Virrius uses the inevitability of Roman army rape as a reason to defect to Hannibal in 216 BCE during the Second Punic War: Virrius no more wants to see the rape of Capuan women, girls and boys than he does his own execution or the sacking of his city.[23] In 204 the Locrians complained about the behaviour of Roman soldiers garrisoned in their town: rape and plunder were rife and relentless.[24]

There were other glimmers of humanity: Livy reports the episode in which a beautiful Spanish girl was brought before Scipio Africanus during the siege of Carthago Nova, presumably to pleasure the commander; Scipio sent her away knowing that she was betrothed to a local man. This civilised and compassionate behaviour evoked obvious gratitude from the girl's family, which in itself resulted in a welcome conscription of 1,400 local men to the Roman cause.

Pausanias records how long before that in the late eighth century Aristomenes the Messenian ambushed a number of virgin dancers of Artemis at Caryae. He had been careful to choose those girls who had the wealthiest and most respected fathers. When he turned the girls over to his men that night for guarding, the men tried to rape the girls. Aristomenes reacted, saying their behaviour was decidedly un-Greek: the men ignored him so he killed the most violently drunk among them. He then ransomed the girls and returned them 'in a state of virginity just as he had taken them'.[25]

But, more usually, the atrocities knew no bounds. An early example of total siege came in 672 BCE when the Romans sacked Alba, destroying all the houses and forcing the whole population out of the smoking town with nowhere to go. The houses they grew up in were gone; their possessions littered the streets and their gods were held captive by the Romans. All the women could do was grieve and wail. Cicero, in his *In Verrem* of 70 BCE, tells of the unbridled lust of Verres in Sicily: 'To how many noble virgins, to how many matrons do you think he offered violence in that foul and obscene lieutenancy? In what town did he set his foot that he did not leave more traces of his rapes and atrocities than he did of his arrival?'[26]

Cicero's observation was undoubtedly reinforced by Verres' lubricious but unsuccessful attempt to rape the virgin daughter of Philodamus in Lampsacus: 'his daughter, who was living with her father because she had not yet got a husband, was a woman of extraordinary beauty, but was also considered exceedingly modest and virtuous. [Verres], when he heard this, was so consumed with desire ... that he said he should like to visit Philodamus immediately'. This behaviour was by no means out of character: Verres' rape of Syracuse was consistent with his soldiers' unjustifiable rape of 'the nobility or the rape of matrons, atrocities which then ... were simply not done, either through hatred of enemies, through military licence, or through the customs of war or the rights of victory'.[27]

In the late Republic Vercingetorix reminded his Gallic countrymen that, dreadful as it was, the depredation of their lands to eliminate food supplies for the Romans was preferable to slaughter and the enslavement of their women and children.

We hear from Tacitus regarding indiscriminate rape, violence, sacrilege and pillage inflicted on Italy by Vitellius' troops in 69 CE: 'Italy was beset by something much worse and atrocious than war: dispersed through towns and colonies Vitellius' troops plundered, raped and polluted the land with their violence and lechery; greed and corruption made them go for the sacred and the profane, the good and the bad', and his armies' unrivalled, un-Roman even, savage licentiousness and rapine later in the year: 'then Vitellius and his army exploded into savagery, rape and lust that knew no equal – the behaviour of foreigners'.[28]

Vitellius' hugely corrupt recruitment campaign among the aggrieved Batavians involved the systematic rape of adolescent boys. The ferocious sack of Cremona by Antonius was notorious for the systematic rape and butchering of women and the elderly on a prodigious scale:

> Forty thousand soldiers burst into Cremona with even more army suppliers and camp followers who were even more corrupted by lust and savagery. Age and dignity provided no protection as they interchanged rape with slaughter and slaughter with rape. Old men and aging women – useless as booty – were dragged into the 'fun'; any grown up girl or

fine-looking man who came along was torn apart at the violent hands of the rapists.

Such was the scale of this atrocity that the public was repelled; it turned the Roman people against the armies of Antonius. Slaves taken at Cremona remained unsold on the slave markets; survivors were surreptitiously ransomed by their relatives.[29] The general disgust may have been exacerbated by the fact that these victims were Italians caught up in a civil war, rather than foreigners or barbarians. This, of course, torpedoes the widely held belief that rape and plunder were always indicative of Roman might and superiority against foreigners. Moreover, a scene from the Column of Marcus Aurelius from around 193 CE clearly shows Roman soldiers looting a defenceless village and assaulting women.

The situation in Rome becomes all the more shocking when it seems that familiarity started to breed contempt. If the historians and rhetoricians are to be believed, the endless descriptions of sacked cities, raped women, pillaged temples and enforced evacuation or enslavement were becoming somewhat tedious. It had, of course, all started many centuries earlier with the sack of Troy – as staple of Greek and Roman school education – and had been relentless ever since. Polybius (2.56.7), writing before *c.*118 BCE, deplored the 'ignoble and womanish' style used by the historian Phylarchus to excite the emotions of his readers, what with all those 'clinging women with their hair dishevelled and their breasts bare ... weeping and lamenting as they are led away into slavery'. Descriptions of women being raped and enslaved would seem already to have become clichéd and commonplace. Sallust, in the first century BCE, complains that 'arms and corpses, gore and grief are all over the place' (*Ad Catilinam*, 51.9); the orator-teacher Quintilian (35–100 CE) gives a master class in the effect of hyperbole, recommending the use of descriptions of women shrieking and mothers fighting desperately to keep hold of their children (*Institutio Oratoria*, 8.3.67–70).

Cities that surrendered rather than being taken by storm received a relative degree of clemency. Livy describes two such cases: Pometia in 502 BCE and Phocaea in 190 BCE. At Cartagena in 209 BCE Scipio stopped the wholesale slaughter (which included slicing dogs and other animals in half) once Mago had surrendered.[30] The clear message from the Romans was that resistance was just not worth the vicious reprisals that would inevitably follow; as far as the Romans were concerned, holding out in a siege was no different from any other form of hostility against the Romans. In 216 BCE savage retribution followed in Capua when the inhabitants locked up a number of Roman citizens in the steaming, airless bath house; they died a terrible, suffocating death according to Livy.[31] Images of the 1756 Nawab of Bengal atrocity, Black Hole of Calcutta, spring to mind.

Heliodorus gives us a fictional account of siege compassion (*Aethiopean Story*, 9.11). When besieged by the Ethiopian army, the people of Cyrene in North Africa realised that resistance was futile so threw themselves on the mercy of the Ethiopians, wailing and carrying branches in supplication, bearing images of their gods. They then sent out their children who, terrified, crawled or toddled over to the lines of the Ethiopians and won their mercy. Whether there is any basis in reality for this cannot be known, but there seems no reason not to believe that there is a germ of truth in there somewhere.

Julius Caesar leaves a seemingly endless catalogue of massacres visited on non-combatants: in 55 BCE when he attacked the Upisi he despatched his cavalry to hunt down the women and children who had fled pell-mell in absolute terror; he casually describes the slaughter of the elderly, the women and the children of Avaricum in 52 BCE in revenge for an earlier Gallic slaughter of Roman citizens in Cenabum; in 51 BCE at Uxellodunum he had the hands of the enemy cut off as a terrible and tangible warning to anyone else contemplating resistance.[32] This mutilation by amputation may seem comparatively merciful compared with execution, but cutting off limbs in the first century BCE was tantamount to a slow death as many of the victims would have soon succumbed to massive loss of blood, sepsis or shock. Even at the end of the last century, the amputation of hands under Saddam Hussein prompted an Amnesty International report in 1996 citing the high mortality of these prisoners due to wound infection.

Caesar also recounts how the women of one surrounded town found themselves in a terrible situation when their menfolk decided to desert them and their children and flee to Vercingetorix. The women got wind of the plan and implored their men not to leave them. But all was in vain, so they made such a clamour and din that it alerted the Romans who realised that something was amiss. The Gaulish men abandoned their plans for their ignominious escape as their route to defection would now be blocked.

In Britannia in 78 CE Agricola started as he meant to go on when he massacred the Ordovices; they had had the temerity to attack one of his cavalry squadrons; he then ruthlessly took the Druid stronghold on the island of Anglesey.[33] While women are not specifically mentioned as victims here, there is no reason to believe that they were not among the casualties.

In 6 CE Judaea became a Roman province, governed by Roman prefects. On one level there was stability – due in large part to an agreement fixed by Augustus whereby the Romans exercised religious tolerance with the Jews while the Jews, or rather the High Priest and the wealthy landowning Sanhedrin, tolerated Roman rule. On another level there was no such satisfaction; many Jews resented the occupation and looked forward to the day when the

Messiah returned to deliver them from the Romans. They marked this anticipation with insurrection. From 6 CE, bands of terrorists, *sicarii*, or knife-men, attacked Roman census officials, melting back into the desert after committing their assaults. These bandits were not only inimical to the Romans, they were the sworn enemies of the Zealots, the opposing faction of Jews who made up the majority of the rebels against Rome. The fragile status quo was, however, upset by the appointment of a series of procurators notable for their corruption and by the rapacity routinely deployed, often with impunity, by the Romans in suppressing the chronic unrest.

Initially, the Romans had been mere spectators to the growing internecine ferment in Judea. Their client king Agrippa II tried to restore peace without success; some of those advocating peace, including the former High Priest Ananias, father of Eleazar ben Hananea, were murdered by the Zealots. The 600-strong Roman garrison tried to prevent further hostilities but were subjected to a siege in their Jerusalem headquarters for their troubles. They were forced to surrender, only allowed free passage on condition they gave up their weapons. The required oaths were taken and the commander Metilius marched out of the former palace of Herod the Great, believing he and his men were on the road to freedom. As soon as the Romans laid down their arms and armour they were brutally attacked by Eleazar and his fellow terrorists. Nonplussed, the Romans frantically appealed to the oaths and the agreement, but to no avail. They were butchered to a man – that is with the exception of Metilius who promised to convert to Judaism, and a humiliating, very un-Roman, circumcision.[34]

In 66 CE the situation deteriorated rapidly and came to a head when Jews in Jerusalem started attacking Roman citizens and the garrison reacted. The soldiers sent in defence were also assaulted: 6,000 Jews were whipped and crucified in reprisal while the Temple was looted in an act of naked iconoclasm. The Jews humiliated Gessius Florus, the inept, venal and corrupt procurator, after his raid; Florus' response was as swift as it was brutal: he released his troops to wreak rapine and death in the upper market in Jerusalem. The result was wholesale slaughter in which Jewish homes were looted and 3,600 men, women and children were slain, the women raped, no doubt, beforehand. Others were crushed to death in the narrow alleyways. This marked the beginning of four years of serial atrocity, culminating in the sack of Jerusalem.[35]

The news of the riots spread, fanning insurrection in other towns and cities: Agrippa II (Herod Agrippa) and his sister Berenice fled Jerusalem to Galilee. After the destruction of the Temple in 70 CE members of the Sicarii left Jerusalem and settled on Masada. Josephus reports that the Sicarii raided nearby Jewish villages including Ein Gedi, where they massacred 700 women and children.[36]

Cestius Gallus, the legate of Syria, advanced on Jerusalem bringing with him something of a reputation for massacring civilians and razing villages, notably Lydda during the feast of Tabernacles. To the rebels this was a red rag to a bull: they attacked the Romans, killing 550 to their 22 casualties. Despite conciliatory attempts to broker a peace by the Romans, the rebels remained obdurate: Gallus took Acre and then Caesarea and Jaffa, where he massacred some 8,400 citizens. He then razed Bezetha and advanced as far as Mount Scopus on the outskirts of Jerusalem, but failed to take the Temple Mount; he halted his advance and prepared to besiege the inner city, but astonishingly, the Romans suddenly withdrew from Jerusalem and headed towards the coast. Gallus had been within an ace of bringing the war to a comparatively bloodless end; this diversion was certainly going to make it much worse for the Jews, as they were to discover to their bloody cost later in 70 CE.

The expected reprisals by the Romans in Jerusalem soon came in the predicable shape of a siege. The siege ended with the sacking of the city and the destruction of its Second Temple. Depictions on the Arch of Titus in the Forum in Rome stand today as testament to the merciless sack of Jerusalem and the Temple. Josephus' description of the most harrowing aspects of the predicted famine could have been written by the epic poet Lucan, noted for his characteristically graphic and melodramatic horror: these are the lengths to which the Romans were prepared to go to avenge Beth Horon:

> children pulled the morsels that their fathers were eating out of their very mouths, and what was still more to be pitied, so did the mothers to their infants; and when those that were most dear were perishing under their hands, they were not ashamed to take from them the very last drops that might preserve their lives: and while they ate after this manner, yet were they not concealed in so doing; but the seditious every where came upon them immediately, and snatched away from them what they had got from others; for when they saw any house shut up, this was to them a signal that the people within had got some food; whereupon they broke open the doors, and ran in, and took pieces of what they were eating almost up out of their very throats, and this by force: the old men, who held their food fast, were beaten; and if the women hid what they had within their hands, their hair was torn out for so doing; nor was there any commiseration shown either to the aged or to the infants, but they lifted up children from the ground as they hung upon the morsels they had got, and shook them down upon the floor. But still they were more barbarously cruel to those that had prevented their coming in, and had actually swallowed down what they were going to seize upon, as if they had been unjustly defrauded of their right. They also invented terrible methods of torture to discover where any food was, such as blocking up the anuses

> of the miserable wretches, and driving sharp stakes into their rectums ... in order to make them confess that he had but one loaf of bread, or that he might discover a handful of barley-meal that was concealed.[37]

Incredible as it may sound, this description is eclipsed by Josephus when he gives an illustration of a woman called Mary who is forced by starvation into cannibalism. She roasts her breastfeeding baby and devours half the body:

> She murdered her son, and then roasted him, and ate half of him, and kept the other half hidden. Evil men soon came in and, smelling the horrid scent of this food, they threatened to cut her throat immediately if she did not show them what food she had been preparing. She replied that she had saved a good portion of it for them, and uncovered what was left of her son. Then they were horrified and astonished, and stood aghast at the sight ... the men went out shaking, never having been so frightened in their lives ... they left the rest of the meat to the mother.[38]

Jerusalem eventually fell to the Romans on 7 September and was systematically destroyed. Josephus records that 1.1 million people were butchered after the siege, most of whom were Jewish; 97,000 were captured and sold into slavery.[39] He sums up the carnage as follows: 'The slaughter inside was even more awful than the spectacle seen from outside. Men and women, old and young, insurgents and priests, combatants and those who wanted mercy, were mowed down in indiscriminate carnage. There was more slain than slayers. The legionaries had to clamber over heaps of dead to proceed with their extermination.'

There was more horrific cannibalism some 300 years later. Ammianus Marcellinus, the Greek historian, is our principal source for the Siege of Amida in 359 CE; he was a Roman commander at the siege and describes events in his *Res Gestae*. The siege occurred when the Sassanids, under Shapur II, besieged the Roman city of Amida in 359 which he had previously lost to the Romans. As the Persian armies advanced, so the Roman farmers were forced to evacuate their lands and seek places of safety such as Amida.[40] Ammianus describes the ferocity of the fighting: '[we were] so crowded together that the bodies of the slaughtered, held up by the crowd, could find nowhere to fall, and that in front of me a soldier with his head cleaved in two, and split into equal halves by a powerful sword stroke, was so crowded on all sides that he stood erect like a stump'.

No surprise, then, that the siege lasted seventy-three days with many disastrous attempts by Shapur to capture the city; siege towers were set on fire by the Romans and plague broke out in Amida. The Persians eventually prevailed and tied up all the men in the amphitheatre where they died. According to the (Christian propagandist) *Chronicle of Pseudo-Joshua*, famine was rife and women were compelled to eat old shoes and stones 'and other horrible things from the streets and squares'.[41] Some resorted to prostitution selling sex for food to the Persians who left the women to die when the food became too scarce. An incredulous Joshua explains how women then resorted to cannibalism, forming bands that kidnapped other women, the elderly and children; these would then be roasted or boiled and eaten. The Persians were alerted to these atrocities by the smell of the roasting flesh and put an end to it by torturing and executing the guilty women.

Nothing shows the suffering of women in war more graphically than the Imperial victory column and victory arch. They are the classical equivalent of propaganda billboards or newsreel, depicting destruction and rapine on a large and explicit scale. We can gaze at the wholesale destruction of the Dacians and the deforestation of their lands on Trajan's Column; the sacrilegious plundering of the Jewish temple on the Arch of Titus. The Arch of Marcus Aurelius gives us newsreel frozen in time, showing the routine abuse and rape of barbarian women at the hands of Roman soldiers. Dacian women are depicted torturing Roman soldiers – something a Roman *matrona* would never dream of doing. The sculptors of Marcus Aurelius also give us a timeless illustration of Dacian women utterly distraught and terrified as they cling on to their children. The clothes of German women are dishevelled, as is their hair; they are dragged along by their hair in the same way that their country is dragged by the Romans into oblivion. The Roman victory column or arch is still with us – not just frozen in time in twenty-first-century Rome but in Picasso's *Guernica* from the last century, and in the footage we see regularly tracing the atrocities and carnage visited on Syria and the women, children and men bombed out and displaced there.

PART THREE

ROME

Chapter 12

Military Women in Roman Legend

We have seen how Livy used women in a military setting as a vehicle for signalling major constitutional reform at Rome. These were by no means the only women to whom Livy ascribes martial qualities.

Rhea Silvia was embroiled in the earliest of Rome's military scenarios. She was the daughter of Numitor, King of Alba Longa, and was descended from Aeneas, founder of Rome. Numitor's younger brother Amulius staged a coup, seized the throne and killed Numitor's son; he then compelled Rhea Silvia to become a Vestal Virgin, effectively consigning her to celibacy for the next thirty years and ensuring that the line of Numitor had no heirs.

Rhea Silvia, however, gave birth to the twins Romulus and Remus claiming the god of war, Mars, as the father. Livy says that she was raped by an unknown man, but 'declared Mars to be the father of her illegitimate offspring, either because she really thought that to be the case, or because it was less shameful to have committed such an offence with a god'. Romulus and Remus, of course, went on to topple Amulius, and reinstate Numitor as King of Alba Longa – the rest is ancient history ...[1]

Hersilia was the wife of Romulus. We learn from Livy's description that she was Rome's first female diplomat and woman military advisor. Romulus was jubilant at his victory over the army of the Antemnates although Hersilia was mobbed by entreaties for clemency from the captured Antemnate women; she implored Romulus to show mercy and take them on as citizens of Rome which would have the advantage of strengthening the state. He was only too happy to do this, thus performing one of the first acts of extending Roman citizenship to the vanquished – a policy that was to serve Rome well for centuries to come.[2]

Titus Tatius (d. 748 BCE) was the Sabine King of Cures, who attacked Rome – aided by the treacherous Tarpeia.[3] After their women had been famously seized and abducted by the Romans, according to Livy, the Sabines declared war on Rome. They were aided in their assault on Rome by Tarpeia who opened the city gates to the Sabines. She had gone outside to fetch water for a sacrifice; she was the daughter of Spurius Tarpeius, the commander of the Roman citadel. Why she betrayed Rome is open to dispute. Some say that she was bribed by Tatius, with the promise of receiving what the Sabine soldiers wore on their shield arms, their left arms. A venal Tarpeia could only see as far as their gold bracelets (*armillae*) and bejewelled rings; the Sabines, unfortunately for her, saw it very differently and crushed her to death with their shields. Others give an aetiological explanation and maintain that the Sabines killed Tarpeia by pushing her off a 25m-high cliff on the Capitol which later took her name (*Rupes Tarpeia* or *Saxum Tarpeium*) and became notorious as a place of terrible execution for murderers, traitors, the mentally and physically disabled. The Latin proverb '*arx Tarpeia Capitoli proxima*', 'The Tarpeian Rock is close to the Capitol', means one's fall from grace can happen very quickly.

The Sabines failed to take the Forum, its gates miraculously protected by boiling jets of water unleashed by Janus. Tarpeia became a byword in the Roman world for greed and treachery: Propertius talks of Tarpeia's 'shameful tomb'.[4]

The Sabine women, by now settled Roman wives and mothers, were in the thick of it again later when they, through some desperate but effective diplomacy, managed to persuade Sabine Tatius and Roman Romulus to bury their differences; this bold, self-deprecating and selfless action led to joint rule by the Romans and Sabines. Livy provides the details:

> [They] went boldly into the midst of the flying missiles with dishevelled hair and torn clothes. Running across the space between the two armies they tried to stop any further fighting and calm the excited passions by appealing to their fathers in the one army, and their husbands in the other. [They begged them] not to bring upon themselves a curse by staining their hands with the blood of a father-in-law or a son-in-law, nor upon their heirs the taint of parricide. 'If,' they cried, 'you are weary of these family ties, these marriage-bonds, then turn your anger upon us; it is we who are the cause of the war, it is we who have wounded and slain our husbands and fathers. Better for us to die than live without one or the other of you, as widows or as orphans.'[5]

Cloelia is described by Livy as *dux* in her brave escapade, an epithet, as we have seen, usually reserved for men; she demonstrated conspicuous bravery and military sagacity. Lars Porsena took some Roman hostages as part of the peace treaty which ended the war between Rome and Clusium in 508 BCE. One of these hostages was a young woman named Cloelia. She managed to flee the Clusian camp on horseback, at the head of a group of Roman girls. According to Valerius Maximus, she then swam across the Tiber back to the Romans. Porsena demanded that she be returned, and the Romans consented. When she arrived back at Clusium, Porsena was so impressed by her bravery that he allowed her go free and to choose half the remaining hostages to be liberated with her. She picked the Roman boys, because she knew that they could continue the war. The Romans gave Cloelia an honour usually reserved for men: an equestrian statue, located at the top of the Via Sacra.[6]

It was women too who averted the Volscian attack on Rome led by Gaius Marcius Coriolanus in 491 BCE. Where envoys and priests had failed to appease Coriolanus, the women of Rome, '*ingens mulierum agmen*', 'a mighty column of women', sought out his mother and his wife, Veturia and Volumnia. They persuaded them to go into the lines with Coriolanus' two young sons and implore him to withdraw: 'where men failed to defend the city with their swords, women may be more successful with their tears and pleas'. Moved by the presence of his loved ones, and by an angry and impassioned Veturia who appealed to his duty as a son and to Rome, Coriolanus relented and withdrew his army.[7] Veturia's success was, no doubt, assured when she reminded Coriolanus how much she had sacrificed for him over the years: providing an education and remaining a *univira*, a one-man woman, after his father's death, thus making her mother, father, nurse, teacher and sister to him. Rome honoured the women by building a temple dedicated to Fortuna.

Chapter 13

Military Women in Roman History

Women were involved in the military security of Rome from very early days, and rose to the occasion during the very first military crisis endured by the city. Servius, in his commentary on the *Aeneid*, tells how in 390 BCE, when the Gauls were besieging the Capitol, women banded together and donated their gold to the treasury, and their hair to make bowstrings for the Roman archers.[1] This, of course, is reminiscent of a similar action by Hellenistic women recorded by Hero of Alexandria. The Gauls blockaded Rome for seven months, finally winning when the Romans surrendered in the face of imminent starvation. Luckily for Rome, the Gauls were more interested in revenge and short-term financial gain than the domination of Italy, or slaying, or enslaving the population; they were accordingly paid off in gold, and promptly returned to their lands in the north.

Some fathers exercised their *ius vitae necisque*, their legal right to decide the fate of their daughters, on moral grounds: in about 642 BCE Horatia, the fiancée of one of the Curiatii, enemies of Rome, made the mistake of mourning his death after he was slain by one of her three brothers, two of whom also died in the fight. The surviving brother, Publius, stabbed Horatia to death, proclaiming that 'any woman who mourns an enemy of Rome will die like this'. Her father sanctioned the sororicide, adding that he would have killed her himself had her brother not.[2]

Valerius Maximus records two Campanian women as *exempla* of gratitude in *Factorum ac dictorum memorabilium libri*, Book 5, Chapter 2. One example comes during the siege of Capua led by Quintus Fulvius Flaccus when these two Campanian women, sympathetic to Rome, spent much time and money to help the city. One was Vestia Oppia, married with a family, who worked hard every day helping the Roman army. The other was Cluvia Facula, a prostitute, who supplied food to the Roman prisoners of war. When they had subdued Capua, the Roman Senate restored the Capuans' freedom and property and, where appropriate, rewarded them. The two women were praised at a Senate meeting in 210 BCE.

At 9.1.8 ext. 13 Valerius tells of Dripetrua (fl. first century BCE, also known as Drypetina) who was the daughter of Laodice and Mithridates VI of Pontus. Dripetrua was born with a double row of teeth (hyperdontia), and lived with the resulting facial deformity. She is noted for staying at her father's side after his defeat by Pompey the Great in the Third Mithridatic War (75–65 BCE), providing, no doubt, considerable moral support. Boccaccio includes her in his De Mulieribus Claris.

Appian records a number of brave and devoted acts by women following the proscriptions imposed after the assassination of Julius Caesar in 44 BCE. The wife of Acilius used her jewellery to bribe and disorientate the soldiers who had come to arrest him, and escaped with him to Sicily; Apuleius' wife threatened to turn him in if he refused to let her escape with him; Antius' wife concealed him in a blanket to effect their escape; and the wife of Rheginus hid him in a sewer and dressed him as a donkey-driving charcoal seller to make good their escape.[3]

Hortensia is perhaps the most famous female orator in the classical world. She was the daughter of Quintus Hortensius Hortalus (114–50 BCE) himself a celebrated speaker, *bon viveur*, friend, rival and sometime colleague of Cicero. Hortensia became something of an academic: she was unusual in that she spent her adolescent education studying rhetoric and the speeches of her father, of Cicero and the Greek orators. Her time was not wasted. In 42 BCE the three members of the Second Triumvirate, Lepidus, Octavian and Mark Antony, were casting around for funds to finance their wars against the assassins of Julius Caesar; the money procured from their campaign of proscriptions was running out. An easy, apparently vulnerable, target was women's wealth: accordingly the triumvirs levied a super tax on the 1,400 most affluent women in Rome. The triumvirs had seriously miscalculated, however. The women were incensed and, in a rare example of public demonstration by *matronae*, they marched on the Forum to vent their anger.

Here Hortensia delivered a passionate and eloquent speech at a *contio*, before the triumvirs. Hortensia demanded to know why women, who had neither say in nor benefit or glory from the wars, should be required to finance those wars. Attempts to disperse the women failed. Hortensia's passion won the day and the taxable quota was radically reduced to 400 women; men with over 100,000 sesterces were obliged to make good the shortfall by loaning money to the Triumvirate; immigrants, *peregrini*, were similarly taxed. Appian says that the speech was so good that one would never know that it had been written and delivered by a woman. Valerius Maximus says that Hortensia's father henceforth 'lived and breathed through the words of his daughter'.

Sempronia, the Catilinarian conspirator, showed bravery and determination but, not surprisingly in the circumstances, she attracted condemnation and vituperation. Sallust writes that she exhibited a boldness worthy of a man; she was well married with children; she was versed in Latin and Greek, accomplished in the lyre and a good dancer, she was an excellent and convivial conversationalist: in short her bravery, social and artistic skills and abilities were above reproach. That said, though, Sallust adds acidly that, marriage and motherhood apart, she displayed none of the qualities expected of the conventional *matrona*: she was impulsive, louche, passionate, a perjurer, an accessory to murder, a liar and a spendthrift. Sempronia to Sallust was a kind of anti-*matrona*: while undoubtedly brave and gifted, she broke the mould and stepped far beyond the traditional boundaries laid down for Roman women. Her support for Catiline and the stigma this brought, her independence of mind and ostentatious social skills would have caused outrage and anger in some quarters.[4]

Suetonius described Porcia as brave because of an incident in her second marriage, to Brutus. Immediately before the assassination of Caesar, it became apparent to Porcia that her husband was unduly distracted and anxious. She determined to share his concerns and proceeded to slash her thigh with a knife, causing substantial loss of blood. Brutus was alarmed; she told him that she was much more than just a sex partner in their marriage, indeed 'a partner in your joys and a partner in your sorrows' – for better or for worse. She, by virtue of her ancestry and her demonstrable bravery, claimed that she was above the usual weakness of women and was worthy of his confidence in anything that might be troubling him. Seeing Porcia's wound, Brutus told Porcia everything. Porcia had proved to him that she could conquer pain and would not, therefore, yield under torture: Brutus could confide in her with absolute confidence. Valerius Maximus concludes that Porcia was assessing for herself how well she would be able to face suicide in the event of Brutus' death. After Caesar's murder Porcia and Brutus parted. Her husband showed how fully he appreciated Porcia's courage: Plutarch says that she was distraught on seeing a painting of Hector leaving Andromache for war; when Brutus heard this he, out of respect, declined to repeat Hector's parting words to Andromache but rephrased it as follows: 'work at your loom and your distaff, and give your commands to your servants. She may not have the strength for the same exploits as are expected of a man, but she has the spirit to fight as nobly for her country as any of us.'

Plutarch's source is Bibulus, their son; it is remarkable because it is the first example we have of an enlightened man enjoining his wife to fight for her country like a man. When Brutus was killed in 43 BCE Porcia committed suicide by swallowing hot coals, an act that Valerius Maximus describes as 'her woman's spirit equal to her father's manly death'.[5]

Mucia Tertia, daughter of Quintus Mucius Scaevola, earned something of a reputation as a diplomat. In 62 BCE Cicero appealed to her to intercede with Metellus Nepos, her kinsman, who had vetoed Cicero's valedictory speech as consul. Later, she was sent to Sicily by Octavian to negotiate with Sextus Pompeius, her son, who was starving Italy with his grain blockade. Mucia was unsuccessful but a later meeting with Pompey that included her daughter-in-law, Scribonia, Sextus's wife, was fruitful and led to the 39 BCE Treaty of Misenum. This ended the blockade and brought Sextus into the Triumvirate. In 31 BCE, after Actium, she persuaded Octavian to spare him.[6]

The impact of war sometimes made *matronae* act bravely en masse. The proposal in 195 BCE to repeal the *Lex Oppia* of 215 BCE evoked conservative male disgust and condemnation when women came out from their homes and demonstrated in the Forum to support the repeal. The law had restricted the use of luxuries by women in the wake of the disastrous Battle of Cannae some twenty years before. It limited the amount of gold women could own and required that all the assets of wards, single women and widows be handed over to the State; the wearing of dresses with purple trim and riding in carriages within Rome or nearby towns was also prohibited, except during religious festivals. The feeling among women was that the law had served its purpose and had run its course. Livy records the speech given by Marcus Porcius Cato and his embarrassment and anathema at what he saw as indecorous behaviour ill-befitting Roman women. To give it some context, Cato, generally no friend of women, had famously paraphrased Themistocles when he asserted: 'All men rule their wives; we rule all men; our wives rule us'. He harps back to the days of the ancestors who permitted women no public activity or commercial dealings without a guardian, and who safeguarded the power of fathers and husbands. Cato is appalled by this populist female action and their demands that they be heard.[7] The repeal, nevertheless, was approved, thanks in no small part to the reasonable and enlightened arguments of Lucius Valerius:

> For a long time our matrons lived by the highest standards of behaviour without any law, what is the risk when it is repealed, that they will give in to luxury? ... Are we to forbid only women to wear purple? When you, a man, can use purple on your clothes, can you not permit the mother of your family to have a purple cloak, and will you let your horse be more finely saddled than your wife is dressed?

Soldiers, Marriage and Baggage Trains

Servius gives us evidence that the Roman attitude to women and the military was more restrictive than the Greek. In his commentary on Virgil's *Aeneid* he indicates that women were never seen in the camps, *castra*, and that camp followers were held in low regard.[8] We know the first statement to be untrue. In 134 BCE Scipio Aemilianus arrived in Numantia to revivify the Roman legions there; one of his first acts was to expel the prostitutes from the camps. To Frontinus, the military strategist, such action appears under his heading *de disciplina*: in his view, clearing out the whores was obviously good for discipline. Camp followers were always disparaged, it seems: Cicero, with an axe to grind and a case to make, describes the followers of Catiline as effeminates – a deep insult to the Roman ego and his *virtus* – who even contemplated taking their wives with them on campaign.[9] Servius is supported by the anonymous author of the rhetorical exercise *Miles Maranus* which was written in the second century CE: discipline demands that women be kept out of the camp.[10] At the end of the Republic, Propertius has a woman in one of his elegies bewail the fact that military camps were off limits to females.[11]

Even if any of that were true, the situation had certainly changed by the early Empire. Military service had an increasing impact on families: many men were absent from home for long periods of time and for much of their careers due to the demands of military service and provincial administration. This would have fostered a degree of independence among wives left behind at home. War widowhood could, paradoxically, have had a beneficial financial advantage for some women who were in a position to inherit the family business or fortune.

The drift towards accompanied postings began when officials started erecting statues in the provinces celebrating their wives; in time, the statues were replaced, at least in the highest echelons, by the real thing: Caecilia Metella led the way when she accompanied her husband, Sulla, to Athens in 86 BCE; in 49 BCE Cornelia went with Pompey to Lesbos and Egypt. In the Empire Agrippina, Plancina and Triaria all bear testament to the fact that women were accompanying their men to the outposts of empire. In 21 CE, Drusus, son of Tiberius, delivered a speech explaining the presence of Livilla, his wife, during his service in the 'remote parts of empire', citing Augustus as having set a precedent with Livia. Livia, though, was also praised for shunning camp life.

It was, however, by no means always a popular development. Aulus Caecina Severus adopts an extreme view, exaggerated in his unsuccessful speech of 21 CE, as recorded by Tacitus; but it contains within it the arguments that no doubt influenced the regulations relating to accompanied postings. Severus always got on with his wife and they had six children together – however, he left her at home for the forty years he was away in the provinces. Why? Because women encourage extravagance in peace time and weakness during war; they are feeble and tire easily; left unrestrained they get angry, they scheme and boss the commanders about; he cites instances of women running patrols and exercises, of how they attract spivs and embrace extortion.

Tiberius took a similarly dim view, complaining that women were taking over the army and that his generals were now redundant. Tacitus would have approved; in the same year he reports a debate in the Senate proposing a ban on governors taking their wives with them to the provinces, and being present at military manoeuvres.[12]

A tragic example of the consequences of a woman infiltrating a barracks occurred in 39 CE during the reign of Caligula. Gaius Calvisius Sabinus had been accused of *maiestas*, treason, in 31 CE after the fall of Sejanus – but he had survived the purge; now he was under suspicion from Caligula. Unfortunately for Sabinus, Cornelia, his wife, was allegedly in the freakish, voyeuristic habit of watching soldiers doing their exercises; less innocently, she also visited sentries at night dressed up in military uniform and offering her body. One such night in 39 CE she was caught in flagrante in the headquarters. Cornelia was accused of entering the camp at night dressed as a soldier, interfering with the guard and committing adultery in the general's headquarters. Her partner in crime was Titus Vinius (12–69 CE) – a general who became very powerful in the reign of Galba, and who was having this daring affair with his commander's wife on what was his very first campaign. Cornelia and Sabinus committed suicide before they could be brought to trial.[13] Propaganda, perhaps, to deter women from fraternising with the military? Vinius' reward was imprisonment by Caligula, but when Caligula was assassinated he was freed. Tacitus called him 'the most worthless of mankind'.[14]

Soldiers were not permitted to marry on service but many, no doubt, did, building relationships with local women and starting families with them.[15] One of the fascinating Vindolanda Tablets from around 100 CE shows that wives of officers clearly did accompany their husbands abroad: Claudia Severa sent an invitation to her sister Lepidina asking her to make her day by coming to her party on 11 September. The body of the letter is written by a scribe but the postscript is written by Claudia and is the oldest example of a woman's handwriting in Latin in existence.[16] These tablets offer us a unique glimpse into military life on the edge of empire, particularly military family life – indisputable proof that such postings were accompanied postings.

Indeed, there is extensive and sound textual tombstone evidence in every Roman province that wives and families accompanied their higher ranking husbands – from commanding offices down to junior officers such as centurions and decurions. In Britannia at Brougham tombstones were erected by wives for their husbands. Yet more conclusive evidence comes from excavated burials which include a child interred in an annexe at Cramond and a murdered woman buried alongside a man under a floor in the Housesteads *vicus*. Significant cremation remains of women, infants and children were also discovered at Brougham, indicating a sizeable non-combatant population. Of the 227 individual remains, twenty-seven were confirmed as female, and thirty-eight infant. Shoes, in sizes which can only fit women, have also been excavated amounting to 14.7 per cent of all shoes excavated in forts in northern Britain. A baby feeding bottle was unearthed at Mumrills and a toy axe at Cramond: infants and children, of course, assume the presence of mothers. Typically, women's jewellery is much in evidence – dress pins, brooches, armlets and bracelets, ear- and hair rings – as are items associated with textile production, although it is quite possible that these were used by men as well as women.

The wife of Aulus Plautius, Pomponia Graecina, *insignis femina* – a distinguished lady according to Tacitus – was a convert to Christianity while with her husband in Britannia. Aulus Plautius was the Roman commander of the invasion force which conquered Britannia in 43 CE and became first governor of the province; Pomponia spent most of her life bravely mourning Julia, a relative of hers and grandaughter of Tiberius', who was executed by Claudius, her maternal uncle. Her public mourning was an outward show of defiance against the Imperial household: after fourteen years of this she was charged in 57 CE for harbouring an *externa superstitio*, a 'foreign superstition' (Christianity). A family court was convened by her husband: she was acquitted and left to get on with her brave act of disobedient mourning for another twenty-six years. The governor Agricola's wife, Domitia Decidiana, was presumably with Agricola in Britannia when Tacitus reports that Agricola was grieving over their dead son who had died the year before in 83 CE.

The marriage ban, then, was originally imposed probably for reasons of *disciplina*, but also because it was difficult to post soldiers from place to place when they had wives in tow. Herodian states that the ban was imposed in the first place because marriage was one of those things 'alien to military discipline and prompt readiness for action'. Severus, the emperor who lifted it, was probably in a weak position because he relied on his armies for support and would have been vulnerable to their demands relating to in-service marriage. His cynical advice to his sons was to enrich the army and despise everyone else. In addition, the ban absolved the State of any responsibility for soldiers' dependants to the extent that it did not need to provide shelter or food, or indeed military protection. In any event a soldier's pay would not have enabled him to support a family until the rise awarded by Severus. The ban kept soldiers focused and it fostered camaraderie undiluted by affection towards and preoccupation with wives and women. This may go some way to explaining why the ban was not revoked when it became standard practice for a soldier to spend most of his service in one garrison (three years minimum, and as long as seven). The children born from the informal relationships that were formed between soldiers and native women were never recognised as legal heirs. Epigraphical evidence shows us that those soldiers who married slave women often made provision for them to be freed when they, the soldiers, died. A bronze military diploma now in the British Museum awarded to a Spanish cavalryman, Gaius Valerius Celsus who served in a Pannonian unit in Britannia in about 100 CE, shows that he and his wife and children received Roman citizenship at the end of his service as an auxiliary. Gaius Julius Caesar Augustus Germanicus (12–41 CE) won his sobriquet, Caligula, from the soldiers who saw him in the little boots he wore around the army camps which he frequented with his mother, Agrippina the Elder, in Germania.[17]

The original aim of the ban was to insulate the soldiery from the normal life going on around them; as armies of occupation they were micro-societies in which marriage initially had no part. Sex, though, obviously always did, and prostitutes would have been much in evidence. D.S. Potter and D.J. Mattingly, in their *Life, Death, and Entertainment in the Roman Empire*, suggest that if the British army's early twentieth-century experience in India is anything to go by then the ratio of prostitutes to soldiers would have been in the region of 1:48 – that amounts to a lot of prostitutes. Suppressed natural sexual energy, of course, goes some way to explaining (but not justifying) the frequency of rape among the military inflicted on subjugated populations – female and male. Boys, like women and girls, were not safe: Tacitus (*Histories*, 4.14) tells how a recruitment campaign among the Batavi on the Rhine during Nero's reign involved dragging off and raping the most handsome of their youth. Soldiers were routinely billeted in civilian homes: we can only speculate on how often and how many families were torn apart by the sexual appetite of their guests.

A discovery in the garrison town of Dura Europos in Syria might suggest that the army laid on entertainment for its troops. A series of inscriptions in a garden there lists the names of a number of male and female entertainers – and of a couple of Roman officials who, presumably, organised it all. While this probably does not indicate a military brothel, it seems very likely that the entertainers would have provided additional services, given what we know about the rather loose morals of dancing girls and flautists. This ties in with Tacitus' acid statement that Syrian legionaries in general were keen to make money wherever they could.[18]

There is precious little detail regarding the logistics surrounding a Roman army; however, Livy does tell us that in 169 BCE the praetor C. Sulpicius took possession of 6,000 togas and 30,000 tunics for his armies in Macedonia. While men certainly did 'work the wool' in the field, it seems inconceivable that women would not have been involved in aspects of what amounted to a huge military textile industry – mending or spinning, for example, which were considered to be particularly feminine crafts.[19]

Women as Military Donors

One particular way in which women contributed to the war effort in the conflicts against Carthage is their invaluable contribution of money or jewellery, or other items, for the war effort. The money would have been used to finance weapons and armour or to fit out ships and make defences. In the Third Punic War, Diodorus reports how the women of Carthage, faced with the extirpation of their city and the end of Punic civilisation, handed in their golden jewellery which was melted down and minted as gold coinage between 149 and 146 BCE.[20]

This was not the first time women provided financial support. During the First Punic War, Carthage found itself under siege in 239 BCE by Libyan mercenaries protesting because their Punic masters were not able to pay them.[21] In an attempt to help their army, Libyan women not only cashed in their jewels but bound themselves by oath to do so. Polybius also tells how in 146 BCE the *poleis* of the Achaean League were threatened by the Romans: the Achaean commander, Diaios, in desperation, released slaves and liberated them in a bid to strengthen his forces; at the same time, the local women were compelled to sell off their jewellery to help the League, but this only led to chaos and further desperation.

Polyaenus describes how the women of Thasos cut off their hair so that it could be used as roping in siege machinery, *mechanemata* – probably torsion catapults.[22] The idea caught on and we have evidence for its use in the *Historia Augusta*, in Lactantius, and Caesar at Salona and at Carthage, Massilia and Rhodos.[23] Financial assistance, and not a little morale, provided by women continued throughout history with its connotations of patriotism, loyalty and self-sacrifice. Perhaps the most famous example is the jewellery donated by Prussian women to finance the fight against Napoleon's armies in 1805: 'The Prussian royal family exhorted its female citizens to hand over their gold and silver jewellery as a personal contribution to the war effort . . . a public attestation of self-denying loyalty and patriotism'.[24]

In 390 BCE, when the Gauls were besieging the Capitol, women banded together and donated their gold and hair to make bowstrings for the Roman archers. After the invention of the catapult in the Hellenistic period, women found for themselves another way in which they might make an important contribution to the war effort: to enable heavy rocks and other missiles to be shot or lobbed long distances there was a need for elastic materials: animal tendons, horse hair – and women's hair was just the job, according to Hero of Alexandria (*c.*10–*c.*70 CE) in his seminal *Treatise on Ranged Weapons* (*Belopoeika*). Nourished with olive oil and plaited, it provided perfect material for torsion on these new weapons of mass destruction. The technical treatise, *Cheiroballistra*, *About Catapults*, is sometimes attributed to Hero.[25]

At the siege of Sinope by Mithridates II in about 250 BCE Polybius records that the Sinopeans received nearly 8 tons of hair and 2 tons of animal tendons from the Rhodians – that must have left many a crop-haired woman and few domestic animals in Rhodes for some time after.[26] In order to use their 2,000 or so catapults when Carthage was under siege by the Romans, the women of the city donated their hair as part of the rush to arm.[27]

Women as Siege Fighters

Unlike their Greek counterparts, the Romans and their allies were not, after the sack of Rome by the Gauls in 387 BCE, compelled to endure many sieges, and so were not so often subject to volleys of lethal roof tiles or other such defensive action from besieged women. However, we know that a few attacks made by the Romans did elicit a belligerent response by fighting women. When the Romans besieged Corioli, a town of the Volsci south of Rome, in 493 BCE, they were met by fierce resistance from 'everyone according to their strength and power', including women. At Fidenae, in 426 BCE, the Fidenates and Veientes joined forces against the Romans, who were under the command of Mamercus Aemilius Mammercinus in his second dictatorship. Mamercus had the hills to his right, the Tiber to his left and occupied a ridge behind, unseen by the enemy. The battle marked a first for the Romans – baptism by fire in hostile enemy use of incendiary tactics; this was the first time they encountered fire-brandishing women who used torches as their only weapons. The mighty enemy column, ablaze with burning torches, instilled sheer terror in the Romans – but after recovering from their initial shock, and being rallied by Mamercus, the Romans seized the enemy torches and turned the tables by attacking the Fidenates with their own brands. The troops on the ridge then attacked, surrounding the Veientes, cutting them down at the Tiber and forcing the Fidenates back to their city, where they surrendered.

An attack on Veii in 396 BCE saw the Romans mine under the city only to emerge from their tunnel to a reception committee which included women hurling tiles from the rooftops.[28] Sallust tells us how, during the north African war with Jugurtha (112–106 BCE), a Roman garrison was wiped out by a combined force of soldiers and women and children at Vacca.[29]

At the siege of Petelia on the coast of Bruttium in the Second Punic War, the Carthaginians apparently faced a combined force of men and women who were 'no less manly than the men' (Appian, 7.5.29). The Roman siege of Carthage saw women working alongside men in the manufacture of munitions; the city became one big arms workshop with men and women working round the clock, turning out all manner of arms and armour: daily production peaked at 100 shields, 300 swords, 1,000 missiles for catapults, 500 arrows and spears and as many catapults as it was possible to make. And then there was that hair, cut off and donated by the women.

Female bravery, generosity and patriotism is evidenced by the actions of Paulina Busa of Canusium (Canosa di Puglia) in 216 BCE after the catastrophe that was Cannae. Stragglers from the battle made their way to Canusium in Apulia where the wealthy Busa fed, clothed and provided them with money; she summoned doctors and medical supplies and took care of the casualties – an early version of Médecins Sans Frontières. Eventually a force of some

10,000 men was assembled which Publius Scipio was able to turn into a viable fighting unit to face Hannibal again. Busa's great loyalty and beneficence was regarded by some with envy, to the extent that the nearby town of Venusia, not to be outdone, themselves recruited a force of cavalry and infantry. Busa was honoured by the senate for her action. To this day a house can still be seen in the Roman city centre of Canosa, known to the locals as belonging to Paulina Busa.

From Plutarch we learn about Praecia – a beauty and a wit – and something of a political and military fixer; she was instrumental in helping Lucius Licinius Lucullus win the governship of Cilicia in 74 BCE, and subsequently the much sought-after command against Mithridates. Praecia, though little more than a prostitute according to Plutarch, wielded great influence and power. She began an affair with the equally influential, and dissolute, Publius Cornelius Cethegus, arch-enemy of Lucullus. She was soon dictating everything Cethegus did: 'nothing important was done in which Cethegus was not involved, and nothing by Cethegus without Praecia'. Lucullus saw his chance and proceeded to insinuate his way successfully into Praecia's affections to the extent that Cethegus was soon backing Lucullus for the governorship. Lucullus won his command and soon dispensed with the services of Praecia and Cethegus.[30]

After the assassination of Julius Caesar, the pivotal family meeting (*consilium*) called by Marcus Junius Brutus and Gaius Cassius Longinus at Antium included three powerful women: Servilia Caepionis (b. *c.*104 BCE, d. after 42 BCE), long-term mistress of Caesar, mother of Brutus and Cato the Younger's half-sister; Porcia Catonis, Brutus' wife and cousin; and Servilia's youngest daughter, Junia Tertia, or Tertulla, (*c.*60 BCE–22 CE). Servilia actually chaired the *consilium* and was party to and influential in one of the most significant sequences of events in Roman history. Cicero tells that she made efforts to have the derisory *curatio frumanti*, the grain commission, removed from a senatorial decree outlining post-assassination appointments for Cassius and Brutus. In 63 BCE, Servilia had caused a scandal in the Senate during a debate on the Catiline conspirators: she sent a love letter to Julius Caesar; her half-brother Cato, who was on the opposite side to Caesar in the debate, accused Caesar of corresponding with the conspirators and, embarrassingly for all, demanded the letter be read out. In 48 BCE Servilia was still influencing Caesar when, after the Battle of Pharsalus, he gave orders for Brutus to be shown clemency even though he had defected to Pompey. Suetonius tells of the fabulously expensive (6 million sesterces) pearl she received from Caesar and how, when she felt that Caesar's interest in her was waning, she offered Tertulla, her daughter, to him. This happened at about the time of an auction at which Caesar sold a number of estates to Servilia at knock-down prices, causing Cicero to quip that they were an even bigger bargain as they had been further discounted by *tertia* – a third/Tertulla. In 43 BCE Servilia chaired another pivotal meeting which was convened to discuss how the conspirators should react to Cicero's plan to ask Brutus to bring his armies to Italy and what they should do about Marcus Aemilius Lepidus, Servilia's son-in-law, who had joined Mark Antony and had been declared an outlaw. P. Servilius Casca (relative and assassin), Labeo (a lawyer), Scaptius (Brutus' agent) and Cicero were also present. Servilia acted also as the eyes and ears in Rome for Cassius and Brutus when they fled the city. Cicero described her as 'a highly capable and dynamic woman'.[31]

The ambitious and assertive Fulvia Flacca Bambula (*c.*83–40 BCE), is most famous for gleefully pricking the tongue of decapitated Cicero with a hairpin: she was exacting indignant revenge because Cicero had insinuated that Mark Antony, her third husband, only married her for her money. The consequence of that indiscretion was Cicero's mutilated head found itself on public display in the Forum after his proscription in 43 BCE.

Fulvia, like many prominent and obtrusive women before her, is likened to a man: 'a woman in body alone' by Velleius Paterculus who evidently regarded her vengeful and gruesome act as unladylike and, by implication, the sort of thing only a man should do.[32] This is Dio's account: 'Fulvia took the head into her hands before it was removed, and after abusing it spitefully and spitting upon it, set it on her knees, opened the mouth, and pulled out the tongue, which she pierced with the pins that she used for her hair, at the same time telling many brutal jokes'.[33]

Pomponia, the widow of Cicero's brother, Quintus Tullius, and sister of Atticus was even more sadistic: when Philologus, the freedman who betrayed the Ciceros, was brought to her she ordered him to cut off strips of his own flesh, cook them and then eat them.[34]

But Fulvia deserves to be remembered for more than her sadism: her military and political achievements are quite remarkable by any standards. Before her relationship with Antony, she had been married to two other powerful men. The first was the notorious Publius Clodius Pulcher in 62 BCE, then in 51 BCE Gaius Scribonius Curio, an ally of Julius Caesar's. Plutarch, in his *Life of Antony*, describes Fulvia as no wool-spinner, no doer of housework, in effect no *matrona*; Fulvia's ambition was to rule over a powerful man, an ambition she was to realise as Mark Antony's third wife. Antony must have been impressed, and her influence must have been significant, because he struck coins bearing her image as a representation of Victory – the first time that Roman coinage had featured a woman. Another possible first was when Antony's supporters renamed the city of Eumenia in Phrygia as Fulviana; subsequent mintings of the coins bear this name.

Fulvia exerted considerable influence on Mark Antony before and after the assassination of Julius Caesar, effectively harnessing the thuggish gangs of former husband Clodius to help Antony against Dolabella. She took an enthusiastically active role in the subsequent bloody proscriptions, accompanying her husband to his military headquarters in Brundisium.[35] She also confounded Cicero's attempts to have Antony declared an enemy of the state in his absence.[36] All the while, Antony was conducting an affair with Lycoris – the accomplished mime artist whom Cornelius Gallus celebrated in his elegies. Such was his attraction to Fulvia, though, that Antony even renounced Lycoris for Fulvia; to add to this, Fulvia's daughter, Clodia Pulcher, married the up and coming young Octavian. According to Dio, Fulvia was one of the most powerful people in Rome at this time and proceeded to fight Antony's corner tirelessly, winning support from veterans when she toured Italy, children in tow.[37] And all of this despite the tensions caused by Antony's latest affair, this time with Cleopatra. Fulvia even helped raise eight legions for Antony to fight against Octavian in the Perusine War in 41–41 BCE.[38]

Velleius Paterculus tells how she was not to get away unscathed from this unwomanly behaviour and impertinent intrusion into the male world of war and battle.[39] Velleius says she 'was creating general confusion by armed violence' and that Octavian's troops tied obscene messages to stones and fired them directly at Fulvia. Two were intended for her clitoris with the unmistakeable suggestion that she was a tribade, a lesbian;[40] Fulvia and Antony were invited to open their arses wide to receive the projectiles. They in turn responded by calling Octavian a 'cock-sucker' and 'wide-arsed', suggesting that he too was open to penetration: the ultimate insult for a free-born man. Martial preserves for us the lascivious epigram that Octavian reputedly composed for Fulvia:

> Because Antony is shagging Glaphyra, Fulvia has decided that my punishment will be that I have to shag her too. Me fuck Fulvia? What if Manius begged me to bugger him? Would I? I think not, if I had any sense. 'Fuck or fight,' she demands. Doesn't she know that I love my prick more than life itself? Let the battle trumpets blow![41]

Octavian is insinuating that the ensuing civil war was caused because Fulvia was put out by his rebuffal.

Fulvia Flacca Bambula was not the only Fulvia who wielded power in the turbulent days of the first century BCE. During the Catiline conspiracy, Gaius Cornelius and Lucius Vargunteius were deputed to assassinate Cicero on 7 November 63 BCE; Quintus Curius, a senator who became one of Cicero's chief informants, warned Cicero of this threat with the help his mistress, Fulvia: by posting guards at his house Cicero avoided the would-be assassins.[42] This Fulvia was not one of those vulnerable women whom Catiline recruited as informants and agents provacateurs in his battle with Cicero – former prostitutes who had fallen on hard times towards the end of their careers, and were thus open to exploitation by the conspirators. This Fulvia, on the other hand, was of good birth; she had learnt of the conspiracy from her indiscreet and tactless lover, Quintus Curius. When she revealed the details to various people, the plot was confounded, and Cicero was voted consul.

The Laudatio Turiae is a first century BCE tombstone engraved with an epitaph, a husband's eulogy of his wife. It provides the gold standard of wifely conduct, erected in Rome by her husband, Quintus Lucretius Vespillo. The *laudatio* celebrates, at some length, the life of a dutiful and exceptional wife and catalogues those qualities expected of the good *matrona: pietas*, wool-working, looking after the household and its gods, modesty of dress and elegance, financial generosity, bravery, loyalty and shrewdness in the face of her husband's enemy, Lepidus, in 46 BCE when he was proscribed. Turia concealed Lucretius in their roof space and convincingly acted out the role of the bereft wife dressing in rags, looking dishevelled and grieving over an apparently lost husband. When Lucretius was in exile she moved in with her mother-in-law and valiantly defended the house against Milo, the renegade former owner who had not reckoned on Turia's bravery and determination.

Octavia the Younger (69–11 BCE) is one of history's first female shuttle diplomats. Her credentials and lineage are impeccable. She was the sister of Gaius Octavius Thurinus, better known as Octavian, and fourth wife of Mark Antony. She was the great-grandmother of Caligula and Agrippina the Younger, maternal grandmother of Claudius and both paternal great-grandmother and maternal great-great grandmother of Nero. Born in Nola, she was the daughter of Gaius Octavius (Octavian's father) and Atia Balba Caesonia (85–43 BCE), the niece of Julius Caesar.

Gaius Octavius died in 59 BCE and Octavia's mother later remarried, to Lucius Marcius Philippus. He was a caring step-father, raising Octavian, Octavia and Octavia the Elder (Octavia's step-sister by Gaius and Ancharia, his first wife), and managing their children's education with Atia. He arranged Octavia's first marriage, to the consul and senator Gaius Claudius Marcellus Minor (88–40 BCE), an opponent of Julius Caesar's. Octavia had three children from the marriage with Marcellus Minor: two daughters, both named Claudia Marcella (the Marcellae), and a son, Marcus Claudius Marcellus (42–23 BCE). Octavia's son, Marcellus, became a favourite of Augustus and was groomed to succeed him before his untimely death at the age of 19; in 25 BCE he had married Julia, Augustus' daughter.

Octavia was Mark Antony's fourth wife; the marriage made Antony and Octavian brothers-in-law, and as such was one of the most politically sensitive unions in a society where political marriages had almost become de rigueur among elite families – Octavia was pregnant with her late husband's third child, Claudia Marcella the Younger. Things went well at the start: Antony minted a coin bearing Octavia's portrait, making her only the second Roman woman to feature on a coin after Fulvia, her immediate predecessor in Antony's affections. The couple travelled together, with Octavia happily residing with her husband in Athens during the winter of 39–38 BCE.[43]

Octavia was a generous and public-spirited woman. Evidence for this comes in a curious incident in 43 BCE when she helped the proscribed Titus Vinius: he had been hidden in a chest by his wife, Tanusia, concealing him in the house of a freedman, claiming that he was dead. This brave woman herself risked death for harbouring a proscribed person. Octavia agreed to help and arranged for Tanusia to deliver the chest to a festival attended by Octavian (without the other triumvirs) where he was sprung from the chest. Octavian was so impressed by the couple's ingenuity and nerve that he pardoned Vinius and Tanusia and promoted the freedman to the equestrian order.[44]

Early signs of her influence over, and diplomacy with, brother and husband came in 37 BCE when she was able to use her unique position as sister and wife to good ends. Octavia travelled with Antony to Italy where he won over a number of Octavian's allies, including Maecenas and Agrippa – the equivalent of Octavian's Home Secretary and Foreign Secretary – pleading the pointlessness of conflict between her two kinsmen. Her intercession led to the Treaty of Tarentum and a five-year extension of the Triumvirate. Other triumphs included persuading Antony to increase the size of the fleet he was despatching to Octavian in the fight against Sextus Pompey off Sicily, and winning a 1,000-strong bodyguard from Octavian to add to the army he was loaning Antony for the Parthian campaign.[45] The *pièce de résistance* came when she nearly arranged the betrothal of Octavian's young daughter, Julia, to Mark Antony's son by Fulvia, M. Antonius Antyllus. The marriage never took place because Antyllus was executed after the Battle of Actium.

Things continued to go well: Octavia had established herself as an influential and respected diplomat, astute enough to capitalise on her unique position as wife and sister to the benefit of both Antony and Octavian, and, ultimately, to a Rome weary with internecine warring.

This, however, did not take into account Antony's unpredictable and mercurial nature when it came to the treatment of his wives in relation to wider political or personal considerations. Just as he had spurned Fulvia, so he snubbed Octavia: their journey together to the Parthian theatre of war was cut short when Antony abruptly sent her back to Rome. Octavia's personal safety was the pretext, but Cleopatra was the real reason.[46] Octavia's unhappiness can only have been heightened when in 36 BCE she learnt that Antony had admitted to the paternity of Cleopatra's children. Octavia remained loyal, though, and won from Octavian permission to take much-needed reinforcements and ordnance to Antony; she got as far as Athens where letters from her husband told her to go home: Octavia meekly responded by asking what she was to do with the men and supplies.

This insult was not lost on Octavian. Antony's treatment of his sister was a growing cause of resentment and he advised Octavia to divorce Antony; she refused and remained the steadfast *matrona*, her main concern being that more enmity might lead to more civil war, despite the fragile peace won by the Treaty of Tarentum.[47] Antony finally divorced Octavia in 32 BCE.

Octavia was a competent diplomat who for many years ensured a degree of non-aggression between her brother and her husband. Despite her loyalty and patience she was treated shabbily by Mark Antony who showed little appreciation of her pacifying influence and competence during one of the most desperate times in Roman history when many others around him were anything but calm.

Vipsania Agrippina II or Agrippina the Elder was born in 14 BCE, daughter of Julia Augusti and Agrippa, and sister of Julia the Younger. She was the second granddaughter of Octavian, now Augustus, and sister-in-law, step-daughter and daughter-in-law of Tiberius. She was the mother of Caligula, maternal second cousin and sister-in-law of Emperor Claudius, and maternal grandmother of Nero. She married the ever popular Germanicus in about 5 CE, when she was 18 years old.

Agrippina was born in Athens, while her mother was accompanying Agrippa on one of his many trips abroad. However, her early childhood was spent under the strict control of Augustus and Livia, with Augustus over-seeing the education of Agrippina, her sister and her brothers – as he had done with her mother. Agrippina's marriage to Germanicus was a happy union, and certainly productive. Three of their children died in childbirth, but six survived: Nero Caesar, Drusus Caesar and Gaius (Caligula), Agrippina the Younger (Julia Agrippina, mother of Nero), Julia Drusilla and Julia Livilla. Germanicus, a scion of both the Julian and Claudian families, had been adopted by Tiberius in 4 CE, and was clearly a contender to succeed Augustus; he was popular with the Roman people, and had the common touch. Agrippina's fecundity (*insigne fecunditate*), her peerless moral rectitude (*praeclara pudicita*) and her strong maternalism made her equally popular with the Roman people.

In 9 CE Rome suffered one of its worst military disasters, when Publius Quinctilius Varus was ambushed by the Roman-trained German chief Arminius in the Teutoburger Wald in between modern Paderborn and Gütersloh; three legions were totally annihilated. The situation on the German frontier would remain tense and unpredictable for years to come.

Tacitus records that soon after Tiberius had succeeded Augustus, Agrip-pina, pregnant with Agrippina Minor, accompanied Germanicus to Germany; he had orders to suppress a mutiny simmering among a number of legions. The legions were disaffected because they had still not been discharged five years after their enlistment, and were suffering in the wake of the

Teutoburger calamity. Moreover, Tiberius was not the most popular of generals or emperors.

At least four legions were implicated in the mutiny in Germany. The situation was exacerbated by poor treatment at the hands of officers and the soldiers' demand for money owed to them in the will of Augustus. They reluctantly swore an oath of allegiance to Tiberius under orders from Germanicus, but the atmosphere remained very tense: officers were being thrown into the Rhine. The pregnant Agrippina and Caligula began to play a significant role. Tacitus describes how the troops valued Agrippina's qualities as a *matrona*; they admired her fertility, her famous chastity and the fact that she took the trouble to visit their camps with Germanicus, even giving birth to Caligula there.[48] Agrippina cleverly exploited her popularity, and her condition. Her presence in the camps initially left the soldiers cowed, but they were soon angrily revived by a demonstration to Germanicus of just how overdue their discharge was: one veteran thrust Germanicus' hand into his edentulous mouth to show how old he was. The rebellion culminated in the legions' demand that Germanicus depose Tiberius. Germanicus was appalled at the suggestion, threatening, or pretending to threaten, to kill himself; despite some efforts to restrain him, other, more cynical, troops told him to go right ahead – one even offered Germanicus his sword, because it was sharper than Germanicus'. Confusion reigned as the mutineers tried to enlist more legions to their cause.

Germanicus resorted to duplicity, rashly pretending that he had received a letter from Tiberius agreeing to discharge the troops who had reached the end of their service, and to double the pay promised by Augustus. Two legions accepted, and two refused. Germanicus managed to find the money required to pay off the troops, who called off the mutiny and swore an oath of allegiance to Tiberius. All seemed well until a delegation arrived from Rome, denying all knowledge of the letter and the terms. The mutiny was back on; Germanicus' standards were removed and, with them, his command of the armies on the Rhine.

It was at this point that Agrippina took firm control of events. It was decided that she and Caligula should leave the Roman camp, along with the other women and children, for the safety of an allied Gallic settlement; perversely, the Gauls were proving more loyal to Rome than the legions. By demonstrating her bravery and loyalty to Rome when she pretended to leave with reluctance, Agrippina evoked shame and pity among the troops and so engineered a turning point in the mutiny. In contrast to Tacitus, however, Dio records that Agrippina and Caligula were held hostage by the mutineers.[49] Whatever the case, the mutiny evaporated and the ringleaders were slain. Germanicus then attacked and defeated the Germans, even gloriously winning back one of the standards that had been lost in the Teutoburger

Wald. Agrippina had cleverly bailed out her husband and probably saved his military career by an astute piece of tactical thinking.

Agrippina, however, had more impressive military expertise to offer. Pliny the Elder (author of the lost *German Wars*), who had served under Germanicus, apparently told how, in 15 CE, Germanicus then made a determined push towards the Elbe. The plan failed when he overreached himself, foolishly advancing as far as the Teutoburger Wald – the very place where Varus and his legions had been ignominiously annihilated by Arminius six years earlier. Arminius re-emerged to haunt Germanicus when he successfully attacked the Romans on all sides with his guerrilla units; one detachment was trapped, evoking memories of Teutoburger Wald. In panic, the Romans rushed to the Rhine Bridge at Castra Vetera (Xantem), aiming to destroy it and prevent the Germans from crossing behind them. Agrippina realised that this would maroon many Roman soldiers on the wrong side of the river, exposing them to the Germans. Accordingly, she took up a position on the bridge, physically preventing its destruction: the pregnant Agrippina stood here, at the head of the bridge, encouraging, resupplying and providing field dressings for the returning legions. Tacitus calls her a great-minded woman (*femina ingens animi*) who assumed the role of a general (*munia ducis*), inspecting the troops and distributing bonuses until Germanicus arrived. Agrippina had saved the day for Rome, and salvaged the career of her husband again.[50]

Agrippina was an influential woman on a number of levels, but her impetuosity and aggressive nature were to be her downfall. She was undoubtedly a faithful army wife, supporting Germanicus and travelling extensively with him. However, conservative Romans would say that she probably overstepped the mark and involved herself in military matters where she had no real business; nevertheless, she won the support of the troops, and no doubt helped her husband's position during the German mutiny. Her later disregard for Germanicus' dying plea – that she curb her fiery temperament and stay away from politics – showed her to be determined to avenge his death and establish the exact circumstances of his demise.

Munatia Plancina was wife of the governor of Syria, Gnaeus Calpurnius Piso, a wild and dissentious man according to Tacitus. She was also a rival of Agrippina's: the couple were accused of poisoning Germanicus, and were backed by Livia, wife of Augustus, as part of her own jealous campaign against Agrippina. In 18 CE Piso and Plancina had had a confrontation with Germanicus and Agrippina in Syria. Livia had asked Plancina to keep Agrippina under control to prevent a repeat of the Rhine bridge episode when Agrippina had assumed the role of a military commander; nevertheless, Planicina took this as an opportunity to involve herself in military matters, taking part in infantry and cavalry exercises and ignoring *decora feminis*, things appropriate for a woman.[51]

Dio Cassius describes a woman who heroically defends another woman's virtue in his record of Pythias, a slave girl who under interrogation and torture stood up for Nero's wife, Octavia, in 63 CE. She defiantly spat in the face of her interrogator, Tigellinus, when he questioned Octavia's virtue, exclaiming: 'my mistress's vagina is cleaner than your mouth'.

Agrippina the Younger

Agrippina the Younger (14–59 CE) was sister of the Emperor Caligula, niece and fourth wife and reputed murderess of the Emperor Claudius, and mother of the Emperor Nero. She shared some of her mother's militaristic tendencies. In Judaea she flexed power and influence when she became involved in a serious uprising there. While ruled by Herod Agrippa – personal friend to Caligula and Claudius – Judaea enjoyed self-rule until Herod's death in 44 CE when it reverted to a province, much to the dismay of the Judaeans. A riot ensued when a Roman soldier, guarding the temple precinct, dropped his trousers and rudely broke wind at worshippers celebrating the Passover – undoubtedly one of the most significant farts in history: as many as 30,000 people died in the subsequent fighting. Even more seriously, in 51 CE a revolt sprang up in Judaea after Galilean pilgrims were murdered by a band of Samaritans; the Roman procurator, Ventidius Cumanus, sided with the aggressors. The situation was complicated by the involvement of Felix, brother of Pallas and procurator of Samaria, a position he gained, no doubt, with the complicity of Agrippina. According to Josephus, the governor of Syria, Ummidius Quadratus, held two inquests, after which a number of Jews and Samaritians were beheaded; he then passed this intractable issue over to Claudius.[52] Claudius came down on the side of the Samaritans but Herod's successor, Agrippa II, appealed to Agrippina who managed to persuade Claudius to take a different, more pro-Jewish view. Three Samaritans were executed and Cumanus was exiled for maladministration; Felix took over as procurator of Judaea, the first freedman to attain such a senior provincial post. Felix repayed his debt to Agrippina, in part, by minting coins in her honour.

In the same year on the other side of the Empire Caratacus, veteran British chieftain and King of the Catuvellauni, led the Silures and Ordovices against Publius Ostorius Scapula. Scapula finally defeated Caratacus, capturing Caratacus' wife and daughter and taking his brothers prisoner. Caratacus himself fled to the Brigantes seeking asylum but, unfortunately for Caratacus, the queen, Cartimandua, had him bound in chains and handed over to the Romans. Caratacus' reputation preceded him: all Rome was anxious to see the man who had defied and irritated Rome for so long. His dignified speech saved his life: Claudius pardoned him, his wife, daughter and brothers. Significantly, Agrippina had seated herself on the podium on a chair next to

Claudius with the result that Caratacus paid her the same respect and homage he paid to the Emperor, thinking nothing of the fact that a woman had equal authority with an emperor of Rome – nothing surprising to a warrior from a country where women chieftains were not unusual: Cartimandua and Boudica are the famous examples. What is remarkable is Agrippina's audacity and her unprecedented position on the platform before the Roman military standards, assuming joint power with the Emperor of Rome. Both Dio and Tacitus gasp at the impudence of it all.

The noblewoman Verulana Gratilla also took centre stage as one of the leading fighters in the defence of the Capitoline against Lucius Vitellius in civil-war-torn 69 CE; Tacitus is careful to point out that, to enable her to do this, she abandoned her family.

The behaviour of Vitellius' second wife, Triaria, was deemed to be similarly inappropriate: Tacitus describes her as behaving with 'unfeminine ferocity'; she went so far as to strap on a sword, revelling in the fighting between the Vitellian and Flavian factions in 68 CE: 'Some accused Triaria, the wife of L. Vitellius, of having armed herself with a soldier's sword, and of having behaved with arrogance and cruelty amid the horrors and massacres of the storming of Tarracina.'[53]

The Vestal Virgins

Tacitus tells how, in that maelstrom of 69 CE, Vestal Virgins were deployed as envoys by an embattled and desperate Vitellius to Marcus Antonius Pius, a supporter of Vespasian. Vitellius was hoping for a ceasefire in which to negotiate peace terms, but Antonius dismissed the Vestals with due honour (*cum honore*) and replied to Vitellius that all deals were off in the wake of the murder of Sabinus and the firing of the Capitol. The involvement of the Vestals here not only demonstrates their role as vital negotiators at times of crisis but also the reverence in which they were held, even amid military chaos. Their sacrosanctity was also clearly evident in their role as intermediaries between Claudius and Messalina. Important state papers and the wills of eminent statesman (such as those belonging to Julius Caesar and Mark Antony) and emperors were kept in their building, the Atrium. In 73 BCE two Vestals had been embroiled in the Catiline conspiracy: Fabia, the half-sister of Terentia, Cicero's wife, was accused of having an affair with Catiline, while Licinia was similarly accused of consorting with Crassus, her cousin. Both were acquitted.

Faustina the Younger

Annia Galeria Faustina Minor or Faustina the Younger (*c.*130–176 CE) was a daughter of Antoninus Pius and Roman Empress Faustina the Elder; she was a Roman empress herself and wife to her first cousin Emperor Marcus Aurelius. Faustina was very popular with the army and was awarded divine honours after her death. Despite this, Dio and the *Augustan History* slur her when they accuse Faustina of ordering deaths by poison and execution; and of instigating the revolt of Avidius Cassius against her husband. The *Augustan History* slanders her with reports of casual adultery with sailors, gladiators and 'men of rank'; however, despite the intrusion of the classical equivalent to the gutter press, Faustina and Aurelius seem to have been devoted to each other.

Faustina accompanied her husband on various military campaigns and enjoyed the devotion and reverence she received from Roman soldiers. Aurelius gave her the title of *Mater Castrorum*, 'Mother of the Camp'. She tried to make a home out of an army camp: between 170 and 175 CE, she travelled to the the north of the Empire, and in 175, she accompanied Aurelius to the east.

Faustina has another, somewhat less glorious, connection with the world of combat: smitten by desire for a gladiator, she finally confessed her passion to her husband, Marcus Aurelius. On consulting the Chaldean magic men the gladiator in question was executed and Faustina was made to bathe in his blood, and then have sex with her husband while still covered in that blood. All passionate thoughts of the gladiator apparently disappeared.

Faustina died in 175 CE, after an accident at the military camp in Halala in Cappadocia. A devastated Aurelius buried her in the Mausoleum of Hadrian in Rome. She was deified: her statue was placed in the Temple of Venus in Rome and a temple was dedicated to her. Halala's name was changed to Faustinopolis and Aurelius opened charity schools for orphan girls called Puellae Faustinianae or 'Girls of Faustina' (Life of Marcus Aurelius, *Historia Augusta*, 26, 4–9). The Baths of Faustina in Miletus bear her name.

Julia Domna

Julia Domna is famous for the encounter reported by Dio between her and the wife of a Caledonian chieftain in Britannia. Julia Domna remarked primly on how the tribal women are somewhat free with their sexual favours; the barbarian woman replied tartly, 'We satisfy our desires in a better way than you Roman women do. We have sex openly with the best men while you are seduced in secret by the worst.' That was probably the end of that particular edge of empire tête-à-tête.

Julia Domna (170–217 CE) was empress and wife of Emperor Lucius Septimius Severus and mother of Emperors Geta and Caracalla; Julia was renowned for her prodigious learning and her extraordinary political influence, and for her expertise in and patronage of philosophy. Julia had her fair share of political enemies, who accused her of treason and adultery, none of which was proven. She and Severus were very close; he insisting she accompany him in the campaign in Britannia from 208. When Severus died in 211 in

Eboracum (York), Julia became mediator between their two sons, Caracalla and Geta, destined to rule as joint emperors. Geta was murdered by Caracalla's soldiers in the same year. Julia too was awarded the title of *Mater Castrorum*, in 195 CE.

Her surviving son Caracalla acceded to the Imperial throne as sole emperor, a tense situation because of his involvement in Geta's murder. Nevertheless, Julia accompanied Caracalla in his campaign against the Parthians in 217. Caracella was assassinated, while Julia committed suicide at Antioch, perhaps because she knew she had breast cancer.

Vitruvia

Vitruvia, also known as Victoria, was prominent in the Roman breakaway realm known as the Gallic Empire in the late third century CE. She was the mother of Victorinus, who ruled as Gallic Emperor until his assassination in 271. In the wake of this, Vitruvia stabilised the Empire using her vast wealth to buy the support of the legions. This enabled her to confirm the elevation of her chosen candidate for emperor, Tetricus I, formerly the governor of Gallia Aquitania.

Vitruvia features in Aurelius Victor's *Liber de Caesaribus*, and in the account of the Thirty Tyrants in the *Historia Augusta*, noted for its tabloid unreliability. In the *Historia Augusta*, she is numbered as one of Rome's Thirty Tyrants, that she herself bore the titles *Mater Castrorum* and Augusta, and minted her own coins.

The Gallic Empire is the name given to a breakaway segment of the Roman Empire that operated de facto as a separate state from 260–74 CE. It came about during the Crisis of the Third Century, established by Postumus in 260 in the wake of barbarian invasions and instability in Rome. At its zenith it included Germania, Gaul, Britannia and (temporarily) Hispania. It was retaken by Aurelian after the Battle of Châlons in 274 CE.

Albia Dominica

Albia Dominica (*c.*337–after 378 CE) was a Roman Augusta, wife to Emperor Valens who ruled from 364–78 CE, as Emperor of the East and Co-Emperor with his brother Valentinian I. Valens died on the field fighting against the Goths at the Battle of Adrianople in 378 CE after which the Goths moved east and attacked Constantinople. This left a power vacuum for the Romans which the Empress Dominica filled: after the death of Valens, Dominica ruled as regent and defended Constantinople against the attacking Goths until Valens' successor, Theodosius I, arrived. To boost her forces, Dominica paid serving soldiers' wages out of the Imperial treasury to any civilian volunteers who were willing to take up arms against the invaders.

Serena

Serena was niece and adoptive daughter of Emperor Theodosius I and wife of Stilicho, whom she married in 384 CE. Stilicho was a German Vandal who rose to the position of leading general and the most powerful man in the Western Roman Empire. He was regent for youthful Honorius, Serena's cousin, and was finally deposed, arrested and executed in 408 CE.

Theodosius had seen that Stilicho could be a useful ally so, to bring him closer and guarantee his loyalty, he arranged the marriage with Serena who gave birth to a son, Eucherius, and two daughters, Maria and Thermantia; they both married Honorius. Serena picked a bride for the poet laureate, Claudian, and looked after Honorius' half-sister, her cousin Galla Placidia. After Stilicho's death, Serena was falsely accused of conspiring with the Visigoths during the siege of Rome and was executed with Galla Placidia's consent.

Claudian's *In Praise of Serena* portrays, somewhat obsequiously, her as the typical army wife, despairing that her husband is leaving for war in Thrace and overjoyed when he eventually returns:

> How you trembled and wept when the cruel bugles summoned your warrior to arms! With a face wet with tears you saw him leave home praying for his safe return after snatching the final hasty kiss from between the bars of his crested helmet's visor. But again what joy when he finally returned, preceded by the clarion of victory and you could hold his still armoured body in your loving arms again! How sweet the long hours of the chaste night when you asked him to tell in safety the story of his battles. When he was away at the wars you didn't comb your shining hair nor wear your usual jewels. You spent your time in worship and in prayer ...[54]

But Serena was not just an unhappy war wife; she put her time on her own to good use, acting as Stilicho's eyes and ears, alert for any sign of opposition or conspiracy against him, maintaining his presence at court *in absentia*. Indeed, Claudian tells us how she once warned him of a 'foul conspiracy' organised by Rufinus, 'who sought means to destroy his master by traitorously stirring up the Getae against Rome'.

Claudian touches on a number of other issues relating to women in war. In his *Bellum Geticum*, he references the unhappy fate of women in war, whatever side they are on: 'He [Alaric] who targeted the women of Rome as victims of his lust has seen his own wives and children led away captive'.[55] He attacks Alaric's wife, mad, *demens*, with greed for Roman booty: '[she] demanded in her madness the jewelled necklaces of Italian matrons for her proud neck and Roman girls for her slaves!' She is in stark contrast to the modesty and frugality, loyalty and chastity displayed by Serena.[56] His personification of Africa bewails the pillage and rape endured by the women of Carthage – discarded by Gildo's troops when they had satisfied their lust and then handed them on to the Moors.[57] Gildo (died 398 CE) was a Berber Roman general in Mauretania who revolted against Honorius but was defeated and committed suicide.

Aelia Galla Placidia (388–450 CE), daughter of the Roman Emperor Theodosius I, was the regent for Emperor Valentinian III from 423 until he attained his majority in 437; she was consort to Ataulf, King of the Goths from 414 until his death in 415, and empress consort to Constantius III in 421 CE.

We have noted how Serena was accused of conspiring with Alaric, 'the whole senate therefore, with Placidia, sister to the emperor, thought it right that she should suffer death'. Thereafter, Galla Placidia became something of a pawn and a victim in the bloody events following the fall of Rome. Before that fall, Placidia was captured by Alaric as recorded by both Jordanes and Marcellinus Comes; she went with the Visigoths from Italy to Gaul in 412 CE. Alaric was succeeded by Ataulf who married Galla Placidia at Narbonne in 414. Ataulf was assassinated in his bath in 415 and Sigeric was installed as the next King of the Visigoths. According to Edward Gibbon in *The History of the Decline and Fall of the Roman Empire*, Sigeric's first act 'was the inhuman murder' of Ataulf's six children from a former marriage, 'whom he tore, without pity, from the feeble arms of a venerable bishop', Sigesar, Bishop of the Goths. Galla Placidia was 'treated with cruel and wanton insult' by being forced to walk more than 12 miles on foot with the other captives driven ahead of Sigeric on horseback.[58] This demeaning sight, however, was one of the factors that caused indignant opponents of Sigeric to assassinate him.

Placidia was given back to Honorius as a clause in the peace treaty between the Goths and Honorius; he then forced her into marriage with Constantius III in 417 CE. The following year she interceded in the succession crisis following the death of Pope Zosimus. Two factions of the Roman clergy had elected their own popes with the obvious result that there were two rival popes, Eulalius and Boniface I whose factions plunged the city into chaos. Placidia and Constantius petitioned the Emperor in favour of Eulalius – the first intervention by an emperor in a papal election. However, it took two synods to finally resolve the issue – Placidia had personally written to the African bishops, summoning them to the second synod. In 421 CE, Constantius was proclaimed an Augustus, becoming co-ruler with the childless Honorius; Placidia was proclaimed an Augusta.

But accusations of incest with Honorius forced Placidia to flee the Western Empire. Olympiodorus of Thebes, a historian used as a source by Zosimus, Sozomen and probably Philostorgius mention incidents of public fondling but Gibbon differs: 'The power of Placidia; and the indecent familiarity of her brother, which might be no more than the symptoms of a childish affection, were universally attributed to incestuous love'.[59]

Placidia and her children arrived in Constantinople, where they were treated with kindness; this was some time after the marriage of Theodosius II to Aelia Eudocia in 421 CE during the festival of the Persian victories – victory celebrations after a Roman–Sassanid War in 421–2. The Empress Eudocia celebrated the war in a poem in heroic metre, according to Socrates of Constantinople.[60] When Honorius died in 423 CE Theodosius dithered over the crucial task of appointing a co-emperor for the west; he eventually prepared the 4-year-old Valentinian III for elevation to the Imperial office and in 424 Valentinian was betrothed to Licinia Eudoxia, his 2-year-old first cousin once removed. She was a daughter of Theodosius II and Aelia Eudocia. During this time Gibbon records that the betrothal was a result of 'the agreement of the three females who governed the Roman world', namely Placidia and her nieces Eudocia and Pulcheria. While Theodosius was procrastinating, the pretender Joannes was appointed as new Western Roman Emperor: Joannes' rule was adopted in Italia, Gaul and Hispania, but not in Africa. Placidia and Valentinian joined the force assembled to depose Joannes. Joannes was captured in Ravenna where his right hand was cut off; he was then mounted on a donkey and paraded ignominiously through the streets, and then beheaded in the Hippodrome of Aquileia.

Valentinian was officially proclaimed emperor of the Western Roman Empire in October 425 CE, but three days after Joannes' death Aetius brought reinforcements for Joannes' army – 60,000 Huns from across the Danube no less. Placidia and Aetius came to an agreement which settled the political landscape of the Western Roman Empire for the next thirty years: the Huns were paid off and went home, while Aetius was awarded the post of *magister militum per Gallias*, commander-in-chief of the Roman army in Gaul. Galla Placidia acted as regent of the Western Roman Empire from 425–37, when Valentinian turned 18.

But not everything had gone smoothly: Bonifacius, general and governor of the Diocese of Africa, was one Placidia's early supporters. However, in 429 CE Procopius tells us how Aetius slyly played the two off against each other, warning Placidia against Bonifacius and advising her to recall him to Rome; at the same time he wrote to Bonifacius, telling him that Placidia was about to recall him for no reason other than to have him assassinated.[61] Bonifacius reacted by allying with the Vandals in Spain who crossed to Libya to join him. Placidia was perplexed by Bonifacius' hostile and uncharacteristic actions; she sent friends to Carthage to intercede with him, at which point he showed them the duplicitous letter from Aetius. The plot was now revealed; Placidia did not move against Aetius but urged Bonifacius to return to Rome.

Bonifacius was now regretting his alliance with the Vandals; but when he tried to persuade them to return to Spain they besieged him at Hippo Regius in Numidia. (St Augustine was Bishop of Hippo and died in this siege.) Unable to take the city, the Vandals eventually raised the siege but the Romans were routed and lost Africa to the Vandals. Bonifacius had meanwhile returned to Rome, where Placidia promoted him to the rank of patrician and made him 'master-general of the Roman armies'. Aetius returned from Gaul with an army of 'barbarians' to be confronted by Bonifacius in the bloody Battle of Ravenna in 432 CE. Bonifacius won the battle, but was wounded and died a few days later. Aetius withdrew to Pannonia.

Placidia, however, was becoming increasingly isolated with many of her allies dead or defected to Aetius and her influence significantly weakened. Atilla the Hun burst into Placidia's life when her daughter, Justa Grata Honoria, asked him in 450 CE to rescue her from a marriage to a Roman senator that the Imperial family, including Placidia, was forcing upon her; she included her engagement ring with the letter. Attila interpreted her missive and the enclosure as a proposal of marriage which he of course accepted, asking for half of the Western Empire as dowry. When Valentinian discovered this, it was Placidia who persuaded him not to kill Honoria. Even though Valentinian wrote to Attila asserting the illegitimacy of the supposed marriage proposal, Attila sent an envoy to Ravenna to proclaim that the proposal had been legitimate, and that he was on his way to claim what was rightfully his. Honoria was hastily married off to Flavius Bassus Herculanus, though Attila continued to press his claim.

Placidia died soon afterwards at Rome in November 450 CE after a life full of military and political intrigue.

Theodora

Theodora (*c.*500–48 CE), the wife of Justinian I and powerful empress of the Byzantine Empire, acted as a virtual co-regent with her husband. Procopius is our main source for her life, but perplexingly he gives us three wildly varying accounts from him in three separate works. *The Wars of Justinian* is complimentary and describes a brave and influential empress; Procopius' *De Aedificiis*, *Buildings of Justinian* is a panegyric which shows Justinian and Theodora as a pious couple.[62] In his *Anekdota* ('unpublished works') or *Secret History*, published a thousand years later, Procopius describes a woman who is vulgar and characterised by unquenchable lust – quite unrecognisable from his descriptions in the earlier works.

Her interest to us is in her involvement in the Nika riots in 532 CE when she urged her hesitant husband with a stirring speech to stand and fight the rebels instead of fleeing.[63] The riots went on for a week in Constantinople and were the most violent in the history of Constantinople, with nearly half the city burned or destroyed and tens of thousands of people killed. They were triggered by what can only be called chariot-racing hooliganism: in 531 CE members of the Blues and Greens factions were arrested on charges of murder in connection with deaths that occurred during rioting after a recent chariot race. Hanging was the sentence and many were hanged but in January 532, two of them, a Blue and a Green, escaped and took refuge in the sanctuary of a church surrounded by an angry mob. Justinian's feeble reaction was to declare that a chariot race would be held on 13 January and commuted the sentences to imprisonment. By race twenty-two, the chanting had changed from 'Blue' or 'Green' to a Νίκα meaning 'Win!' or 'Conquer!', and the crowds began to storm the palace, which for the next five days was under siege. Inevitably, taxation became the issue and a rallying cry: a terrified Justinian considered fleeing, that is until Theodora advised otherwise saying, 'Those who have worn the crown should never survive its loss. Never will I see the day when I am not saluted as empress.' Although an escape route across the sea beckoned, Theodora insisted that she would stay in the city, quoting an ancient saying, 'Royalty is a fine burial shroud', or 'the Purple makes a fine winding sheet'.[64] Justinian pulled himself together and the rebels were eventually killed. Procopius says that about 30,000 rioters were slaughtered and the senators who had supported the riot were exiled. He rebuilt Constantinople including the Hagia Sophia.

Antonina

Flavius Belisarius (*c.*505–65 CE) was a successful general and faithful supporter of Justinian, largely responsible for suppressing the Nika riots. His wife, the duplicitous Antonina (*c.*484–after 565 CE), brought with her a reputation even worse than Theodora's, her long-standing friend. Among other scandals, she was alleged to have had an affair with her adopted son, Theodosius. Sex for Antonina often took place in front of the slaves with Antonina 'a slave to her lust'; even when Belisarius caught them in the act he was unwilling to believe what he was seeing with his own eyes. Antonina was a schemer and a fixer of the first order, playing a prominent role in the downfalls of Pope Silverius and John the Cappadocian, 'making Silverius appear a pro-Gothic traitor' and implicating John 'in a conspiracy to gain the throne': 'Having first disposed of Silverius … she first cut out all their tongues, and then cut [the bodies] up bit by bit, threw the pieces into sacks, and then, without further ado, dumped them in the sea'.[65]

Antonina zealously followed Belisarius on campaign so that she, according to Procopius, could exert control over Belisarius: 'she had taken care to travel all over the world with him'.[66] Her first foray into his military world was during the Vandalic War (533–4 CE), during which, while crossing the Adriatic Sea, the water supplies of the navy were polluted. But not on Belisarius' own ship: Antonina had stored their water in jars in a darkened room protecting them from contamination.

She was instrumental at Ad Decimum in North Africa when 'Belisarius left his wife and the barricaded camp to the infantry, and advanced with all the horsemen ... But on the following day the infantry with the wife of Belisarius came up and we all proceeded together on the road to Carthage'.[67]

When in 537 CE Belisarius sent Antonina to Naples, allegedly for her own safety, she helped Procopius, then secretary to Belisarius, to raise a fleet which was used to transport grain and reinforcements to Rome through Ostia. She may also have been implicated in the death of Constantinus in late 537 when she persuaded Belisarius to kill him.

Procopius bluntly sums up Antonina's dubious backgound: both her father and grandfather were charioteers:

> her mother was one of the prostitutes attached to the theatre ... [Antonina] having in her early years lived a lewd sort of a life and having become dissolute in character. She not only consorted a lot with the cheap sorcerers who surrounded her parents, but also acquired knowledge of what she needed to know from them. She later became the wife of Belisarius, after having been the mother of many children.[68]

After joining her husband in the Lazic War, the cuckolded Belisarius eventually arrested Antonina on evidence provided by bedchamber servants; but he was unable to bring himself to exact punishment – due, according to Procopius, to Antonina's skilful use of the black arts. The informants were deemed to be lying: Antonina exacted what seems to have been her favourite punishment: she had their tongues cut out and their bodies chopped up; the body parts were dumped in the sea. Theodora eventually restored Theodosius to a grateful Antonina, but not before she had whipped Photius, her son, half to death for concealing him. Unfortunately, Theodosius died soon after of dysentery.[69]

When Justinian caught the plague it obviously prompted discussions about his succession. Belisarius swore to oppose any emperor chosen without his consent. Theodora took offence at this and had them recalled to Constantinople where Belisarius was relieved of his command. Procopius credits Antonina with getting him restored to Imperial favour. However, she prevented him from getting his old job back – *magister militum per orientem*: chief of the eastern army. Belisarius was sent back to the Gothic War and Antonina went with him. She was at Portus in 546 CE, Croton in late 547 and Hydruntum in 548. She was then posted to Constantinople to muster reinforcements for the Gothic War. When she arrived she found that Theodora had died and convinced Justinian to recall Belisarius.

Chapter 14

Foreign Women Fighters

During the Republic and in the early days of the Empire, Rome had to deal with a number of foreign queens, influential noblewomen and at least one pirate who all demonstrated impressive military and leadership skills.

The Women of Tegea

In a battle in the war of the Tegeatans against the Lacedaemonian King Charillus in the eighth century CE, the women of Tegea ambushed their enemy; this ambuscade led to victory so the women, therefore, celebrated the victory on their own, and excluded the men from the sacrificial feast. This, according to Pausanias (8.48), is the origin of this surname of Ares – Gynaecothoenas, 'the god feasted by women'.

Hippo

To Valerius Maximus, the Greek Hippo (fl. fifth century BCE) was a paragon of feminine virtue, up there with Roman Lucretia, Verginia and Cornelia. In his *Facta et Dicta Memorabilia* (*Unforgettable Words and Deeds*) he tells us that when Hippo was kidnapped by an enemy fleet, she determined to put her chastity above her life, and threw herself into the sea to drown. Her body was washed up on the shore of Erythrae where a tomb was built for her which survived in his own day.

Youtab

Youtab (Persian: یوتاب, fl. fourth century BCE) was the sister of Ariobarzanes, Satrap of Persis. She is famous for fighting alongside her brother against Alexander the Great at the Battle of the Persian Gate in 330 BCE. Apparently, she fought ferociously before dying in battle excelling in hand to hand fighting; even unarmed tribal refugees joined the fight.

Arsinoe III Philopator

Arsinoe III was Queen of Egypt from 220–204 BCE, and a daughter of Ptolemy III and Berenice II; she was married to her brother, Ptolemy IV, and took an active part in the government and military activity of the country. She rode at the head of infantry and cavalry to fight Antiochus the Great at the Battle of Raphia (Gaza) in 217 BCE. With the battle going badly, she stood before the troops and urged them to fight to defend their families. She also promised a bonus of two minas of gold to each of them if they won the battle, which they did. In 204 BCE she was murdered in a palace coup, soon after the death of her husband.

Teuta – Woman Pirate

Teuta was the queen regent of the Ardiaei tribe in Illyria; she reigned from *c.*231–227 BCE when she succeeded her alcoholic husband Agron (250–230 BCE) acting as regent for her young stepson Pinnes. Polybius, somewhat disparagingly, says that she ruled 'by women's reasoning'. The trouble started when Teuta backed the piratical raids of her subjects against neighbouring states (Polybius, 2.4.1).

The situation began to escalate when Teuta captured and later fortified Dyrrachium and Phoenice; while off the coast of Onchesmos, her ships raided a flotilla of Roman merchant vessels after which Teuta's forces extended their piratical operations further southward into the Ionian Sea, defeating the combined Achaean and Aetolian fleet in the Battle of Paxos and capturing the island of Corcyra. This enabled Teuta to attack the crucial trade routes between mainland Greece and the Greek cities in Magna Graecia. Teuta had now become the 'terror of the Adriatic Sea'.

Rome was naturally concerned about its trade and with the protection of that trade. In 230 BCE the Romans protested to Queen Teuta about the piracy which she openly condoned. Queen Teuta glibly told the ambassadors that according to Ilyrian law, piracy was a legal trade and that her government had no right to interfere in what was effectively private enterprise and that 'it was never the custom of royalty to prevent any advantage its subjects could get from the sea'. One of the envoys allegedly replied that Rome would make it her business to introduce better law among the Illyrians as 'we Romans have an excellent custom of punishing private wrongs by public revenge'.[1] Teuta then imprudently sanctioned the murder of one Roman envoy, Coruncanius, and imprisoned the other. The Romans reacted on a massive scale when they mobilised and set sail with an army of 20,000 troops, 200 cavalry and the entire Roman fleet of 200 ships, under the command of consuls Lucius Postumius Albinus and Gnaeus Fulvius Centumalus. This was the first time Roman armies had crossed the Adriatic. Significantly, they set up Demetrius of Pharos as a client king to challenge Teuta's power. Demetrius had previously enjoyed a similar position under Teuta, and was himself renowned as a pirate. The Romans took Corcyra, Apollonia, Epidamnus and Pharos, and finally laid siege to Scodra, Teuta's capital city. She surrendered ignominiously in 227 BCE, and was subjected to restrictions on military and naval activity, most crucially not to sail outside a restriction zone south of Lissa, while her territory was confined to the region around Scodra. However, the Romans did not quash the Illyrians but set up a protectorate instead. This meant that the Illyrians, as *amici* (friends), remained free, unoccupied and untaxed, but had a moral obligation to show gratitude to Rome in the shape of practical support as required.

Sophonisba

Sophonisba was the daughter of Hasdrubal Gisco; she is remembered for her bravery and loyalty to Carthage when she poisoned herself rather than be humiliated in a Roman triumph. Renowned for her beauty, Sophonisba had been betrothed to King Massinissa, leader of the Massylii (or eastern) Numidians. But when Massinissa allied himself to Rome he severed his links with Carthage and was replaced in Hasdrubal's favour by Syphax, King of the Masaesyli (or western Numidians). Sophonisba was part of the package.

Unfortunately, Syphax was defeated and captured in 203 BCE by Massinissa and Scipio Africanus at Bagradas. Masinissa then married Sophonisba but Scipio insisted that he give her up so that she could be taken to Rome and feature in Scipio's triumph. Masinissa feared the Romans more than he loved Sophonisba, admitting to her that he was not able to free her from captivity or the Romans. He asked her to die like a true Carthaginian princess: she then cooly drank the poison he offered her. Her story can be found in Polybius (14.4ff.); Livy (30.12.11–15.11), Diodorus (27.7), Appian (*Punica*, 27–8), and Dio (Zonaras, 9.11).

The Kandakes of Kush

Kandake or Kentake was the title bestowed on queens of the ancient Kingdom of Kush in the Nile Valley. It is a derivative of Candace, a Meroitic language term for 'queen' or 'queen-mother'. Pliny (*Natural History*, 6.35) records that the 'Queen of the Ethiopians' bore the title Candace.

The Kandakes of Kush were powerful women rulers; they all ruled as queens but the following are the ones whom we know had a specific militaristic dimension to their rule. In the hierarchy of Kush, the mothers would rule and then install their sons as rulers, but they also deposed their own sons too and could order a son-king to commit suicide to terminate his rule.

According to the much embellished *Alexander Romance*, Candace of Meroë engaged Alexander the Great in battle, although it is doubtful if Alexander ever attacked Nubia or advanced further south than Siwa in Egypt. Apparently, when Alexander tried to take her lands in 332 BCE, she arranged her armies strategically to confront him and seated herself on a war elephant when he approached. Alexander, seeing the strength of her armies, apparently decided to withdraw from Nubia and headed for Egypt instead.

Shanakdakheto is the earliest known ruling African queen of ancient Nubia and reigned from about 170–150 BCE, ruling with full power, without a king, in the Meroë Empire.

Amanirenas was a queen of the Meroitic Kingdom of Kush from about 40–10 BCE. She is notable for her role leading Kushite armies against the Romans in a war from 27–22 BCE. According to Strabo (17.1.54) and Dio (54.5), taking advantage of the absence of Aelius Gallus, prefect of Egypt in 24 BCE, Amanirenas and Akinidad defeated Roman forces at Syene (Aswan) and Philae, and drove the Jews from Elephantine Island; they returned to Kush with prisoners and loot, including several statues of Augustus. Augustus destroyed the city of Napata, Amanirenas' capital, in retaliation. After this initial victory, the Kushites were expelled from Egypt by Gaius Petronius enabling the Romans to establish a new frontier with a garrison in Qasr Ibrim (Primis). Amanirenas was described as brave, and blind in one eye.

Amanikhatashan was Queen of Kush from *c.*62–*c.*85 CE); her pyramid is at Meroe in the Sudan. Amanikhatashan is noted for having sent her cavalry to support Titus during the Great Jewish Revolt in 70 CE.

Cleopatra II

Cleopatra II, Ptolemy VI and their brother, Ptolemy VIII, co-ruled Egypt from *c.*171–164 BCE. In 169, Antiochus IV of Syria invaded Egypt; Ptolemy VI Philometor joined up with Antiochus IV outside Alexandria. Ptolemy VI was crowned in Memphis and ruled with Cleopatra II. Cleopatra II married her other brother, Ptolemy VIII Euergetes II, in 145 BCE.

Cleopatra II later led a rebellion against Ptolemy VIII in 131 BCE, and drove him and Cleopatra III, her daughter, out of Egypt. Ptolemy VIII had his son murdered, dismembered and his head, hands and feet sent to Cleopatra II in Alexandria as a birthday present. Cleopatra II ruled Egypt from 130–127 BCE when she was forced to flee to Syria. Cleopatra and Ptolemy VIII settled their differences in 124 BCE. After this she ruled jointly with her brother and daughter until 116 BCE when Ptolemy died, leaving the kingdom to Cleopatra III. Cleopatra II herself died soon after.

The Bracari Women

The Bracari were a belligerent Celtic tribe of Gallaecia in the northwest of modern Portugal. In 138 BCE Sextus Junius Brutus found that in Lusitania the women were 'fighting and dying with the men with such bravery that they did not cry out even when being slaughtered'. Appian wrote that they were a very warlike people. According to him, the Bracari women warriors fought defending their town, 'never turning, never never showing their backs, or uttering a cry', preferring death to captivity.

'Doom Monster': Cleopatra VII

Cleopatra VII grew into a shrewd military strategist, a highly successful diplomat and a competent military commander; she was co-ruler of Egypt with her father, Ptolemy XII Auletes, and later with her brother-husbands, Ptolemy XIII and Ptolemy XIV. She followed in a tradition of Egyptian women warriors which included the Pharaoh Hatchepsut who fought in several battles during her younger years; Nefertiti, wife of the Pharaoh Akhenaten, depicted as attacking enemies, and Ahhotep, wife of Seqenenre Tao II, who was believed to have been in command of the army while her son Ahmose I was still young.

Cleopatra VII Philopator (69–30 BCE) was a formidable foreign woman who had a significant impact on Rome both militarily and politically, and who changed the course of Roman history. She was born into the Ptolemaic dynasty, the Macedonian family that ruled Egypt after the death of Alexander in 323 BCE.

When her father, Ptolemy XII Auletes, began funding a temple-building programme by re-establishing trade with India, the horn of Africa and re-exports to the rest of the Mediterranean world, his new wealth attracted the attention of the Triumvirate: Pompey, Caesar and Crassus, and this culminated in Auletes assisting Pompey in his highly lucrative victory over Mithridates VI. A coup by Cleopatra V and a daughter, Berenice IV, in 58 BCE forced Auletes', and Cleopatra's, exile to Rome via Rhodes and Athens. In 55 BCE, after much bribery and intrigue between Auletes and the Romans alternately playing and paying Berenice off, Ptolemy XII was restored to the throne – but at a price: it was with the support of the Roman general Aulus Gabinius and Mark Antony, sent to Egypt by Pompey. When Auletes killed Berenice and her followers, Cleopatra, aged 14, was elevated to joint ruler with her father. Ptolemy died in March 51 BCE midst an ominous partial solar eclipse; he expired weighed down with debt, leaving Cleopatra and her brother, now Ptolemy XIII, joint monarchs.

Perceptively, Cleopatra set about winning back the support of her disgruntled people. She learned the local Egyptian language and spoke it, the first and only Ptolemy in 300 years to do so. With prudent and skilful handling of the economy, combined with tax increases – tolerated by her subjects due to her growing popularity – Cleopatra began to turn around the ailing country she had inherited. Meanwhile, in Italy, Julius Caesar crossed the River Rubicon in 49 BCE, while the armies of Pompey retreated south before requesting military support from Egypt. Cleopatra responded by sending a detachment of the Gabiniani and warships.

Later that year relations between brother and sister broke down irreparably with Ptolemy virtually ostracised and Cleopatra flying in the face of the Ptolemaic tradition that made female rulers subserviant to male. In 48 BCE Cleopatra incurred the wrath of the Gabiniani; when they killed the sons of the Roman governor of Syria, Marcus Calpurnius Bibulus, she handed the murderers over to Bibulus. This led to Cleopatra's fall from power: she fled

Egypt with her sister, Arsinoe, while Ptolemy XIII assumed sole rule. Pompey ingratiated himself with Ptolemy.

That same year the decisive Battle of Pharsalus saw Pompey defeated by Caesar. It also marked the start of relations between Caesar and Cleopatra, impatient to win her country back from Ptolemy. However, things started to go wrong between Pompey and Egypt when Pompey sought sanctuary in Alexandria with the remnants of his army, looking to win support from his former ally. The 14-year-old Ptolemy XIII haughtily watched Pompey's arrival from a specially erected throne in the harbour; Pompey was stabbed in the back and beheaded in front of his wife and children; his body was burnt, his head embalmed. This atrocity was nothing more than a half-baked attempt by Ptolemy to curry favour with Caesar, but it backfired. Caesar sailed to Egypt to collect an outstanding debt owed to him by Auletes, only to be presented with Pompey's head; he was incandescent and upset. Pompey, although an enemy, was nevertheless still a Roman consul and the widower of Caesar's daughter, Julia, who had died in childbirth giving birth to Pompey's son, Caesar's grandson. Caesar reacted by taking Alexandria and setting himself up as mediator between Ptolemy and Cleopatra, an inevitable consequence of which was contact with Cleopatra.

Cleopatra was clever enough to interpret the rancour between Caesar and her brother as an opportunity; her envoys had achieved little or nothing, so she determined to meet Caesar face to face to plead her case. Cleopatra was only too conscious of her innate sophistication and sexuality: Cassius Dio tells us that she 'reposed in her beauty all her claims to the throne'. Caesar agreed to a meeting, despite Ptolemy's attempts to thwart it.[2] Arriving regally at the palace in the royal boat (Ptolemy had blocked all the roads), Lucan says that she bribed the guards to unchain the harbour; Plutarch, improbably, insists that she, somewhat less regally, inveigled her way in, rolled up in a carpet carried by Apollodorus the Sicilian, her faithful acolyte. According to Plutarch it is Cleopatra's ingenuity here, her 'provocative impudence', that enraptured Caesar.[3] He cannot, though, have failed to be impressed at the same time by the physical appearance of Cleopatra: Lucan, with typical hyperbole, describes her as: 'wearing the spoils of the Red Sea, treasures in her hair and around her neck, great pains taken on her refinement'.[4] Nine months later Cleopatra gave birth to their son, Ptolemy Caesar, nicknamed Caesarion, 'little Caesar'. Caesar's paternity is doubted and was denied by him, possibly pending changes to the laws of marriage which Caesar promised in order to have his 'marriage' to Cleopatra officially recognised. Critically, he opted to name Octavian, his grandnephew, as his heir and main beneficiary, despite Cleopatra's predictable insistence on Caesarion.

Caesar now backed Cleopatra's claim to the throne but was besieged in Alexandria by Achillas and Pothinus, allies of Ptolemy. Mithridates of

Pergamum raised the siege allowing Caesar to defeat Ptolemy's army at the Battle of the Nile at which Ptolemy XIII drowned. Caesar restored Cleopatra to her throne, with her younger brother Ptolemy XIV as co-ruler. Arsinoe, who had defected to Achillas, was taken prisoner, marched through Rome in disgrace, banished to the Temple of Artemis in Ephesus and later executed by Cleopatra and Mark Antony on the temple steps in an act of sacrilege. Caesar and Cleopatra married according to Egyptian rites – a polygamous union that initially went unrecognised in Rome as Caesar was still married to Calpurnia Pisonis, his third wife; what is more, it was still illegal for a Roman citizen to marry a foreigner. When Caesar left Egypt some weeks later to prosecute the Civil War he left behind him three legions (the XXVIIth, XXXVIIth and XXXIXth) under the command of Rufio, as much to keep Cleopatra under observation as to ensure the safety and security of Egypt, his new ally.[5]

Already, at the age of just 21, Cleopatra was starting to exert significant influence on Rome. In 46 BCE Cleopatra, Ptolemy XIV and Caesarion arrived in Rome, staying for two years in one of Caesar's houses on the fashionable Janiculum Hill[6] – an ill-judged move on Caesar's part as Roman society was beginning to remember that Caesar was, scandalously, still married to Calpurnia. Cleopatra had left her most trusted advisors behind to govern the country in her absence.

Caesar could not have been more indiscreet and inappropriately honorific when he went so far as to erect a golden statue of Cleopatra as Isis in the Temple of Venus Genetrix, the ancestor of his gens, situated in the prestigious new Forum Julium.[7] The five-fold shock of a statue in honour of a living person, a woman at that, a foreigner and a queen and the religious association with Roman Venus, all in the middle of Rome, was incalculable.

Cleopatra was still in Rome when Caesar was assassinated on 15 March 44 BCE.[8] She and her retinue then returned to Egypt, travelling via Cyprus to re-establish Egyptian control there; Cyprus had been restored to her by Caesar in 48 BCE. Cicero, and many other Republicans besides, despised the foreign queen, Horace's *fatale montrum*, and were glad to see the back of her.

Young Ptolemy XIV died soon after – some allege that he was poisoned by Cleopatra; whatever, she was anxious that her son by Caesar, Caesarion, be a key player in her dealings with Rome; accordingly, she made Caesarion her co-regent and successor, bestowing on him the titles Ptolemy XV Caesar Theos Philopator Philometor (Father – and mother – loving God). The Civil War raged on and Cleopatra sided with the Caesarian faction led by Mark Antony and Octavian, against the assassins of Caesar under Marcus Junius Brutus and Gaius Cassius Longinus. In 43 BCE, she allied with the leader of the Caesarian party in the East, the fickle Publius Cornelius Dolabella, who recognised Caesarion as her co-ruler. But soon after, Dolabella was defeated in Laodicea by Cassius and took his own life.[9]

In 41 BCE Cleopatra finally met with Mark Antony in Cilicia after a number of previous requests had gone ignored; ostensibly, Antony wanted to question the queen about the money she had allegedly paid to Cassius, and to win her support in his planned campaign against Parthia. Cleopatra arrived for the rendezvous in a blaze of astonishing extravagance and ceremony on the sumptuous royal barge, 'a get-up that beggared description', Plutarch spluttered: he describes her magical boat as having gold prows, purple sails, silver oars and a gold-studded canopy. Cleopatra's ostentatious display was so compelling that Antony, her second world leader, was so besotted that he spent the winter of 41–40 BCE with her in Alexandria. Appian tartly describes the total seduction of Mark Antony: he was captivated by her beauty, he was enthralled by her as if he were a young man, even though he was 40.

Antony attracted censure for dallying in Alexandria. Rumours of lavish entertaining percolated back to Rome and were exaggerated, excoriated and scandalised by a pious Cicero in his *Philippics*: Antony's house rang with the clamour of drunkards, the streets sloshed with wine; Antony was a booze-soaked, sex-sodden wreck of a man. Cleopatra's reputation suffered equally by association. The accusations of drunken debauchery persisted for years: according to Horace and Propertius, spokes-poets for the Octavian-Augustan regime, Cleopatra was ever befuddled by Mareotic wine, her voice slurring with a wine-soaked tongue.[10]

It was at this point that Cleopatra and Antony ordered the execution of Cleopatra's sister Arsinoe, carried out, as we have seen, sacrilegiously and scandalously on the steps of the Temple of Artemis; apart from her fatal defection, Arsinoe was, it seems, a pretender to Cleopatra's throne. Cleopatra also had Sarapion, her top general in Cyprus, killed for his collusion with Cassius; Appian concludes that now Antony was putty in Cleopatra's hands: 'whatever Cleopatra wished it was done, regardless of the laws, human or divine'.[11] Cleopatra agreed the reinforcements Antony needed for his campaign against Parthia.

Cleopatra gave birth to twins on 25 December 40 BCE; the father was Mark Antony and they were named Alexander Helios and Cleopatra Selene II. There then followed a gap in the relationship. In 36 BCE, Antony returned to Alexandria on his way to wage his war with the Parthians; the old relationship was renewed and Antony remained in the Egyptian capital, marrying Cleopatra, even though he was at the time married to Octavia Minor, Octavian's sister. He and Cleopatra had another child, Ptolemy Philadelphus, during their stay in Antioch. Lucrative territories were signed over to Cleopatra to bolster the substantial natural resources and crops to which she already had access: Crete, Cyrene, Cilicia, Phoenicia and part of Arabia corresponding to modern Jordan were all ceded to Egypt, as were parts of Syria and troublesome Judaea, ruled by Herod the Great (74–4 BCE). Cedar from

Lebanon for shipbuilding; the rare and priceless Balm of Gilead (*commiphora gileadensis* and *pistacia lentiscus*) from around Herod's Jericho, a key constituent in popular medicines and fine perfumes; Dead Sea bitumen, an essential ingredient in embalming fluid – all of this highly lucrative trade and produce came Cleopatra's way by the trireme load.

In 34 BCE another opulent and dazzling ceremony – again with Cleopatra symbolising Isis – celebrated the carving up of the Egyptian Empire, the so-called 'Donations of Alexandria'. Cleopatra became Queen of Kings over Egypt while Caesarion took Cyprus as King of Kings; Alexander Helios and Philadelphus became Kings East (Armenia, Media and Parthia) and West (Phoenicia, Syria and Cilicia) of the Euphrates respectively while Cleopatra Selene II was made Queen of Cyrene and Libya. Cleopatra added Philopatris, lover of her country, to her titles. Caesarion was depicted as the Egyptian god Horus while Cleopatra called herself *Nea Isis*.

Meanwhile, the increasingly fragile relationship between the triumvirates Octavian and Mark Antony finally shattered in 33 BCE. Octavian had the Senate vote a war against Egypt culminating in 31 BCE in the decisive naval action off Actium. Cleopatra took flight with her fleet, soon to be followed by Antony. Octavian then invaded Egypt; Antony's armies deserted to Octavian on 1 August 30 BCE. Antony had no choice but to take his own life. Cleopatra committed suicide soon after.

Cleopatra was a powerful, successful, ruthless and influential woman who thrived at a time when Rome and the Mediterranean world were vulnerable and in turmoil, when political and military intrigue and activity were at their most volatile and explosive. Cleopatra moved in the highest circles within the Mediterranean states, forging alliances and pursuing affairs with two of the most powerful men of the day. Cleopatra was of course never popular in Rome. In her relationships with Julius Caesar and Mark Antony she was viewed with considerable malice and suspicion, a queen who evoked memories of the regal kings of early Rome and their dreadful tyranny. Octavian, not surprisingly, capitalised on Cleopatra's alleged reputation in a bid to slur and diminish her. He sacrilegiously took Antony's will from the safekeeping of the Vestal Virgins and read it out to the Senate, but not before he had carefully selected the extracts most likely to discredit Antony: he focused on Antony's state funeral preparations and the 'repatriation' of his body to Egypt; other damaging revelations included the news that he had left Cleopatra the 200,000 volumes deposited in the libraries at Pergamum. Furthermore, it was revealed that at a banquet he had stood up and annointed Cleopatra's feet as a bet, and he had allowed the Ephesians to salute Cleopatra as their queen; he would often receive love-letters from her on onyx or crystal, and read them while seated in tribunals when he was supposed to be dispensing justice. Once, when the famous orator Furnius was speaking, he saw Cleopatra outside being carried through the forum on a litter – he abandoned the trial and escorted her on her way.

Octavian humiliated Antony and Cleopatra when he officially declared war on the queen and relieved Antony of the powers he had feebly ceded to a woman. He alleged that Antony's mind was stunted by drugs, insinuating that Cleopatra was a witch; the war would be fought against a motley force comprising Mardian the eunuch, Potheinus, Iras, Cleopatra's hairdresser, and Charmian, her lady-in-waiting – for they it was who were in charge in the Antony and Cleopatra camp. Octavian's propaganda machine was highly effective, painting Cleopatra as the enemy of the state before and after Actium; the slurs were to achieve mythic proportions in the fiercely nationalist poems of Virgil, Horace and Propertius. The more Octavian 'bigged up' Cleopatra and the threat she posed to a paranoid Roman state, then the bigger his achievement when he eventually defeated her and extinguished that threat.

Those poets at the end of the Republic and in the early Principate sang as one in their contempt and derision. Virgil is perhaps the most restrained but, by the same token, the most penetrating. In Book 4 of the *Aeneid*, Dido, the woman who led the 'Carthaginians' out of Phoenicia to Libya as *dux*, is betrayed by Aeneas, sidelined by his dedication to the mission to found Rome. Dido is incandescent with rage and in a pit of misery and despair: 'you'll pay the price, you traitor, and I will hear about it – the news will reach me deep down among the deadmen'.[12]

Virgil's audience would have appreciated the poignancy of Dido's threats: Rome did indeed 'pay the price' for Aeneas' duplicity, it paid with the devastating, near cataclysmic Punic Wars. Indeed, the Dido episode resonated uncomfortably with the political upheaval caused so recently by Cleopatra, a foreign queen eerily reminiscent of Dido, whose facility for global powerplay would be likened to the unnatural skills of Dido as sorceress. More graphically, Cleopatra takes centre-stage on Aeneas' highly symbolic shield *ekphrasis*, which describes the armour given to him by Venus to fight for the foundation of Rome and the genesis of its proud history. She is called *regina* – 'the queen' – twice, still a dirty word so long after the fall of the monarchy; her gods are monsters, *monstra*, which are ranged against the Roman pantheon; her defeat and her flight to Egypt are thrown into full relief.[13]

Horace excoriates the 'mad queen' – *regina demens* – in his famous poem celebrating her defeat at Actium: 'now the time has come to drink, get up and dance and feast!' Horace was sure that Cleopatra was plotting the death and destruction of Roman rule; her retinue is made up of eunuchs: squalid, diseased and drunk; she too is inebriated: she is a doom monster – *fatale monstrum*. The poem echoes sentiments first introduced in the *Epodes*: written soon after Actium, Horace sneers at the Roman soldiers under Antony and

Cleopatra – they are enslaved by a woman and answer to eunuchs, they have all gone soft with their un-Roman use of the *canopium*, the mosquito net.[14]

Propertius is no less hostile. Cleopatra informs a poem in which she is compared with Cynthia: Cynthia is the poet's *domina*; she dominated him just as Cleopatra dominated Antony; Propertius is enslaved by a woman, just as Antony was. Calling a mistress *domina*, and calling their love for that *domina*, *servitium amoris* were two of the outrageous ways in which the love poets controversially turned the usual man-woman relationship on its head: in the real world the man was always the dominant partner, a real Roman would never be a slave to love. Insinuating Cleopatra in the poem embroils her in the decadent, un-Roman world of the love poets where women ruled, men were slaves, were love-sick or languished, *exclusus*, locked-out on the doorstep, were cuckolded as pathetic lovers whose only combat experience, unlike a real Roman, was in the war of love, *militia amoris*. Propertius, of course, was one of the champions of this roué lifestyle; his portrait of Cleopatra here is probably one of the first instances of her as an embodiment of woman's domination of men, a motif that has endured down the years.

Cleopatra comes at the end of a list of powerful women Propertius uses to justify Cynthia's domineering behaviour, but it is with Cleopatra that he piles on the invective when the poem elides into a celebration of Augustus. Chiming with Horace, we hear that she plotted to subjugate Rome, her marriage was sordid, Egypt a place of deceit and disaster; she is a whore of a queen, *meretrix regina*, she tried to supplant Roman gods with Egyptian, to influence the Tiber with the Nile before her death by asp. She has stained the reputation of the Macedonian house; she is 'shagged out' (*trita*) by sex with her own slaves – a cardinal sin and a shameful crime which would have resonated with Propertius' audience: centuries before, Romans would remember that Lucretia lost her life at the mere prospect of such a heinous crime. This triumphalist tone is resumed by Propertius in his celebratory poem exulting at the victory of the Battle of Actium, and the defeat of Cleopatra.[15]

Cleopatra turns up, predictably, in Juvenal's list of effeminate pathics – grieving after Actium, *maesta*.[16] Martial satirises her death.[17] In the epic *de Bello Civili*, Lucan's Cleopatra is evil: her beauty is sluttish (*incesta*) – a swipe at her marriage to her brother – her make-up is vulgarly overdone and her jewels and clothing blingy; the banquet she lays on for Caesar is opulent beyond anything then known at Rome, and is described at length as the last word in *luxus*.[18] In the words of Pothinus as he tries to foment war against Caesar, she is *inpia* – unholy, devoid of *pietas* – as she scuttles between husbands; she is selling off Rome and, witch-like, seduces Caesar with potions; their affair is *incestus*, their love *obscaena* and impious. Michael Grant called it all 'one of the most terrible outbursts of hatred in history'.[19]

Cleopatra, too, came to embody everything the Roman man suspected, feared or despised about difficult and flamboyant women. She was certainly gifted, highly intelligent, assertive, ambitious and conspicuous: after all, she was a successful head of state, a respected military commander and a capable strategist; she was a leader of men and she was ruler of a rich and bountiful country; she infatuated and manipulated two of Rome's leading generals and most powerful politicians. Quite simply, her actions helped form the shape of the Roman constitution and the establishment of the Roman empire after the Battle of Actium.

Cleopatra was the anti-*matrona*, the anti-Roman. A few decades after her death the emperors' women, of course, were able to assume a public prominence and political, dynastic profiles far and away above anything possible to women in the Republic. Only a foreign woman, Cleopatra, came close to the power wielded by the Julias, Livias, Agrippinas and the Messalinas by virtue of their position at the heart of government and in the beds of the men running those governments.

Hypsicratea

This is a story of undying love in a woman who could not bear to be without her husband, even to the point of following him into battle armed to the teeth. Hypsicratea (fl. 63 BCE) was a Queen of Pontus, ruling over a number of states with Mithridates VI.

Evidently, she was so devoted to Mithridates that she dressed up in men's clothes, cropped her hair, learned combat skills and followed him into exile. She rode with him into battle, putting down rebellions and fighting tooth and nail against the Romans – with axe, lance, sword and bow and arrow.

During Mithridates' defeat by Pompey, Hypsicratea escaped from Pompey's blockade and was one of only three comrades who remained by the king's side. According to Plutarch (*Pompey*, 32.8), she was 'always manly [*androdes*] and extremely bold, the king consequently liked to call her Hypsicrates ... she neither flagged physically over the distances they ran nor did she tire of tending the body and horse of the king, until they came to a place called Sinor, which was full of the king's coins and treasures'.

Cartimandua

In 43 CE, after a century of Roman indecision and hesitation with aborted invasions led by Julius Caesar and Caligula, the Emperor Claudius invaded Britain, ostensibly to help the client king Verica of the Atrebates who had been exiled after a revolt by the Catuvellauni.[20]

When we think of fighting women in the early days of Roman Britain, we automatically think of Boudica; Boudica, however, was preceded by another British warrior woman: Cartimandua, queen of the Brigantes tribe.

The area corresponding to modern-day Wales proved particularly intractable to the Romans with fierce resistance coming from the Silures, Ordovices and Deceangli. Caratacus and his guerrilla attacks were a real problem; the Brigantes and the Iceni were also proving troublesome. In 51 CE things came to a head when Publius Ostorius Scapula defeated the Silures under Caratacus. He fled to the Brigantes, but their Queen Cartimandua treacherously handed him over him to the Romans, whence he was sent to Rome, a reluctant star in Claudius' triumph. The Silures, however, would still not lie down and Cartimandua's former husband Venutius, a hater of the name of Rome, *Romani nominis odium*, took over the mantle of leader of British resistance against Rome.[21]

Cartimandua had probably been leader of the Brigantes well before Claudius invaded and may well be one of the eleven monarchs who surrendered to Rome without a fight, as depicted on Claudius' triumphal arch.[22] She was now considerably wealthy, due to her shrewd support of Rome and her betrayal of Caratacus; however, Venutius attacked the Brigantes, only failing to defeat them when the Romans sent Caesius Nasica and the IXth *Hispana*, to reinforce Cartimandua.[23]

Cartimandua had married Vellocatus, former armour-bearer to Venutius, and installed him as her king. Although Tacitus records that Cartimandua was a good ally of Rome's and acknowledges her *nobilitas*, he savages her in much the same way as the Augustan poets had destroyed Cleopatra. He describes her treatment of Caratacus as *per dolum*, evily-intended, and highlights her wealth and luxurious lifestyle; her liaison with Vellocatus is adulterous and driven by rage and lust: *pro adultero libido reginae et saevitia*, it is a disgraceful act, *flagitium*, which shook the very foundations of her dynasty. She is a queen, *regina* – as we know, a dirty word in Rome – and is duplicitous in her behaviour towards Venutius and his family: *callidisque Cartimandua artibus*; Tacitus disparages the power that she wields as being ignominious to men. The next powerful woman, the next British queen the Romans encountered, was to prove even more belligerent and somewhat less compliant.

Boudica

Aulus Plautius was appointed commander-in-chief and governor of the province. The invasion force comprised four legions: the IXth *Hispana*, the IInd *Augusta*, the XIVth *Gemina* and the XXth *Valeria Victrix*, supported by 20,000 auxiliary troops from Thrace and Batavia in Germany. After sixteen days, according to Dio, Cogidubnus was installed as a client king and Claudius returned to Rome, his victory celebrated with a triumphal arch.

Plautius proceeded to subdue Britannia, advancing inland. The IXth moved through the northeastern territories of the Catuvellauni into the

Coritani lands in modern Leicestershire and Lincolnshire, establishing fortresses near the borders with the Iceni and near the borders with the Brigantes, both client states.

Tacitus describes the in-fighting and divisiveness that was characteristic of the British tribes and which was ultimately responsible for their conquest: *ita singuli pugnant, universi vincuntur* – 'and so they fight individually, and all are conquered'. Nevertheless, they are generally compliant and toe the line, so long as their trust is not abused, *iniuriae absint*.

One of the ways in which local tribes could engender peaceful co-existence with Rome was to bequeath to the Romans their lands on the death of the king or queen. Prasutagus, prosperous King of the Iceni – inhabiting what is roughly today's East Anglia – did just that, citing Nero as heir but with an additional clause naming his daughters as co-heirs. The Iceni had been on friendly terms since the early days of the invasion but on the king's death in 60 CE the Romans chose to ignore the small print in the king's will, subdued the kingdom and plundered it. The Iceni's status as *civitas peregrine*, 'foreign state', was annulled. Perhaps it was naive of the Iceni to expect an extension of the special relationship after Prasutagus' death but the aftermath of the Romans' arrogance was shocking, brutal and highly provocative: Prasutagus' daughters were raped, *filiae stupro violatae sunt*, Queen Boudica, his wife, was flogged, the family was treated no better than slaves and his Roman creditors called in their loans, loans that the Iceni thought were gifts.[24] Boudica was humiliated and outraged.

Tacitus describes how trouble between the Britons and Rome had been brewing for some time; deplorable breach of trust, the high-handed treatment of Prasutagus' family and the growing burden of occupation provided the spark for revolution. We hear of *servitus*, *iniuriae*, *subiecti* and *contumelia*: slavery, injustice, subjugation and abuse. The Romans are greedy and rapacious cowards; they are driven by avarice and riotous living – *avaritia* and *luxuria*. The Iceni and other tribes had been disarmed by Ostorius Scapula, *legatus* from 57–62 CE, a cause of great resentment to the Iceni who believed they were a client kingdom and therefore immune from such treatment. The Roman veterans in Camulodunum (Colchester) behaved with contempt and arrogance towards the natives. The Britons had had enough: they planned a violent response.

Tacitus describes Boudica as a woman of royal and noble birth, pointing out that it was, unlike in Rome, not unusual for a woman to lead a British army. The only other description we have of Boudica comes from Cassius Dio writing at the end of the second century BCE; he vividly describes her as follows after the Iceni's victory over the Romans:

> all this ruin was brought upon the Romans by a woman, a fact which in itself caused them the greatest shame. . . . But the person who was chiefly

> instrumental in rousing the natives and persuading them to fight the Romans, the person who was thought worthy to be their leader and who directed the conduct of the entire war, was Bouduica, a Briton woman of the royal family and possessed of greater intelligence than often belongs to women. ... In stature she was very tall, in appearance most terrifying, in the glance of her eye most fierce, and her voice was harsh; a great mass of the tawniest hair fell to her hips; around her neck was a large golden necklace; and she wore a tunic of divers colours over which a thick mantle was fastened with a brooch. This was her invariable attire. She now grasped a spear to aid her in terrifying all beholders'.[25]

It is impossible to tell how accurate a representation of the woman this is. Dio would have been careful to build her up – the bigger the woman the bigger Rome's success in defeating her. He echoes the typical Roman man's anxiety at being outclassed by a mere woman and grudgingly accepts that she was intelligent, to Romans an unwelcome quality for a woman, and a capable strategist, something else to be feared. Physically she was big, frightening, a bit common and gaudily attired – in common with the elegant and classy Cleopatra, she was another unwelcome antithesis to the discreet, unobtrusive, compliant and modestly attired traditional Roman *matrona*.

Gaius Suetonius Paulinus, the Roman governor, was preoccupied with trying to take Mona, Anglesey in order to eradicate the exiled Druid community there and end the island's status as a haven for disaffected refugees. Druidism was feared by the Romans, not least because of its reputation for focusing opposition to Roman rule and the mystery surrounding the arcane rites; Druids were also known for their veneration of the human head which manifested itself in routine decapitation of corpses after battles. This head-hunting can be seen depicted on Trajan's Column. According to Tacitus, the Druids had a reputation for 'soaking their altars in the blood of prisoners and using human entrails in their divination'. Equally disturbing for the Romans was the fact that the Druids deployed women in their ranks: Paulinus' soldiers lined up opposite an armed pack among which were 'women dressed in black robes with dishevelled hair like Furies, brandishing torches. Next to them were the Druids, their hands raised to the skies, screaming fearsome curses'. The Romans were at first paralysed with fear but then attacked, slaughtered all before them and hacked down the sacred groves.[26]

The time was ripe for British rebellion: the Iceni under Boudica marched on Camolodunum and, with its Temple of Claudius, a citadel symbolic of oppressive Roman rule and the seat of their servitude: *sedes servitutis*. The Iceni were joined by the disaffected Trinovantes, the tribe that had been displaced and enslaved to make way for the *colonia*. The omens were not good for the Romans:

> the statue of Victory in Camulodunum crashed to the ground, supine as if in flight; lamentations could be heard though no mortal man had uttered the words or the groan; hysterical women chanted impending doom, 'at night there was heard to issue from the senate-house foreign jargon mingled with laughter, and from the theatre outcries and a ghost town on the Thames was seen to be in ruins and the Channel turned blood red; shapes like bodies were washed up'.

More crucially, Camuoldunum was not fortified and was largely undefended – Suetonius Paulinus had fatefully posted the XXth legion to the Welsh borders. The procurator was found wanting: when the beleagured Roman inhabitants clamoured for reinforcements Catus Decianus sent a mere 200 auxiliary troops.

The *colonia* was sacked and the temple fell after two days; the *saevitia*, savagery, of Boudica's forces was uncompromising. The IXth legion under Petillius Cerealis rushed to relieve the defenders but was annihilated. Catus Decianus fled to Gaul. Suetonius reached Londinium – an important but undefended trading port – he calculated that it was impossible to defend with the forces at his disposal. Londinium was abandoned and those left behind were slaughtered in the carnage that ensued. Tacitus paints a wretched picture of the doomed city, and of Suetonius' dilemma:

> Alarmed by this disaster and by the fury of the province which he had goaded into war by his rapacity, the procurator Catus crossed over into Gaul. Suetonius, however, with wonderful resolution, marched amidst a hostile population to Londinium, which, though undistinguished by the name of a colony, was much frequented by a number of merchants and trading vessels. Uncertain whether he should choose it as a seat of war, as he looked round on his scanty force of soldiers, and remembered with what a serious warning the rashness of Petilius had been punished, he resolved to save the province at the cost of a single town. Nor did the tears and weeping of the people, as they implored his aid, deter him from giving the signal of departure and receiving into his army all who would go with him. Those who were chained to the spot by the weakness of their sex, or the infirmity of age, or the attractions of the place, were cut off by the enemy.

Excavations have revealed a thick red layer of burnt detritus covering coins and pottery dating from before 60 CE.

Euphoric and drunk on their easy successes, the Britons then devastated Verulamium (St Albans), a stronghold of the pro-Roman Catuvellauni. According to (an exaggerating) Tacitus, up to 80,000 men, women and children were slain in the orgy of destruction visited on the three towns of

Camulodunum, Londinium and Verulamium by Boudica's forces. The Britons were not in the habit of taking prisoners: they had no interest in selling slaves and they showed no quarter, the only options being slaughter, hanging, burning alive and crucifixion – *caedes*, *patibula*, *ignes*, *cruces*. Dio's graphic account tells us that the noblest women were impaled on sharpened spikes the length of their bodies and that their breasts were hacked off and sewn onto their mouths, 'to the accompaniment of sacrifices, feasts, and lewd behaviour' sacrilegiously performed in sacred places, like the groves of Andraste, a British goddess of victory.[27]

Suetonius hurriedly assembled a force of around 10,000 men and prepared for battle, the Battle of Watling Street. These soldiers were massively outnumbered by Boudica's 230,000 – no doubt a huge exaggeration but Boudica did have substantial superiority; as Dio says, even if the Romans were lined up one deep, they would not have reached the end of Boudica's line. So casual and confident was Boudica's army of victory that women were allowed to spectate in wagons on the edge of the battlefield. Boudica herself rallied her troops from a chariot, her raped daughters standing before her: in a rousing speech recorded by Tacitus she declared that there was nothing unusual about a woman commander among the British and invoked her lineage from mighty men; *solitum quidem Britannis feminarum ductu bellare testabatur*. She was fighting for her kingdom, her freedom and to avenge her battered body and her violated daughters: *contrectata filiarum pudicitia*. She was intent on wiping out the *stuprum* inflicted on her daughters and restoring their violated *pudicitia*. The Romans slay the elderly and rape virgins: *ut non corpora, ne senectam quidem aut virginitatem impollutam relinquant*. The Romans are cowards but the Iceni and she, a woman, will defeat them. She concludes by affirming her resolve, the resolve of a woman – men can live in servitude (if that is what they want): *id mulieri destinatum: viverent viri et servirent*.[28] Boudica certainly will not. Paulinus' speech to his men takes up the gender theme when he disparagingly declares that there are more women in the ranks of the Britons than men. Suetonius, the historian, that is, suggests that the Iceni army comprised more women than men.

Tacitus plays down the enormity of the very thought of a woman with a successful military career; a woman with the temerity to do battle with men – nothing could be more un-Roman, exemplified only by two wives of the damned-from-memory and disgraced Mark Antony: Fulvia Furia Bibacula and that other exotic, foreign queen, Cleopatra VII.

There was to be no victory; Boudica was soundly defeated. The Britons were hampered by their poor manoeuvrability and inexperience of disciplined open-field tactics. The narrowness of the battlefield restricted the numbers Boudica could deploy at any one time thus diluting her numerical advantage.

The Britons were felled in their droves by the Roman *pilae*, spears, which rained down on them.

Even women and animals were slain that terrible day; Boudica's retreating soldiers were hampered by the wagons nearby, full of spectators. According to Tacitus (exaggerating again), 80,000 Britons died to the Romans' loss of 400. No doubt there was more rape; it is not known what happened to Boudica's daughters.[29] Boudica committed suicide by poisoning: *Boudica vitam veneno finivit*. Dio disputes this, or at least paraphrases the detail out of the same story, and claims that Boudica fell ill and died, and was buried at great expense and with full military honours.

If Boudica was loathed by some Romans in the years following her death, then the verdict of St Gildas (*c.*500–70 CE), or *Gildas Sapiens*, a sixth-century British monk, probably outdoes them for vitriol. He calls her 'a treacherous lioness' who 'butchered the governors who had been left to give fuller voice and strength to the endeavours of Roman rule' in his searing polemic *De Excidio et Conquestu Britanniae*, a history of the Britons before and during the Saxon period. The tract includes a description of the Roman occupation and the 'Groans of the Britons', in which the Britons request military assistance from the departed Roman army.

Andraste was, according to Dio Cassius, an Icenic war goddess invoked by Boudica in her battles against the Romans:

> 'Let us, therefore, oppose [the Romans], trusting boldly to good fortune. Let us show them that they are hares and foxes trying to rule over dogs and wolves'. When she [Boudica] had finished speaking, she employed a type of divination, letting a hare escape from the fold of her dress; and since it ran on what they considered the auspicious side, the whole crowd shouted with pleasure, and Boudica, raising her hand toward heaven, said: 'thank you, Andraste, and call upon you as woman speaking to woman ... I beg you for victory and preservation of liberty'.[30]

Cimbrian Women

Strabo is our main source for this belligerent tribe of women.[31] Strabo and Tacitus say the Cimbri and Teutones left Frisia and Jutland in the second century BCE when the sea began to encroach, possibly through climate change. After years of wandering along the Danube and the Rhine they found themselves on the borders of Italy near Noricum (modern Austria and part of Slovenia) where they defeated the Taurisci in 113 BCE. The Taurisci were allies of Rome and appealed for help. The Cimbri and Teutones made a conciliatory offer to retreat, but were duplicitously attacked by a Roman army under Gnaeus Papirius Carbo at Noreia (modern Ljubljana); the Cimbri won convincingly but stopped short of advancing into Italy; the Romans were saved from total annihilation by a storm. Carbo committed suicide by drinking vitriol (*atramentum sutorium*, or sulphuric acid).

In 105 BCE for the first time since the Second Punic War Rome was under serious threat from a foreign power: the Cimbri and the Teutones were back with reinforcements. An army was levied, led by Cnaeus Mallius Maximus; the diplomacy offered by the Cimbri was rejected while Mallius' army became increasingly ill-disciplined and disputatious. The consul Q. Servilius Caepio, meanwhile, was not obeying orders and failed to support Mallius: a disturbing example of class rivalry in which the *novus homo*, Mallius, was snubbed by the aristocratic Caepio. A skirmish opened the hostilities in which Marcus Aurelius Scaurus and his cavalry were attacked in their camp. Scaurus was captured and brought before the King of the Cimbri, Boiorix, whom he arrogantly advised to withdraw before inevitable annihilation by the Romans. Boiorix responded by having Scaurus roasted alive slowly in a wicker cage – a truly horrible death which he apparently endured with consummate dignity. This atrocity is, of course, familiar to us: a similar fate befell Royal Jordanian Air Force pilot Lieutenant Moath Youssef al-Kasasbeh, murdered by Islamic State in 2015.

The Romans, though in disarray, attacked at Arausio (Orange) and were routed in what was the worst disaster since Cannae, with the loss of 80,000 men. Typically, Orosius exaggerates when he claims that only ten Romans survived, a nod perhaps to the fact that the camp followers and servants were also butchered. Plutarch graphically records that the soil from the battlefield was made so fertile by corpses and body parts that they produced *magna copia*, high yields, for years to come. Rome panicked, and the *terror cimbricus* became a byword for the disaster as comparisons with Gauls outside of the gates of Rome loomed large in the collective psyche. Desperate measures were taken when, contrary to the Roman constitution, Gaius Marius was elected consul and supreme commander for five consecutive years (104–100 BCE). At Aquae Sextiae, the Romans won two battles; at the Battle of Vercellae, in 101 BCE, Cimbri suffered a devastating defeat and their two chieftains, Lugius and Boiorix, died on the field, while the other chieftains Caesorix and Claodicus were captured. Before killing themselves the women killed their children to avoid slavery.

Strabo graphically describes for us the gruesome role played by women in this bellicose tribe as they habitually followed their men, with their children, in carts:

> Their wives, who would accompany them on their expeditions, were attended by priestesses who were prophetesses; these were grey-haired, clad in white, with flaxen cloaks fastened on with clasps, girt with girdles of bronze, and bare-footed. With sword in hand these priestesses would

> meet with the prisoners of war throughout the camp, and having first crowned them with wreaths would lead them to a brazen vessel of about twenty amphorae; and they had a raised platform which the priestess would mount, and then, bending over the kettle, would cut the throat of each prisoner after he had been lifted up; and from the blood that poured forth into the vessel some of the priestesses would draw a prophecy, while still others would split open the body and from an inspection of the entrails would utter a prophecy of victory for their own people; and during the battles they would beat on the hides that were stretched over the wicker-bodies of the wagons and in this way produce an unearthly noise.[32]

The Cimbrian women joined their men in battle, forming a line with their wagons and fought with poles and lances, as well as staves, stones and swords. Whenever the women saw that defeat was imminent, they killed their children and committed suicide rather than be taken prisoner.

Evidence for Cimbrian ritualistic sacrifice comes with the Haraldskær Woman discovered in Jutland in 1835. Noosemarks and skin piercing were evident; she had been thrown into a bog rather than buried or cremated which was more usual. The Gundestrup cauldron, excavated in Himmerland, may well be a sacrificial vessel like the one described in Strabo's text.

The Ambrones

Psychological warfare and ferocious female combat feature in a confrontation before and during the Battle of Aquae Sextiae. While the Cimbri were descending on Noricum the Ambrones and Teutones split off and marched through Liguria against Marius, who had set up camp on the Rhône. Plutarch gives the numbers advancing on Italy as 300,000 armed fighting men, and many more women and children. The Teutones and Ambrones attacked the Maran camp; they were repulsed so they surrounded the camp and encircled it for six days, repeatedly asking the Romans if they had any messages for their wives because they, the Ambrones, would very soon be joining those wives (Plutarch, *Marius*, 18.2). Marius eventually won the day; Plutarch says that he took 100,000 prisoners, something of an exaggeration no doubt. But before that the Ambrones took terrible losses: 'and the river was filled with their blood and their dead bodies ... the Romans kept slaying them'. The Ambrones' women joined in, fighting ferociously (Plutarch, *Marius*, 19.6–7; cf. Livy, *Periochae*, Book 68):

> Here the women met them, swords and axes in their hands, and with hideous shrieks of rage tried to drive back fugitives and pursuers alike, the fugitives as traitors, and the pursuers as foes; they mixed themselves up with the combatants, with bare hands tore away the shields of the

Romans or grasped their swords, and endured wounds and mutilation, their fierce spirits unvanquished to the end.

German Women of War

Tacitus gives us some fascinating details in Books 7 and 8 of his *Germania*, written in about 98 CE, relating to the role of German women in war and battle. He points out that a major contribution to German military success lay in the fact that their fighting units were not made up of warriors randomly formed into motly units, but rather are 'composed of men from one family and that their households go with them to the battle, and the shrieks of their women and the wailings of their children ring in their ears'. The effect of this is that 'Each man feels bound to play the hero before such an audience and to earn their most coveted praise'. The wounded German soldier brings his wounds 'to his mother and to his wife he brings his wounds; and they do not shrink from counting them, nor from searching them, while they carry food to the fighters and give them encouragement'. In Book 8 Tacitus adds that 'their traditions show that more than once, when a German line was wavering on the point of crumbling it was the women who rallied it, urgently urging the men to fight on, baring their breasts and crying out that their captivity was imminent'.

Women then played a key role in motivating their warrior men, goading them with the prospect of capture by the enemy. These women knew also that their men hated captivity for their women far more than for themselves – aware, no doubt, of the servitude and rape that awaited them. The Romans capitalised on this fact: Tacitus says 'as a matter of fact, we always obtain the firmest hold over those states which are compelled to include amongst the hostages they send us some girls of noble birth'. He goes on to assert that the Germans even ascribe to women 'a certain inspiration and power of prophecy'; they actually listen to what their women say and pay heed to their prophecies. Prophecies would presumably have included forecasts relating to the outcome of battles, and perhaps even strategy, thus giving their women a crucial role in military planning.

The Sitones, Tacitus says 'are conterminous with those of the Swedes, whom they resemble in all respects with only one point of difference: they are ruled by a woman'. This would assume some sort of position of responsibility in military matters.[33] Tacitus adds acidly, 'So far, they [the Germans] fall not merely below the position of free-men, but even beneath that of slaves'.

He adds an interesting footnote when he references two women, one called Veleda, a prisoner in Rome in the days of Vespasian, and one Aurinia (also known as Albruna).[34] In the second half of the 1st century CE Veleda, a priestess of the Bructeri, was regarded as a goddess by most tribes in central Germany and was accordingly very influential. She lived in a tower on the

Lippe River, near modern-day Lippstadt. The inhabitants of the Roman settlement of Colonia Claudia Ara Agrippinensium (Cologne) accepted her arbitration in a conflict with the Tencteri, an unfederated tribe of Germany. Her finest hour, though, came during the Batavian rebellion of 69–70 CE, led by the Romanised Batavian chieftain Gaius Julius Civilis, when she correctly predicted the initial successes of the rebels against the Romans.

The Batavian leader Civilis originally raised his army as an ally of Vespasian in 69 CE, but when he saw the weakened condition of the legions in Romanised Germany he openly rebelled. Veleda may have had a hand in inciting the rebellion which, in early 70, was joined by Julius Classicus and Julius Tutor, leaders of the Treviri, all Roman citizens like Civilis. The Roman garrison at Novaesium (now Neuss) capitulated without a fight, as did Castra Vetera (near modern Xanten). The commander of the Roman garrison, Munius Lupercus, was ordered to Veleda, but he was killed en route, evidently in an ambush. The praetorian trireme was captured, and rowed up the Lippe as a gift to Veleda.[35]

It took nine Roman legions under Gaius Licinius Mucianus to end the rebellion. The Romans treated the rebels leniently in order to reconcile them to Roman rule and military service. Veleda was left to go free for several years. In 77 CE the Romans either captured her, perhaps as a hostage, or offered her asylum. Whatever, she enjoyed international fame: a Greek epigram has been found at Ardea, south of Rome, that satirises her prophetic powers.[36] Veleda may have later acted for Rome by negotiating the acceptance of a pro-Roman king by the Bructeri in 83 CE.

Thusnelda (b. *c.*10 BCE) was a Germanic noblewoman, the daughter of the Cheruscan Prince Segestes. Although her father had intended her for another, she was captured by Arminius who made her pregnant. This was the same Arminius who annihilated three Roman legions at the Battle of Teutoburg Forest in 9 CE. In 15 CE Thusnelda was surrendered to Germanicus by her father in gratitude to Germanicus for driving off Arminius' besieging forces. In 17 CE Thusnelda and her son were paraded as prized trophies in the triumph granted to Germanicus.[37]

Ganna was a German prophet and priestess of the Semnones tribe. She was politically active, and acted as a diplomat and representative for the Semnones in negotiations with the Emperor Domitian (r. 81–96 CE).

Fritigil (or Fritigils), Queen of the Marcomanni, is the last known ruler of the Germanic peoples who were, in the mid-fourth century, in Pannonia. Fritigil corresponded with Ambrose of Milan regarding the conversion of her people to Christianity. According to the *Notitia Dignitatum*, she convinced her husband to submit to Roman authority and the tribe fell under the power of a tribune.

Amalafrida was the daughter of Theodemir, King of the Ostrogoths, and sister of Theodoric the Great. In 500 CE, to further consolidate his authority over the Vandals, Theodoric arranged a marriage alliance with Thrasamund, King of the Vandals and Amalafrida. She brought not only a very large dowry, but also 5,000 Gothic troops.

When Thrasamund died, his successor Hilderic issued orders for the return of all the Catholic bishops from exile, and Boniface, a devout advocate of orthodoxy, bishop of the African Church. In response Amalfrida headed a revolt, brought in the Moors as reinforcements and did battle at Capsa in Libya. In 523 Amalafrida was beaten, Hilderic had her arrested and imprisoned in a successful bid to overthrow Ostrogothic hegemony; he also had her Gothic troops slaughtered. She died in prison.

On Celtic and Gallic women in war there is ample evidence to prove their natural bellicosity and bravery. Dio Cassius: 'the Romans pursued the Celts to their wagons and fought with their women.' Diodorus Siculus: 'The women of the Gauls are not only like men in their great stature, but they are a match for them in courage as well.' (5.32). Ammianus Marcellinus (*Res Gestae*, p. 80):

> A whole band of foreigners will be unable to cope with one [Gaul] in a fight, if he calls in his wife, she is stronger than he by far and has flashing eyes; least of all when she swells her neck and gnashes her teeth, and poising her huge white arms, begins to rain blows mingled with kicks, like shots discharged by the twisted cords of a catapult.

Caesar, in his *De Bello Gallico*, describes the fighting spirit of Gallic women and their behaviour in military situations, particularly sieges:

> [2.13] The women and children appeared on the wall with arms outstretched in the usual manner of entreaty and begged for peace.
> [7.14] You may think these measures harsh and cruel, but you must admit that it would be a still harsher fate to have your wives and children carried off into slavery and be killed yourselves – which is what will happen if you are conquered.
> [7.28] They were exasperated by the massacre of Romans at Cenabum and the labour of the siege and spared neither old men nor women nor children.
> [7.47] The married women threw clothes and money down from the ramparts and leaning over with bared breasts and outstretched hands begged the Romans to spare them, and not to kill the women and children [at the siege of Gergovia].
> [7.48] The women who a moment before had been holding out their hands in supplication to the Romans began to appeal to their own folk,

leaning over the wall with their hair flung loose in Gallic fashion, and bringing out the children for the men to see.
[7.66] The cavalry cried that they should all swear a solemn oath not to allow any man who had not ridden twice through the enemy's column to enter his home again or to see his wife, children or parents.
[7.78] So the Mandubian population, who had received the other Gauls into their town, were forced to leave it with their wives and children. They came up to the Roman fortifications and with tears begged the soldiers to take them as slaves and relieve their hunger; but Caesar posted guards on the ramparts with orders to refuse them admission.

Zenobia

'I am a queen; and as long as I live I will reign'

Attributed to Zenobia by Gibbon

Julia Aurelia Zenobia (240–*c.*275 CE) was a queen of the Palmyrene Empire in Syria and is famous for her revolt against the Roman Empire. She became queen as the second wife of King Septimius Odaenathus, following Odaenathus' death in 267. Zenobia was nothing if not industrious and ambitious: by 269 she had expanded her empire by conquering Roman-held Egypt; when the Roman prefect, Tenagino Probus, tried to recover the territory, he was beheaded for his troubles. Zenobia ruled over Egypt as queen until 271, when she was defeated and paraded as a hostage in Rome by the Emperor Aurelian.

Zenobia's father, Amr ibn al-Ẓarib, was the sheikh of the Amlaqi; when members of the rival Tanukh tribal confederation killed him, Zenobia became chief of these nomadic Amlaqis, cutting her teeth in leadership skills over men, according to the Arabic version of her story told by the ninth-century Al-Tabari. Unsurprisingly, Zenobia claimed descent both from Dido, Queen of Carthage and Cleopatra VII of Egypt. The historian Callinicus dedicated a ten-book history of Alexandria to a 'Cleopatra', that is to Zenobia.

The picture we are left with gives us the impression of a notable woman. Some of it comes from the notoriously unreliable *Historia Augusta* and has been paraphrased for eternity by Gibbon in his *Decline and Fall of the Roman Empire*.[38] Gibbon says of her: 'Zenobia is perhaps the only female whose superior genius broke through the servile indolence imposed on her sex by the climate and manners of Asia ... Zenobia was esteemed the most lovely as well as the most heroic of her sex'. The *Scriptores Historiae Augustae* is a collection of biographies from the fourth century CE which details the Roman emperors from 117–284. Though the *Scriptores* was apparently authored by six men, only two, Trebellius Pollio and Flavius Vopiscus, are credited with the period of the queen's rule.[39]

Zenobia was reputedly beautiful, with a dark complexion, white teeth, and black eyes – more beautiful even than Cleopatra, although she differed markedly from Cleopatra by virtue of her reputation for chastity. Zenobia only had sex for the procreation of children and, after her marriage, refused to sleep with her husband except for that purpose. Apparently, she carried herself like a man, riding, hunting and enjoying, responsibly, a drink with her officers; she became known as a 'Warrior Queen' and would routinely march 3 or 4 miles with her infantry. 'Her voice was clear and like a man's; her sternness, when required, was that of a tyrant, her clemency . . . that of a good emperor'.[40] Highly educated and fluent in Greek, Aramaic and Egyptian, with a smattering of Latin, she hosted literary salons and surrounded herself with philosophers and poets, the most famous of these being the sophist Cassius Longinus. Comparisons could be drawn with aspects of Cleopatra and with Cornelia from the second century BCE, mother of the Gracchi and paragon of feminine virtue, equally famous for her coteries.

Amid the chaos that was the Imperial Crisis, Zenobia conquered Anatolia as far east as Ancyra (Ankara) and Chalcedon, followed by Syria, Judaea and Lebanon; in so doing she snatched vital trade routes from the Romans, severing grain supplies to the Empire and causing a bread shortage in Rome. Zenobia controversially issued coins – those potent symbols of Roman *imperium* and propaganda – depicting her likeness and others with that of her son, Vaballathus; this may well have provoked the Romans, although the coins do acknowledge Rome's sovereignty, for example, with an image of Vaballathus on one side and Aurelian on the other as joint rulers of Egypt. Others describe her title, *Augusta*, and show her with a diadem on a crescent; the reverse depicts Juno Regina, holding a *patera* in her right hand, a sceptre in her left, a peacock at her feet and a shining star to the right. She also negotiated trade agreements with the Sassanid Persians, and annexed territories to her empire without consulting Rome or considering Rome's interests. By 271 CE she ruled over an empire that stretched from modern-day Iraq, across Turkey and down through Egypt.

The Emperor Aurelian (r. 270–5 CE) initiated a mighty campaign to restore the Roman Empire in 272–3. He destroyed every town and city loyal to Zenobia until he reached Tyana, home of the famous philosopher Apollonius of Tyana whom Aurelian admired. Aurelian's initial reaction to the fact that he was locked out from there did not bode well for the inhabitants: he declared: 'In this city, I will not leave even a dog alive'. Aurelian then dreamed that Apollonius appeared and advised him to show clemency if he wanted victory, and so Aurelian spared the city and marched on. Other cities then recognised the good sense in surrendering to an emperor who was merciful rather than incurring his wrath by resisting. After Tyana, no more cities opposed him.

Aurelian and Zenobia clashed at Daphne near Antioch; Zenobia's forces were crushed. She was then besieged at Emesa, site of her treasury; Zenobia and her son, Vaballathus, escaped by camel assisted by the Sassanids, but she was captured on the Euphrates; those remaining Palmyrenes who refused to surrender were executed. Zenobia blamed her advisors for her disloyalty to Rome, particularly Cassius Longinus, who was executed. Aurelian, despite it all, showed respect for Zenobia but he was undoubtedly embarrassed by the fact that a woman had overrun one-third of his Empire. He sought to mitigate this in a letter: 'Those who speak with contempt of the war I am waging against a woman, are ignorant both of the character and power of Zenobia. It is impossible to enumerate her warlike preparations of stones, of arrows, and of every type of missile, weapons and military engines'. Adding in other correspondence, 'the Romans are saying that I am merely waging a war with a woman', and, in a letter to the Senate: 'I have heard, Conscript Fathers, that men are reproaching me for having performed an unmanly deed in leading Zenobia to triumph'.[41]

This male diffidence, arrogance even, alleged by Aurelian goes all the way back at least to Virgil and beyond when the poet has Aeneas articulate how little consequence there was in defeating a mere woman, Helen of Troy in this case: 'There's no great glory in a woman's punishment, and such a conquest wins no praise'.[42]

During the siege there was, apparently, an exchange of letters between Aurelian and Zenobia: 'I bid you surrender, promising that your lives shall be spared.' She replied, 'You demand my surrender as though you were not aware that Cleopatra preferred to die a queen rather than remain alive.'

Palmyra was sacked after a second revolt. Aurelian, in a letter to Bassus, one of his staff officers, bewailed the fact, in an all too familiar refrain, that 'The swords of the soldiers should not proceed further. ... We have not spared the women, we have slain the children, we have butchered the old men, we have destroyed the peasants'.

Fittingly, perhaps, the accounts of Zenobia's death are as confusing as those concerning Cleopatra's death. According to Zosimus, she and her son died in the Bosporus en route to Rome.[43] The *Historia Augusta* has it that in 274 CE, Zenobia reportedly starred in golden chains in Aurelian's triumph in Rome; ironically, free bread was handed out to the crowds.

According to Vopiscus:

> It was a most brilliant spectacle. Chariots, wild beasts, tigers, leopards, elephants, prisoners, and gladiators paraded through the streets. Each group was labelled with a placard identifying captives and booty from sixteen conquered nations for the spectators. One placard identified Odainat's chariot, another that of Zenobia. But, as she had often walked with her soldiers on foot, Zenobia did not ride that fateful day. Rather, she walked, without a placard, though the expectant crowd had no

> trouble recognizing her, 'adorned with gems so huge that she laboured under the weight of her jewelry'.

Pollio adds:

> This woman, courageous though she was, halted often, saying that she could not bear the weight of the gems. Furthermore, her feet were bound with shackles of gold and her hands with golden fetters, and even on her neck she wore a chain of gold, the weight of which was borne by a Persian buffoon.

Some say she died soon after through illness, hunger strike or beheading. An alternative outcome is that Aurelian, taken by her beauty and dignity, gifted her a smart villa in Tibur (modern Tivoli) where she lived out her life in luxury and became a leading philosopher, socialite and Roman *matrona*. Zonaras claims she was taken back to Rome, never appeared in any triumph and was not paraded through the streets in chains; instead, she married a wealthy Roman husband, while Aurelian wed one of her daughters.[44] In Al-Tabari's account, Zenobia murdered a tribal chief named Jadhima on their wedding night. When his nephew sought revenge he pursued her to Palmyra where she escaped on a camel and fled to the Euphrates. She had earlier ordered a tunnel to be dug beneath the river in case she needed to escape and was just entering this tunnel when she was caught. She then either killed herself by drinking poison or was executed. A possible descendant of Zenobia is St Zenobius of Florence, a Christian bishop from the fifth century. Zenobia has influenced later monarchs such as Catherine the Great of Russia (1729–96 CE), who compared herself to Zenobia and her court to the one at Palmyra.

To Arab historians such as Al-Tabari, Zenobia was a tribal queen of Arab, rather than Greek, descent, whose original name was Zaynab, or al-Zabba.[45] According to *The New Yorker*, among Muslims she is 'a herald of the Islamic conquests that came four centuries later', a view shared by the Syrian Assad regime which has featured Zenobia on its currency, and 'which also resonates within radical Islamic circles'. Palmyra is celebrated as the place where Zenobia stood up to the Roman Emperor. Indeed, ISIS fighters, after seizing Palmyra, released a video showing the temples and colonnades at the ruins, a UNESCO World Heritage site, intact. 'Concerning the historical city, we will preserve it,' an ISIS commander, Abu Laith al-Saudi, told a Syrian radio station. 'What we will do is pulverize the statues the miscreants used to pray to.'

The rest is (modern) history. Ironically, the atrocity perpetrated by Aurelius presaged the atrocities perpetrated by ISIS who 'have killed scores of civilians near Palmyra and executed soldiers in its ancient amphitheatre, in order to make yet another grotesque video documenting a new age of barbarism. In a region once ruled by a strong-willed queen, women who don't bend to ISIS's narrow beliefs may be sold into sex slavery'.[46]

Mavia

Mavia (r. 375–425 CE) (Arabic: ماوية) was a pugnacious Arab warrior-queen, who ruled over a confederation of semi-nomadic Arabs in southern Syria. She led her troops in rebellion against Roman rule, riding at the head of her army into Phoenicia and Palestine. Having repeatedly defeated the Romans, the Romans finally struck a truce with her on conditions stipulated by Mavia when she reached the borders with Egypt. The Romans later called on her for assistance when under attack by the Goths, to which she responded by despatching cavalry.

Mavia was considered 'the most powerful woman in the late antique Arab world after Zenobia'.[47] She and the Tanukhids forsook Aleppo for the desert, leaving the Romans wrong-footed without a standing target to attack. Mavia's highly mobile units, using classic guerilla warfare tactics, mounted numerous raids and thwarted Roman attempts to subdue the revolt – an early version in some respects of the nimble and irrepressible British Long Range Desert Group which operated in north Africa in the Second World War. Mavia defeated the first attempt to quell her and when a second attempt, led by no less than the Roman military commander of the East himself, was sent out to meet Mavia's forces in open battle, she defeated this too. Mavia led her armies from the front and proved to be not only a very astute political leader but also a highly competent field tactician. Her army, deploying Roman battlefield strategy and tactics and their own traditional fighting methods, had a massive impact; her agile cavalry units armed with long lances were used with deadly effect. Valens could do nothing, other than surrender. Rufinus sums it up:

> Mavia, queen of the Saracens, had begun to convulse the villages and towns on the border of Palestine and Arabia with a violent war and to ravage the neighbouring provinces. After she had worn down the Roman army in several battles, had felled a great many, and had put the remainder to flight, she was asked to make peace, which she did on the condition already declared: that a certain monk Moses be ordained bishop for her people.[48]

Mavia successfully negotiated to regain the Tanukhs' allied status and the privileges they had enjoyed before Julian's reign. At the end of the war Mavia's daughter, Princess Chasidat, was married to a devout Nicene commander in Rome's army, Victor, to cement the alliance. Mavia had delivered to the Arabs a just peace. Things, however, started to go wrong: as part of the peace treaty, Mavia sent her troops to Thrace to help the Romans in the fight against the Goths. Her forces proved ineffective while the Goths pushed the

Romans back to Constantinople, killing Valens, the Emperor, in the process. Mavia's forces returned home dejected and depleted in number. Theodosius I, the new Emperor, favoured the Goths over the Arabs who felt betrayed; they mounted another revolt in 383 CE which was quickly suppressed. In 425 CE Mavia died in Anasartha, east of Aleppo where there is an inscription recording her passing.

Chapter 15

Women and War in Roman Epic

The women described in the wars of Virgilian and later epic embrace a number of functions, derived in part from their roles in Greek epic, and no doubt non-extant earlier Roman epic. They are *casus belli*, like Helen and Lavinia; they are victims of war like Dido (also a *casus*); they are combatants like Camilla and the Amazon Penthesilea; or they stoke the fires of war and impel it forward, as with witches, goddesses and Furies.

Camilla

Camilla is the woman-warrior par excellence – effective, successful, competent, a leader of men and women and a reliable second-in-command. She only dies when, in a momentary lapse of battlefield concentration, she is distracted by the spoils of war. Virgil suggests that she was intent on plundering the corpse of Chloreus either to dedicate Trojan weapons in a temple, or else to have them to wear herself. Either way, the temptation is to dismiss or diminish Camilla by comparing her to an Amazon: while there are undeniable parallels, both are *bellatrices* and both expose a breast, for example, and although Virgil calls her 'Amazon', Camilla is her own woman-warrior and she should be treated as such. Homogenising her does her little justice.[1]

We first meet her as the final warrior in the parade of Italian allies in Book 7 of the *Aeneid*, a privileged position; she is a *bellatrix*, a woman-warrior.[2] Not for her womanly wool-working or baskets of wool – Camilla's forte was battling hard and speeding, fleet-footed, after the enemy. Young men and *matronae* were agog in admiration at her splendour as she rode by. Her position at the end of Virgil's procession emphasises her martial expertise and closes the catalogue on a high note, a morale booster for the people.

She reappears in Book 11 in the thick of battle, but before that a short biography describes her early childhood as a semi-feral daughter in a one-parent family living and hunting in the woods, trained by her father.[3] The shadow of war informs her life right from the start when her father ties her to a spear and shoots her over a river to escape a pursuing foe. Camilla thus becomes a weapon of war, a metaphor for war.[4] Her beauty, however, would remain undiminished as later she was much in demand as a wife for many a mother's son.[5]

Her battlefield prowess is second to none: 'She rains down from her hand volleys of pliant shafts, or whirls with tireless arm a stout battle-axe; her shoulder bears Diana's sounding arms and golden bow. Sometimes retreating when forced to flee, this maiden shoots arrows with a rearward-pointing bow as she flies.' Her comrades are hand-picked, armed to the teeth, just like Amazons:

> Around her move her chosen peers, Larina, virgin brave, Tarpeia, brandishing an axe of bronze, and Tulla, virgins from Italy whom the divine Camilla chose to share her glory, each a faithful servant in days of peace or war. The maids of Thrace ride along Thermodon's frozen flood, and fight with emblazoned Amazonian arms around Hippolyta; or when Penthesilea returns in triumphal chariot amid shrill shouting, and all her host of women clash the moon-shaped shield in the air.[6]

Lavinia

Lavinia was the second wife of Aeneas, founder of Rome. When she was widowed by Aeneas's death, Lavinia took on the role of regent until Ascanius, her stepson, was old enough to rule independently. Indeed, when he did finally assume full regal responsibility, one of his first acts was to found a new settlement in the Alban Hills, leaving Lavinia to rule Lavinium. Livy tells us that '*tanta indoles in Lavinia erat*', that she was a very gifted woman. Obviously she was exactly that if she was able to govern in such a bellicose and unpredictable environment. Significantly, Lavinia, a capable and strong woman, was instrumental in the very birth of Rome and joins that small, exclusive club which numbers Argia, Lucretia and Verginia in its membership of women who influenced the constitution and political history of Rome.

As a young woman, Lavinia had always been a desirable match; when she was old enough to marry, *matura viro* (7.53), she was pursued by Turnus no less, Aeneas' arch enemy. However, her father, King Latinus, had been warned by the gods that a Latin husband was not an option and that she should marry a foreign warrior. The prophecy was reinforced when Lavinia's hair burst into flames and the seer predicted great glory for her but terrible wars for Italy – the very same *magnum bellum* announced by the Cumaean Sibyl to Aeneas in the Underworld in Book 6.

Dido

Dido is both a tragic victim of war and a potent catalyst of future war.[7] Betrayed by Aeneas and cruelly sidelined by his dedication to his mission to fight for and found Rome, Dido is incandescent with rage and in a pit of misery and despair: 'she rages, out of her mind, and rushes through the city, mad as a Bacchant'. She confronts the treacherous Aeneas and promises to haunt him for eternity in a threat reminiscent of a *defixio*: 'When I'm gone I'll follow you into the black fire of Hell, when icy death draws out the spirit from my limbs; my ghost will be everywhere; you'll pay the price, you traitor, and I will hear about it – the news will reach me deep down among the deadmen.'[8]

Dido had seen the future and the future frightened her; when she made offerings to the gods on the incense-burning altars, the milk turned black and the wine she poured congealed into an obscene gore. '*Horrendus dictu*': 'shocking to say it', she resolved to take her own life. She can hear the ghost of Sychaeus, her late husband, and a solitary owl wails its song of death; Dido recalls the warnings of the *pious* priests who predicted that her relationship with Aeneas would all end in tears. She instructs her sister, Anna, to build a pyre and enlist the services of the remarkable Massylian priestess; she will wipe out the memory of Aeneas with her spells; his things will burn on the pyre. By implication, Aeneas is *impious*; he is subjected to a kind of *damnatio memoriae* in which he is erased from Dido's memory, as befits a traitor.

As already noted, Virgil's audience would have understood only too well the potency of Dido's threats: Rome did indeed pay the price for Aeneas' duplicity, with three devastating Punic Wars one of which was near terminal for Rome; Aeneas was indeed haunted by Dido, in their frosty meeting in the Underworld. The Dido episode resonates uncomfortably with the political upheaval caused so recently to Virgil's time by Cleopatra, a foreign queen eerily reminiscent of Dido, whose facility for global power-play would be viewed as comparable to the unnatural skills of a sorceress.

Amata

Queen Amata was Lavinia's mother, wife to King Latinus and a source of great irritation to Aeneas. He, of course, was the foreign warrior who was ordained to marry Lavinia, but this was never to be with the blessing of Amata. She is possessed by Allecto, Juno's furious stooge, and vehemently opposes the match, preferring instead the native Latin Turnus for a son-in-law. When she concealed Lavinia in the woods and incited the Latin women, Amata triggered the war between the Trojans and the Etruscan King Mezentius who was allied with Turnus. When she learns of Turnus' death at the hands of Aeneas she hangs herself.[9]

Juturna

Juturna was a goddess of fountains, wells and springs. As devoted sister to Turnus, she, on Juno's bidding, supported him against Aeneas by giving him back his sword after he dropped it in battle, as well as whisking him away from the jaws of death when it looked very likely he would be killed by Aeneas.[10]

Alecto

Alecto (the 'implacable' or 'unceasing anger') is one of the Erinyes, or Furies. Her credentials for evil are impeccable: Hesiod says that she was the daughter of Gaea fertilised by the blood spilled from Uranus when Cronus castrated him; she is the sister of Tisiphone (Vengeance) and Megaera (Jealousy). Her job is to punish crimes such as anger, especially if they are against other people; in this way, she functions like Nemesis who castigates crimes against the gods.

In the *Aeneid* she is ordered by Juno in Book 7 to prevent the Trojans from currying favour with King Latinus either by marriage or militarily; her mission is to wreak havoc among the Trojans and bring about their defeat. To do this, Alecto possesses Queen Amata, who incites all of the Latin mothers to riot against the Trojans. She disguises herself as Juno's priestess, Calybe, and comes to Turnus in a dream, persuading him to begin the war against the Trojans. Turnus ridicules her so she attacks Turnus with a flaming torch, causing his blood to 'boil with the passion for war'. Alecto cannot be sated in her war-mongering and so asks Juno if she can provoke more strife by sucking bordering towns in to the conflict. Juno replies archly that she will manage the rest of the war herself.[11]

Juno

Juno bears Aeneas and the Trojans a monumental grudge which manifests itself throughout the poem as she tries ceaselessly to thwart Aeneas' destiny to found Rome. She tries to confound him at every opportunity. Juno had remained implacable after the insult meted out to her when Paris overlooked her in the divine beauty contest in favour of Aphrodite. Juno favours Carthage and Dido, both slighted and doomed to destruction by Aeneas and his descendants, and she backs Turnus in the wars against Aeneas in Italy. She is, then, a prime and avid mover of war, a major dynamic force in stoking the *arma* and confounding the *vir* that are the subjects of Virgil's poem.

Right from the start of his odyssey, Aeneas' progress is dogged by this persistent and determined Juno. She makes a dramatic entrance into the action when she persuades Aeolus to raise a storm to blow him, literally and figuratively, off course (1.50ff.). She conspires to have Dido detain him in Carthage and she arranges for the scuttling of the Trojan fleet (5.680ff.) to try and get a marooned Aeneas to found his homeland in Sicily. Luckily, Jupiter and rain save the day. By Book 7 she has accepted Fate and restates her determination to at least delay Aeneas (313–16). As we have seen, she possesses Amata through Alecto and turns her against Aeneas and his desire to marry Lavinia; and she inspires Juturna to save the life of Turnus, her brother and Aeneas' enemy.

Callisto

The virginal nymph Callisto was an acolyte of the goddess Artemis – virginity was a prerequisite for cult membership – so Zeus disguised himself as Artemis in order to lure Callisto into the woods to rape her. Artemis discovered that her supposedly virginal follower was pregnant; she turned Callisto into a bear and set her loose in the forest, where she gave birth to Arcas. Zeus hid Arcas away and never revealed the true identity of his mother – with fatal consequences: Arcas went out hunting one day and shot a bear dead, that bear was Callisto, his mother, no less. According to Ovid, Callisto, like Camilla, had no interest in wool-working; neither was she interested in having her hair done – thus eschewing one key matronly emblem and repudiating the feminine practice of making oneself look nice.[12] Instead, Ovid calls her a *miles*, a woman soldier who arms herself with bow and arrow and a soldier of Phoebe. Callisto was, then, a kind of female *miles amoris*, a soldier of love. Sadly, her military characteristics failed to prevent her from being raped.

Atalanta

Ovid's Atalanta is a huntress and a virgin who is doomed never to lose her virginity even though she is loved by Meleager. She excels in the hunt against the Calydonian boar and is involved in battles when on the *Argo* as the only woman passenger with Meleager, Jason and the Argonauts as crew. Atalanta sustains an injury while fighting in a battle at Colchis. Apollonius, however, says that Atalanta was never on the *Argo* because Jason had concerns over having a woman on board with an otherwise all-male crew.[13]

We meet her again in Statius' *Thebaid* where she is 'pacifying groves' with her bow.[14] Later in the poem she is a victim of war. Grief-stricken, she promises her son, Parthenopaeus, as he departs for war, that when he is older she will give him *bella* and *ferrum*, wars and weapons, and will not detain or deter him with a mother's weeping.[15]

Erictho

Witches and other female chthonic agents were stock characters in ancient epic, with, it seems, each poet vying with their predecessors to produce the most terrifying and abhorrent cabalistic characters. Erictho is a witch surpassing even Seneca's Medea in her repulsiveness; Erictho is the satanic witch-queen of all witches. She is literally an 'epic' witch. Her role in Lucan's (39–65 CE) *De Bello Civili* is to prophesy the outcome of the civil war: by providing Sextus Pompey with this information she influences the course of the war.[16]

Sextus Pompey, driven by fear, wants to know the future; he rejects conventional forms of divination, electing instead to deploy the ungodly, 'the mysteries of the furious enchantress'. Being in Thessaly he is local to the world's most dreadul witches and their *herbae nocentes*, pernicious herbs; when the witches here cast their spells even the gods above pause to take note, and can sometimes be persuaded to do the witches' will. Sextus Pompey is anxious to hear the outcome of the imminent Battle of Pharsalus in 48 BCE.

Wild Erictho takes even their evil excesses to new extremes: she communes with the dead and is expert in all things eschatological; where she goes contagion follows; she buries the living and brings the dead back to life; she snatches burning babies from their pyres for occult research and experimentation and assaults the corpses of the dead, scooping out eyeballs and gnawing at their nails. She tears flesh from corpses crucified on crosses, harvests the black putrid congealed gore suppurating from the limbs of the decaying; she steals the meat ripped off putrefying bodies by rapacious wolves. She is a serial murderess, performing crude caesarean sections on pregnant women whenever a baby is required for the pyre; she rips the faces off young boys; at funerals she opens the mouths of the dead with her teeth, bites their tongues and thereby communicates with Hell.

She is a liminal figure living on the outskirts of civilised society and inhabiting 'graveyards, gibbets, and the battlefields generously supplied by civil war'; she uses body parts in her grisly magic spells. 'When the dead are confined in a sarcophagus . . . then she eagerly rages every limb. She plunges her hand into the eyes, delights at digging out the congealed eyeballs, and gnaws the pallid nails on a desiccated hand' (6.538–43).

She is an expert necromancer; when she surveys corpses in a battlefield, 'If she had tried to raise up the entire army on the field to revert to war, the laws of Erebus would have yielded, and a host – pulled from the Stygian Avernus by her terrible power – would have gone to war' (6.633–6). It is precisely for these reasons that she is sought out by Sextus Pompeius. Erictho is just the witch for him; Erictho would be a fine teacher for any of the black arts, but it is in reanimation that she excels. Erichtho complies with Pompey's request to know the future and the outcome of Pharsalus; she wanders amidst an old battlefield and seeks out a cadaver with 'uninjured tissues of a stiffened lung' (6.630). She cleanses the corpse's organs, and fills the body with a potion consisting of, among other things, a mixture of warm blood, 'lunar poison' and 'everything that nature wickedly bears' so as to bring the dead body back to life (6.667–71). The spirit is summoned, but initially refuses to return to its former body (6.721–9). She then promptly threatens the whole universe by promising to summon 'that god at whose dread name earth trembles' (6.744–6). The corpse is now successfully brought back to life and delivers a profoundly bleak description of a civil war in the Underworld, and an ambiguous prophecy about the fate that awaits Pompey and his family.

Erictho may well have been the inspiration for Mary Shelley's *Frankenstein* some 1,750 years later: she would have been familiar with the episode through her husband, the poet Shelley, who was a great admirer of Lucan.

Asbyte

Silius Italicus (*c.*28–*c.*103 CE) is the author of the 12,202-line epic *Punica*, the longest surviving poem in Latin from antiquity. Book 2 opens with Hannibal dismissing Roman envoys from Saguntum and addressing his troops with a threat to Rome. The siege of Saguntum continues, during which the warrior princess Asbyte is killed by Theron, who in turn is slain by Hannibal and mutilated. A Camilla-type figure, Asbyte hails from Libya and, like other warrior women, she is never much bothered by traditional women's work; she is a huntress and a virgin. She is bold, *audax*, and has a band of sisters to follow her in battle – only some of whom are virgins; she has the characteristics of an Amazon, including the bare right breast.[17] Her arrows whizz into the citadel she is attacking while one of her comrades, Harpe, saves her from certain death when she stands to face the flight of an arrow loosed by Mopsus and in mid-shout takes it in her open mouth from where it passes right through her. The battle raged on but Asbyte's time on Silius' blood-soaked battlefield is short: the warrior maiden, *belligera virgo*, is targeted by Theron; she escapes one attack but returns to the fray, at which point she is tipped out of her chariot and:

> As Asbyte tried to flee from the fight, he [Theron] sprang up to stop her, and smashed her with his club between her two temples; he spattered the shiny wheels and the reins, tangled up by the terrified horses, with the brains that gushed from her broken skull ... he then cut off the head of the maiden when she rolled out of her chariot. His rage was still not sated, for he fixed her head on a long pike for all to see, and made men carry it in front of the Punic army.[18]

Argia

Like Polyxena in Seneca's *Troades*, Argia exhibits shades of masculine militarism as portrayed in Statius' epic *Thebaid*, written between 80 and 92 CE. Statius' narrative is, of course, familiar to us from Greek tragedy, in particular from the *Phoenissae* and the *Suppliants* by Euripides. We pick up the story where Eteocles and Polynices have slain each other in battle and their mother, a distraught Jocasta, commits suicide. Argia is the wife of Polynices and graduates in the action of the poem from deserted, *relicta*, army wife to heroine and female warrior.

Argia was a catalyst for war well before she knew it herself. It all began with her marriage to Polynices, which Statius describes as *semina belli*, seeds of war, contrived by Jupiter to enable him to launch the Seven Against Thebes. Argia is soon abandoned as Polynices heads off to his unjust war; at first she is fearful but soon is asking her father, Adrastus, to bring on the war.[19]

The crucial moment for Argia comes in the final book of the epic when Creon, recently installed as the new tyrant of Thebes, defies all convention and the rules of war and religion by denying Argia the right to bury Polynices. Meanwhile, the Argive women, acting as ambassadors, appealed to Theseus, King of Athens, for military aid; Argia leaves for Thebes to resolve the issues over Polynices' burial despite the possibility of a cruel death at the hands of Creon.

Argia moves from being one of the inferior sex, men being the better sex, *melior sexus*, to a *matrona virilis*, a matron with strong, masculine qualities, who has deserted her gender: *sexu relictu*. Her *virtus* is definitely not a woman's – *non feminea* – and she later 'cultivates a sudden passion for courage that is not a woman's courage': *non femineae subitum virtutis amorem colligit Argia*.[20]

Women generally can exhibit *virtus*, despite its explicit connotations of manliness (*vir*) and traditional male attributes of strength and bravery, as well as of virtue. From Seneca we hear about the conspicuous valour of Cornelia and Rutilia, described as *conspecta virtus*; in the *Ad Marciam* he spells out his belief that women are just as capable of displaying *virtutes* as men.[21] Argia's *virtus* brings her up against Creon. Her decision to defy Creon's orders is a huge undertaking – *immane opus* – it is an *arte dolum* – a crafty strategy and it is driven by chaste love and *pietas*, a duty to her son carried out in the face of the *fulmina regni*, the thunderbolts of the king, a blood-stained king, an abominable king. The three times repetition of the word king would have resonated with Statius' audience, still sensitively mindful of the monarchy from the early days of Rome. Argia is, therefore, linked with Lucretia and Verginia – two virtuous *matronae* who stood up to tyranny and who, in so doing, paved the way for the Roman Republic and Roman democracy. But she is better than that even: she is more than a match for the Amazons, those celebrated female warriors who became, for many, the role models for the fighting woman in classical times. Belligerent woman as she is, Argia's chutzpah is, nevertheless, founded on sound *matrona* values, as we see when she poignantly recognises Polynices' body on the battlefield by the clothes which she herself had woven for him.[22]

The Lemnian Women

A Lemnian deed or crime is defined as a deed or crime of exceptional barbarity and cruelty. The phrase originates from two shocking massacres perpetrated by the Lemnians: one was the extermination of all the men and male children on the island by their outraged women; the other was the infanticide by the men of all the children born of Athenian parents in Lemnos. The epic poet Valerius Flaccus (d. *c.*90 CE) describes the first for us, leaning heavily on a long literary tradition. The day arrived when the Thracian men were routed by the women in this internecine war: *Thracas qui fuderat armis*; ominously, the returning Thracian boats approached Lemnos laden with the spoils of war, *praemia belli*, which included livestock, and foreign women.[23] Venus – angry that the Lemnians did not pay her honour – fired up Rumour to spread the word that the Lemnian warriors were bringing back Thracian women to share their beds; Venus aimed to inflict pain on and incite the Thracian women into a frenzy. Rumour ran from house to house, comparing the local Lemnian women's beauty, their patient chastity, their wool-working as good *matronae* – comparing them with these foreign imports with painted faces. The poet describes the Lemnians as Penelopes and Lucretias, paragons of female virtue, while the Thracians are depicted as whores. Further rumours have it that the Lemnian women will be exiled, their places taken by these Thracians: the Lemnians are further incensed. They feign a welcoming homecoming for their husbands and then retire with their husbands to eat and make love.[24] Venus then takes on the role of commander and incites revenge: covered in blood she rushes into a house clutching a still pulsating decapitated head; Venus is the first avenger: she forcibly arms the women with swords and they fall on their once-loved husbands, many of whom are paralysed by fear at the butchery going on all around them, perpetrated by wives, mothers and daughters. Blood flows in the bedrooms and gurgles and foams in the chests of the slain Lemnian men; the women add to the atrocity when they torch their houses and block the exits.

In the *Thebaid* of Statius (45–96 CE), Hypsipile's description of the carnage is even more gruesome than that of Valerius:

> And we made our way through the deserted streets of the city, concealed in the dark, finding everywhere the heaped corpses from the night's massacre, where cruel twilight had seen them slain in the sacred groves. Here were faces pressed to beds, sword-hilts erect in wounded breasts, broken fragments of huge spears, knife-rent clothes among corpses, upturned wine-bowls, entrails drenched in blood, and bloody wine pouring over the wine cups out of severed throats.

Statius even mentions some of the women by name. In the third century BCE Apollonius Rhodius had given us more precise reasons why Lemnos came to be ruled by women:

> Led by Jason, the Argonauts first sailed to Lemnos which was then totally devoid of men and ruled by Hypsipyle, daughter of Thoas. This

> came about as follows: the women of Lemnos had neglected the cult of Aphrodite. As a punishment, the goddess afflicted them with a foul smell. The men found this repellent and took up with captive women from nearby Thrace. The Lemnian women were outraged at this humiliation and proceeded to murder their fathers and husbands.[25]

Apollonius tells us of the Lemnians' subsequent belligerence and rejection of traditionally female activities:

> The Lemnian women found it easier to look after cattle, put on a suit of bronze, and plough the earth for corn rather than to devote themselves, as they had done before, to the tasks of which Athene is the patroness [handicrafts]. Nevertheless, they lived in dire dread of the Thracians ... so when they saw the Argo rowing up to the island, they immediately got ready for war ... Hypsipyle joined them, dressed in her father Thoas' armour.[26]

Hypsipile, the leader, advised conciliation and provisioning the Argonauts on their ships in order to keep them at bay. Venerable old Polyxo added her wisdom and foresight, born as it was of age and experience; 'provisioning', it seems, must also include procreation:

> Hypsipyle is right. We must accommodate these strangers ... there are many troubles worse than war that you will have to confront as time goes by. When the older ones among us have died off, how are you younger women going to face the miseries of old age without children? Will the oxen yoke themselves? Will they go out into the fields and drag the ploughshare through the unyielding fallow? Will they watch the changing seasons and harvest at the right time?[27]

Valerius Flaccus states it more bluntly: 'Venus herself wishes us to join our bodies with theirs, while our wombs still remain strong and we are not beyond childbearing age'.[28]

Many years earlier, in the *Libation Bearers*, Aeschylus had put the criminal actions of the Lemnian women into some sort of context, providing a benchmark for female atrocity against male: 'Indeed the Lemnian holds first place among evils as stories go: it has long been told with groans as an abominable calamity. Men compare each new horror to Lemnian troubles; and because of a woeful deed abhorred by the gods a race has disappeared, cast out in infamy from among mortals.'[29]

The androcide committed by the Lemnian women has led to a race without men, thus introducing the need for the women to adopt traditionally male functions: working the fields instead of the wool and defending their homes in war, rather than managing the *oikos* in domestic peace. Avoiding extinction was, as with the Amazons, a chronic problem, eased here, on the advice of the wisest of Lemnian heads, by the chance visit of the Argonauts.

* * *

The Roman novel too featured women caught up in the shadowy and disturbing world of war; the following example is from Heliodorus.

The Old Woman of Bessa

Heliodorus of Emesa, in the third century CE, describes a necromancy conducted by a witch, an old woman of Bessa, in his *Aethiopica*. Calasiris, a priest of Isis, and Charicleia, the heroine of the novel, come across the aftermath of a battle between the Persians and the Egyptians, a battlefield strewn with corpses. The only living soul is an elderly Egyptian woman mourning her dead son; she invites the couple to spend the night there with her. During the night Charicleia witnesses a shocking scene: the old woman digs a trench, lights two pyres on either side and places the body of her son between them. She pours libations into the trench and throws in a male effigy made from dough. Shaking, and in a trance, the old woman cuts her arm with a sword and drips her blood into the trench uttering wild and exotic prayers to the moon. After some magic she chants into her son's ear and makes him stand up; she then questions him about the fate of her other son, his brother.

The corpse at first says nothing but, because his mother persists, rebukes her for sinning against nature and breaking the law when she should have been organising his burial. He reveals not only that his brother is dead but that she too will soon die violently because of her life of unlawful practice. Before collapsing again, the corpse reveals the awful truth that the necromancy had been witnessed not only by a priest 'beloved by the gods' but also by a young girl who has travelled to the ends of the earth looking for her lover. A happy outcome is promised for both; the old mother is outraged by this intrusion and, while pursuing Calasiris and Charicleia, is fatally impaled on a spear.[30] The old woman exemplifies the victim of war: in this case a mother bereft of her two sons slain in battle; she is so desperate for news of the one who is missing that she resorts to the black arts in a necromancy with the other.

Chapter 16

Women and War and *Militia Amoris*

The Roman love poets would have us believe that they were anti-war because of the distressing effects war and long service overseas had on wives and mistresses. Propertius chastises Postumus for going off to Parthia in 21 BCE to help recover the standards lost by Rome at the Battle of Carrhae in 53 BCE. The poet deplores the fact that, as a consequence of Postumus' posting, Aelia Galla is reduced to tears and that Postumus has ignored her repeated requests for him to stay, a scenario that must have been repeated countless thousands of times in Roman relationships over the years.[1] Propertius concludes his tirade by condemning Postumus' greed and wishing him dead for preferring 'arms to the faithful marriage bed'. He sympathises with Galla and the anxiety that will eat away at her, beside herself with worry about groundless rumours, and fearful that Postumus will be killed and come home in a funeral urn – ancient Rome's answer to the body bag.[2] Propertius shows Galla considerable respect when he compares her situation to that of the patient and faithful Penelope who endured twenty years of waiting for her hero, Odysseus, to return from his war, and after some serious Mediterranean carousing and horizontal collaboration chez Circe and Calypso. Penelope and Galla are the direct opposite of the capricious women Propertius usually consorted with.[3]

Tibullus echoes Propertius' concerns when he asserts that no amount of gold or emeralds is worth reducing a woman to tears.[4] He personalises this generalisation (1.3) when he describes Delia, one of his regular mistresses, as weeping and worrying about him on an imagined posting overseas.[5] The irony of it is, though, that it is precisely his military service that would furnish the luxurious goods desired both by Delia and Nemesis, his other mistress.[6] The Gallas of the elegists' world had a very different take on war and the spoils of war to the venal Delias and Nemesises.

The boot is on the other foot when Propertius finds himself trying to deter Cynthia from following a rival lover to icy Illyria. He emphasises the uncomfortable voyage and the wintry weather she will have to endure; to her however, her lover is worth such hardships, or at least the munificent gifts which he brings are. In the end, though, Propertius prevails and Cynthia stays in Rome with him.[7] Propertius is lucky it seems: the elegist Gallus was obviously less persuasive and bemoans the fact that Lycoris, his mistress, is crossing the snowy Alps and the frozen Rhine without him, no doubt in the company of a soldier lover.[8] That Cynthia still has at least one eye on the spoils of empire is confirmed in 2.16 where she is rebuked by Propertius for preferring the *maxima praeda* offered by a praetor returning from Illyria – it may be 'great booty' to Cynthia but for the poet it represents his *maxima cura*, his 'biggest worry'. Propertius would have it that Cynthia cares little about a lover's official status or his *cursus honorum* – his worth for her is determined by what is in his money bags. For Cynthia, war is an opportunity and the more she plunders returning heroes, the more they have to go back overseas to replenish their spoils of war.[9]

Ovid, too, despises the increasing venality he observes in his mistress, comparing her to a whore; presumably she is grasping imported booty as well – booty that, in this case, comes from Drepane (Trapani) in western Sicily.[10] He can barely disguise his scorn in another poem in which his mistress sports a wig made from a German woman's hair: his woman ponders whether it is actually her being admired or some Sygambrian woman.[11]

The mistresses of the elegiac poets, and many prostitutes and their meretricious bawds were themselves born out of Roman Imperial expansion. Some of these women will have shrewdly emigrated to Rome and to other Roman cities on the tide of eastern immigration and imports as Rome expanded around the Mediterranean basin; many others will have been shipped to Rome and Italy as slaves – taken prisoner by pirates or by Roman armies as the Empire sucked in more foreign states. The former, who probably came by choice, may well have enjoyed a cosmopolitan and relatively affluent lifestyle like the Corinnas, Delias and Cynthias; the far more numerous slaves, on the other hand, were destined to spend the rest of their lives in servitude with the prospect of unremitting sexual and physical abuse. Other immigrants still will have sold their bodies under the guise of dancers and musicians at dinner parties, or else walked the streets plying their trade under arches or in cemeteries, in brothels or inns. Rome's wars clearly had a wide range of effects on different women right across Roman society, depending on their individual circumstances. A privileged few will have welcomed expansionism and the endless flow of luxury goods that it brought, for they, like Cynthia or Delia, were the lucky recipients. Ovid, for one, knows that Rome is full of foreign women eligible as mistresses: quantity and diversity are not a problem.[12] Not only were the denizens of the love poets' demi-monde the product of Roman conquests, the very poetry that celebrated them was too: without conquest there were no such mistresses, without mistresses there was no love poetry.

Propertius' bawd, Acanthis, trains her charges not to be disappointed by soldiers and sailors, reminding them that it is only the gold they bring with them that counts.[13]

Militia Amoris and the *Miles Amoris*

By the middle of the first century BCE a number of male, and some female, poets were able to eschew the traditional *mos maiorum* (the way their ancestors did things), rejecting the *cursus honorum* (sequence of offices), for a life of *otium* (leisure) during which they would while away their time penning poetry and pursuing the objects of their languid affection. Horrified traditionalists regarded this otiose lifestyle as frivolous, effeminate and decidedly un-Roman. The poets allowed themselves to be dominated by their women (their *dominae*), and even to be enslaved by them in the slavery of love, *servitium amoris*; their *cursus honorum* and military career was as *militia amoris*, a soldier of love; they languished, locked-out and rejected, on the doorstep of a capricious woman, *exclusus amator*. Good Roman *militia* was being supplanted by weak, lovelorn 'soldiers', who were under the control of, and dominated by, women rather than by centurions.

The 'war of love' stood in direct and defiant contrast to traditional military service, *militia*, to the *cursus honorum*, the ladder of civic and military service, to *virtus*, bravery and manhood – badges of Roman-ness, and to the *mores* of the day in the early Roman Empire. The soldier of love, *miles amoris*, was a social outcast – what we might call, anachronistically, something of a Bohemian. Propertius, Tibullus and Ovid and others cliché and chorus 'make love not war' in direct opposition to what was expected of them as men and citizens of Imperial, expansionist and bellicose Rome: instead of donning greaves and shields and spending the best part of their lives guarding or acquiring far-flung foreign parts, they assumed the mantle of pacifist and poet, versifying in a life of *otium* in a cosmopolitan Rome. Anagrammatically, they had joined the army of *Amor*, not the army of *Roma*. The soldier of love was fighting endless battles with his mistress in a war of love, the objective of his pseudo-military service. She is the enemy, the other half of the conflict, the *sine qua non* of the poet's military service and of the battles he was fighting; she is his *casus belli*.

There are at least a hundred examples of *militia amoris* – love expressed in militaristic terms – in Roman elegy and lyric poetry, and three poems are devoted exclusively to it; the device appears in Propertius, Tibullus and Ovid. It originates in Greek poetry and makes its first extant appearance in Sappho's work, where the poet views her attempts to win over an uninterested girl as a battle and enlists the support of Aphrodite as an ally. For Anacreon, being in love is like being at war. Theognis describes his failure to win the object of his attentions as a victory for the other side and his efforts as warfare. Sophocles and Euripides both have gods and goddesses fighting wars or battles. In Menander's *Perikeiromene* the attempts to free Glycera from the house of a neighbour are couched in terms of a military assault.[14] In the *Greek Anthology*, Hedylus describes items of clothing from the raped Aglaonice as 'spoils'.[15]

In extant Roman literature Plautus is responsible for introducing the motif in Latin and, in so doing, probably signifies how ubiquitous it was in Greek New Comedy. In the *Persa* Paegnium retorts to Sophoclidisca, 'Still, that warfare is waged [*militia militatur*] much more successfully by spirit than by weight', when she suggests that he is not old, or even heavy enough, for love (231). In *Truculentus*, Astaphium says that he is a soldier under his mistress' command (229).

In Roman erotic poetry Catullus picks up the Hedylus metaphor when he describes Berenice's clothing as spoils (66.13f.); he uses *militia amoris* language in *Carmen*, 37: 'For my girl, who has fled from my embrace, she whom I loved as none will be loved, for whom I fought ferocious fights' (l. 13). Presumably, the girl could not care less about Catullus' skirmishes: she is now being made love to by all the 'piddling back-alley fornicators' and in particular by one Egnatius who brushes his teeth in Spanish urine. Propertius has a similar problem with Cynthia and vows to 'pitch his camp, *castra*, some where else' (4.8.55) because of her infidelity. In the face of violent and hysterical protests from Cynthia, when he thinks about going off travelling with Tullus (1.6.29–30), Propertius exclaims, 'I'm not naturally suited to glory or for arms; Love's is the only warfare which the Fates have planned for me'. He must serve under Venus (4.1.37). A victory over Cynthia is likened to a triumph (2.14.23ff.) complete with booty, an instance of how the elegists deconstructed the triumph motif; Ovid adopts it when he describes himself as booty (*Amores*, 1.2.19f.); the analogy is developed at *Amores*, 1.7.35f. with a lively, albeit mocking, description of Ovid parading as a conquering hero in glory behind his one humiliated captive, a dishevelled girl. It is, though, with *Amores*, 1.2.19ff. that the *militia amoris* motif is most extensively drawn with numerous military references aimed at explaining the debilitating effects of unrequited love, Cupid's latest victim: surrender, military action, armistices, unarmed enemies, chariots, prisoners whipped into line, pomp, gold are all there.

Propertius introduces mythology into the device when he associates himself with such famous battle heroes as Achilles and wild Hector (2.22.34). Paris fought his biggest battles on Helen's knee (3.8.32). Sex with Cynthia are lengthy *Iliads*, the ultimate war story in Propertius' day (2.1.13). Ovid redresses the balance between the sexes in the battle of the sexes in the *Ars Amatoria* (3.1–4) which, in Books 1–2, had been advice for men chasing women. Women, he refers to mock-heroically, as Amazons and Penthesilea, and these are the subject of his tuition in Book 3. By 'arming' women he, nevertheless, realises he is effectively betraying his own sex, his comrades in arms (3.667f.).

In the opening poem of his collection, Tibullus assures Delia that he too has no care for glory in a few lines that manage to combine two similar motifs with *militia amoris* – the slave of love, *servitium amoris*, and the locked-out lover, *paraclausithyron* or *exclusus amator* (1.1.53; 75):

> It's right for you, Messalla, to campaign by land and sea and that the front of your house exhibits the spoils of enemies: I am a captive bound in the chains of a lovely girl; I sit as a doorman before her unyielding doors. I care not for glory, Delia dear; I only want to be with you, and I will pray that people will call me slob and idler ... here I am brave centurion and legionary. Go away, trumpets and ensigns! Take wounds to the greedy men, and take them wealth.

For a brief moment in 2.6 Tibullus capitulates and tells Macer, a military friend whom he chastises for going off to war, that he too is off to do battle. If Amor looks after soldiers, Tibullus will be a soldier, *miles*, headed for the camps, the *castra*: goodbye sex and goodbye girls! Tibullus is strong, *vires*, and he loves the sound of that trumpet. But the slamming of the door in his face brings him back to reality and silences his bragging and boastful words (7–12). In 2.3.33, Cupid has set up camp, *castra*, in his house, a notion repeated by Ovid, *castra Cupido* (*Amores*, 1.9.1).

Ovid has the most elaborate use of the metaphor of lover as soldier in *Amores* 1.9. Here, in over forty-six lines, he runs through the gamut of military life in the field and makes a comparison with life as a lover, trying to convince his readers that lovers are far from lazy: on the contrary, physically and psychologically they compete with soldiers. Every lover is in arms and Cupid is camped out in the field: '*militat omnis amans, et habet sua castra Cupido*'. He reprises this in the *Ars Amatoria* (2.233–42), his version of *The Joy of Sex*, where he asserts that love is a kind of warfare, '*militia species amor est*', and that military skills are the same as those required to win over a woman. In *Amores* 2.12 he compares his conquest, his night of love with Corinna, to a military victory, reinforcing the metaphor by citing warring women who have shaped history – Helen of Troy, Lavinia and the Sabine women. Lovers betray and open up the gates to the enemy (*Ars Amatoria*, 3.577f.). His once-flaccid penis is now erect, keen to go on campaign and get on the job (*Amores*, 3.7.68). Ovid is a war veteran (*Ars Amatoria*, 3.559f. and *Remedia Amoris*, 4), ready to retire from service (*Amores*, 2.9a.23f.) but happy to train new recruits (*Ars Amatoria*, 1.36). Giving advice on the remedies for love is like supplying weapons, *arma*, to the combatants (*Remedia Amoris*, 50).

Women themselves are portrayed as victors in the war of love, reducing the poet to humilating and abject defeat; but the poets, typically and masochistically, derive a perverse pleasure from this. We see it in Propertius (1.6.29f.) and in Ovid (*Amores*, 1.9.43f.). Subserviance in capitulation to the woman is vital (*Ars Amatoria*, 2.175; 2.197; 2.539f.; 3.565f. and 3.752f.) thus bringing the lover's defeat in the war of love ever closer to slavery of love, *servitium amoris*. Just as in the real world of war, defeat in battle often leads to servitude.

Chapter 17

Military Tendencies in Women in Seneca's *Troades*

Seneca the Younger (*c.*4 BCE–65 CE) was statesman, philosopher and dramatist: his works include philosophical essays, 124 letters dealing with moral issues, nine tragedies and a disputed satire. All of his works are imbued with Stoic philosophical doctrine, one of the tenets of which was the need to control one's emotions.[1] Seneca explores this and other Stoic doctrine when, in a number of his tragedies, he reprises plays written by Aeschylus, Euripides and Sophocles dealing with the Greeks when they got home after the Trojan War. The *Troades* – *Trojan Women* – is one such play which focuses on the trials and tribulations of the prisoner of war and displaced Trojan women further encumbered by the devastating news that Polyxena – daughter of Agamemnon – and Astyanax, son of Hector and Andromache – are to be sacrificed as a precondition of the Greeks' return to their homeland.

At the beginning of the play, Hecuba describes the fall of Troy in graphic detail to emphasise the greatness of that fall but then urges the other women to resume their lamentations, encouraging them to strip to the waist to enable them to beat their breasts all the more effectively.[2] The immodesty involved in such a display of semi-nudity and the violent self-harm Hecuba encourages is deemed insignificant, given the dire situation the women are in: disoriented captives pending a future life of enslavement and serial physical and sexual violence.[3] She orders the women to lament for both Hector and Priam; in the case of Priam she enumerates all the demeaning symbols and realities of defeat he, as a man, has avoided – he will not be just another item of booty, he will not be enslaved and fettered, he will not be paraded as a trophy in the Greek equivalent of a Roman triumph. In so doing, Hecuba delineates the very indignities and humiliations which await the Troades. The lamentations here are the culmination of ten years of grief and lament – a kind of crescendo to the fighting engaged in by their men; the rousing lamentations by these women represent the final vocalisation of the civic entity that was Troy; as such it demonstrates their indomitable spirit in the face of the extermination of their homeland.[4]

Andromache, already devastated by the death of Hector, has a particularly intractable and trenchant problem which involves the fate of Astyanax, her son. She attempts to save him from death by lying when challenged by the wily Ulysses, not just out of maternal love but so that Astyanax can fulfil his obligation to extend the family line and, one day, rule over a renascent Troy – the natural reaction of any Trojan war widow.[5] Ulysses and the Greeks, of course, are concerned about the prospect of a renascent Hector for different reasons, namely, the renewed conflict that he could bring.[6] Ulysses is unimpressed by Andromache's prevarication and threatens torture to elicit the truth, not believing that he has fled or died.

Ultimately, it is not grief or fear of torture but uncontrolled fear for Astyanax that is Andromache's undoing that, when found, Astyanax will be hurled from the battlements to his death. What was usually a place of relative safety for non-combatants in a teichoscopy would become the place of her son's murder. Andromache loses all control of her emotions at the prospect.[7] Ulysses continues to pile on the unbearable agony when he asserts that, because they cannot find Astyanax they cannot kill Astyanax, so Hector's tomb will be desecrated and destroyed to compensate and his ashes will be stolen and scattered: Hector's tomb is where Andromache has hidden Astyanax. She resolves this dilemma as a warrior's wife would, demonstrating good strategic, visionary sense and opting to save the one whom the Greeks fear more – Astyanax as resurgent Hector.[8] How? By physically, and heroically, standing in Ulysses' way defending the ashes, *socia*, all described in unmistakably military terms.[9]

In this, Seneca's stoic version of events, Andromache loses and Ulysses wins because Andromache loses control of her emotions and Ulysses retains control of his. Andromache is dignified and appeals to Ulysses' sensitivity and sympathy, but while Ulysses is sympathetic he is, nevertheless, obdurate.[10]

Astyanax and Polyxena both meet their respective fates: Astyanax is hurled from the battlements symbolising the last vestige of Trojan power; Polyxena is sacrificed in a ceremony masquerading as a marriage in which Seneca describes her as an *audax virago* – a bold, manlike woman who demonstrates her anger even on the point of death when she is more akin to a warrior than to a mere woman.[11] The Trojan women are then shared out among their Greek victors.

PART FOUR

WARRIOR WOMEN IN THE ARTS AND ENTERTAINMENT

Chapter 18

Military Women in the Visual Arts

We have already noted how Roman emperors celebrated their military victories with arches, columns and triumphal processions. These arches and columns graphically depict, among other trappings of conquest, proud, civilised Romans slaughtering 'barbarians' – men, women and children – and raping their women. They are the propaganda posters of the day, reminding Romans and others as they gaze at them just how powerful and unforgiving the Roman war machine was, and what will befall any nation or tribe which has the temerity to challenge that power. Defeated enemy leaders and subjugated women took pride of place in the lavish, booty-fuelled triumphs laid on by exulting victors in the late Republic and the Empire: the humiliation and ignominy for these men and women must have been palpable; the crowds clamouring in the streets and forums certainly loved it.

The museums of the world are awash with vases painted with scenes depicting snapshots from everyday Greek and Roman life. As war was a constant in both cultures, it is not surprising that we see today numerous vases depicting aspects of war and military life.

Warriors were leaving home for war all the time which, of course, explains why this scenario is so common in Greek black and red figure vases, typically showing the warrior flanked by an elderly man and a woman (father and mother or wife), sometimes holding hands, with libations being poured. The scene is commonplace and is indicative of the general support a family, a household, *oikos*, gave to military service to defend the *polis*; it also clearly demonstrates how important a part women played in this endorsement. The army mother and wife have a crucial role here, a role that is just as important as their involvement in the religious observances relating to burial and funerals, and the handing down from generation to generation of the responsibility to fight to protect *polis*, household, wife, mother and father and extended family.

But the mother and wife of the warrior have a further role to play here. Just as often she is portrayed as handing over and holding the weapons and armour of her departing son or husband. She thus becomes inextricably involved in her man's army service and the conflict for which he is departing. She is not just an army mother or wife, she is a war mother or war wife. Helmets and shields are routinely passed over and held in imitation of Thetis who helped arm her son Achilles in Homer's *Iliad*. This particular scenario appears on an amphora now in the Boston Museum of Fine Arts with Thetis and Nereids carrying items of armour: shield, breastplate, greaves and helmet. It, and many others, exemplifies women's key role in Greek society: producers of male children whom they will prepare for military service up to the day of departure. Another famous example, by Euthymides, shows Hector donning his armour, flanked by Priam, his father, and Hecuba, his mother, handing him his helmet.

The scenario extends into the world of the gods, illustrated by an amphora in the Capitoline Museum in Rome, where Athena, guardian goddess of Athens, stands by as a hoplite arms himself: Athena points to his helmet on the ground; an old man holds his sword. With Amazons, of course, the gendering disappears as all the characters are female; an amphora now on display in the Ashmolean Museum in Oxford shows this. Further blurring of roles occurs here because all the characters are armed and all the characters are warriors – both givers and recipients.

Victory, Nike, can be seen resplendent on the parapet on the Temple of Athena Nike on the acropolis in Athens, on the Paionios of Mende's statue of Victory at Olympia and as the Winged Victory of Samothrace now in the Louvre, originally on a fountain in the sanctuary of the Cabeiri.[1] This has been described as 'the greatest masterpiece of Hellenistic sculpture'.[2]

Goddesses are shown doing their bit against the race of Giants in Gigantomachic scenes. Take for example Athena laying low a Giant on the pediment of the Old Temple of Athena on the acropolis in Athens. The Siphnian Treasury frieze at Delphi depicts Athena, Artemis and Cybele; Athena and Artemis are joined by Nike, Hecate, Nyx and others on the Great Altar of Pergamon.

Amazonomachy scenes obviously show Amazons; for example, Antiope being hauled off by Theseus on the west pediment of the temple at Eretria. Amazon battles appear on the friezes of temples of Apollo Epikourios at Bassae and of Artemis Leukophryene at Magnesia-on-the Maeander, and on the tomb of King Mausolus at Halicarnassus. The slaying of Penthesilea by Achilles is a common motif on Attic vases.

Greek and Roman pottery display for us the whole panoply of Greek and Roman life: that fateful day, common as it was, when warrior-husband or son departs for the front, is one of the many scenes we can observe. Belligerent goddesses and Amazons, who populated enduring myths, found their way onto some of the most impressive architecture, statuary and ceramics still on show today.

Chapter 19

Women as Gladiators

It may all have started when female sword fighters performed at funerals in the very early days of Rome. There may also be some connection between women participating in chariot racing and women gladiators. The Heraean Games were pivotal: they were a four-yearly female sports event dedicated to Hera and founded by the legendary Queen Hippodameia; they would later become a template for the Olympics and continued for centuries until surpressed by the Christians. Apart from the usual foot races, javelin throwing and so on, the games included chariot races and the chariot races included female chariot races. According to Pausanias' *Description of Greece* (5.15.1–6), Hippodameia assembled a group known as the 'Sixteen Women' to organise the Heraean Games, during which the women competitors, incidentally, wore men's clothes. A first-century CE inscription from Delphi records that two young women competed in races, possibly those at the Sebasta festival in Naples in the Empire; in Domitian's reign there were races for women at the Capitoline Games in Rome in 86 CE.

When we think of gladiators now we tend to imagine men, thanks largely to some epic Hollywood films. However, women were not uncommon competitors in the amphitheatres around the Roman world, playing out the phoney fights and grappling in close combat, much to the delight and sexual titillation of the audiences: the nearest the modern world has come to it is probably female wrestling. The women were usually warm-up acts providing light relief in between the top-of-the-bill, crowd-pleasing gruesome and gory acts.

Women gladiators might share their particular stage, for example, with an elephant walking on a tightrope – as at games arranged by Nero in honour of his mother, Agripinna the Younger, whom he had recently murdered. Tacitus is outraged by female participation: 'Many ladies of distinction, however, and senators, disgraced themselves by appearing in the amphitheatre'.[1] The fact that these were rich women and had no need of manumission or celebrity suggests that they did it for the adrenalin, the sexual high. Dio tells of another spectacle in which Nero, entertaining King Tiridates I of Armenia, put on a gladiatorial show featuring Ethiopian men, women and children.[2] Petronius describes a woman who fights from a chariot booked for a gladiatorial show at a festival.[3]

Women gladiators were just one of many variations on a theme put into the arena to keep the baying crowds entertained. In the hundred-day games staged by Titus, women competed in a battle between cranes and one between four elephants – just a handful of the 9,000 beasts slaughtered in one single day, 'and women took part in despatching them'.[4] They must surely have participated in Trajan's games in 108 CE which lasted 123 days and in which 'eleven thousand or so animals both wild and tame were killed and ten thousand gladiators fought'.[5] Martial, in his *De Spectaculis*, describes women battling in the arena, one dressed as Venus. Another, as a *venatrix*, animal hunter, subdues a lion: 'Caesar, we now have seen such things done by women's courage', he marvels. Statius describes in his *Silvae* 'the sex untrained in weapons recklessly dares men's fights! You would think a band of Amazons was battling by the River Tanais'.

The Romans would appear to have held their show dogs in higher regard than women fighters. While some men were incredulous at the thought and sight of women fighting in the arena, Roman audiences seem to have had no qualms about domestic dogs. In one of his epigrams, Martial tells the sad story of the hunting dog, *venatrix*, Lydia, who was brought up amid the *amphitheatrales*, trainers at the amphitheatres; she loved her job and was loyal to Dexter, her trainer. Age did not wither her; rather she was killed by a lightning-quick goring from a huge, slavering wild boar. Lydia had no complaints; she could not have asked for a nobler death.[6]

Domitian put on 'hunts of wild beasts, gladiatorial shows at night by the light of torches, and not only combats between men but between women as well', and, Dio adds, 'sometimes he would pit dwarfs and women against each other'.[7] Statius sums it all up with: 'Women untrained to the *rudis* take their stand, daring, how recklessly, virile battles!'[8] The *rudis* was the wooden sword given to a gladiator when he was freed after a series of conspicuous victories. Martial praises Titus for showing women fighting like Hercules. Juvenal sardonically describes 'Mevia', hunting wild boars in the arena 'holding her spear, breasts exposed'. Elsewhere he was more scathing:

> How shameful is a woman wearing a helmet, who shuns femininity and loves brute force. ... If a sale is held of your wife's effects, how proud you will be of her belt and arm-pads and plumes, and her half-length left-leg shin-guard! Or, if instead she prefers a different form of combat, how pleased you'll be when the girl you love sells off her greaves! ... Hear her grunt while she practises thrusts from the trainer, wilting under the weight of the helmet.[9]

Nicolaus of Damascus mentions women gladiators.[10] Nero dealt with annoying senators by threatening to have their wives thrown into the arena to do combat. Marriage guidance at its best?

In addition to performing, women were also, if the men are to be believed, seduced by the sheer sexuality exuded by some male gladiators. We have seen how Faustina, wife of Marcus Aurelius, was smitten by a gladiator and finally confessed her passion to her husband. The upshot was that the gladiator was slain and Faustina was made to bathe in his blood, and then have sex with her husband still covered in blood. Much earlier Juvenal highlights this *libido* for gladiators in his tirade against women (6.103–12) which in turn chimes with Ovid's assertion that some elite women were partial to 'a bit of rough', and with Petronius in his *Satyricon* (126); he has Chrysis describe how some well-to-do women burn with desire for men of the lower orders:

> There are some women, you see, whose lust is triggered only at the sight of slaves or messenger boys with their tunics belted right up high. Gladiators in the arena, a mule driver covered with dust, an actor in the shameful exposure of this performance – that's what it takes to get some females heated up.

These women even go so far as to lick the wounds of the flogged. Some women bribed guards to allow them access to the gladiator billets. Excavations in the armoury of the Pompeii gladiatorial barracks unearthed eighteen skeletons in two rooms, presumably of gladiators; but they were not alone. There was also the bones of a woman wearing gold and expensive jewellery, and an emerald-studded necklace; she clearly was not there just to serve the rations …

The gladiator, usually a slave and inhabiting the lowest level of Roman society, was seen as something of a fascinating paradox with magical qualities. His blood was used as a remedy for impotence, an aphrodisiac and any sensible bride would have her hair parted by a spear to ensure a fertile married life – ideally one that had been dipped in the blood of a defeated and dead gladiator. Medical authorities had it that drinking a gladiator's blood or eating his liver cured epileptics. Only gladiators, it seems, were given vasectomies.

Despite their lowly station, some became celebrities and were depicted in mosaics and sculptures, on lamps and tombstones: graffiti was scrawled by them and about them: 'Celadus the Thracian, thrice victor and thrice crowned, the young girls' heart-throb' and 'Crescens the Netter of young girls by night'. Gladiators were all the rage. But, even in victory, a gladiator remained what he was: *infamis* and a slave, unable to escape his ranking alongside criminals, whores, actors, dancers and similar so-called dregs. The paradox and irony was not lost on Tertullian (*De Spectaculus*, 22–3): 'Men surrender them their souls, and women their bodies, too …'. Archaeological evidence for the sexualisation of gladiators has been found in the shape of a multitude of objects depicting phalluses: a phallus-shaped terracotta gladiatorial helmet; a stone relief from Beneventum, showing a heavily armed gladiator in combat with a huge penis. The very word *gladius*, sword, carries unmistakeable sexual connotations and is sometimes slang for penis. The famous bronze figurine from Pompeii shows a menacing gladiator using his sword to fend off a dog-like beast which is growing out of his huge erect penis. Five bells are suspended from his body: every woman's ideal doorbell?

Juvenal (6.82–103) relates with contempt the sorry tale of:

> Eppia, the senator's wife, [who] ran off with a gladiator … And what were the youthful charms which captivated Eppia? What did she see in him to allow herself to be called 'a she-Gladiator'? Her dear Sergius had already begun to shave; a wounded arm gave promise of military discharge, and there were sundry deformities in his face: a scar caused by the helmet, a huge wen upon his nose, a nasty humour always trickling from his eye. But then he was a gladiator! It is this that transforms these fellows into Hyacinths! It was this that she preferred to children and to country, to sister and to husband. What these women love is the sword.

In 11 CE an attempt to ban senators and women performing in amphitheatres and on the stage was unsuccessful.[11] The law decreed that 'no female of free birth of less than twenty years of age and for no male of free birth of less than twenty-five years of age to pledge himself as a gladiator or hire out his services'. In 19 CE the Tabula Larinus prohibited the gladiatorial recruitment of daughters, granddaughters and great-granddaughters of senators or of knights, under the age of 20.

Nearly 200 years later in 200 CE Septimius Severus barred any female from fighting in the arena, μονομαχεῖν, because, as Dio reports, 'women took part, vying with one another most fiercely, with the result that jokes were made about other very distinguished women as well. Therefore it was henceforth forbidden for any freeborn woman, no matter what her origin, to fight in single combat'.[12] This came about after Severus' visit to the Antiochene Olympic Games where he would have seen traditional Greek female athletics. His attempt to impress the mob in Rome with a similar extravaganza was met with derision from the crowds in the Colisseum.

In September 2000, the Museum of London announced that they had discovered the grave of a female gladiator, from the first century CE, in Southwark – the first ever to be found, although it is hotly disputed that it is actually a woman gladiator. A shard of pottery has been discovered with the inscription '*VERECVNDA LVDIA LVCIUS GLADIATOR*', 'Verecunda the woman gladiator, Lucius the gladiator', but this may just be the paraphernalia typical of a woman who was married to or a mistress of a gladiator.

The British Museum has a first–second-century CE marble relief commemorating the release, *missio*, from service of two female gladiators, with 'stage names' Amazon and Achillia. It was found in Halicarnassus, modern Bodrum in Turkey. They are armed with swords and shields, and are advancing towards each other to attack. The gladiatrix on the right has lost her head – damaged, not decapitated. They are standing on a platform, and below on each side a spectator can be seen. It is inscribed above and on the platform in Greek with the two names and the word *apeluthesan*, telling us the the fight was a draw. They have the same equipment as male gladiators, but without helmets and are heavily armed with a greave, loin cloths and belt; they carry a mid-sized rectangular shield and a dagger in their right hand which is protected by the *manica* (arm protection); the *galege* (armlets) of both women are at their backs on the floor; their hair is cropped in the style of a slave and their breasts are bare. Such a spectacle must have been prestigious for it to be commemorated in this way.[13]

An epigraph from Ostia praises Hostilianus as the first to 'arm women', *mulieres*, in the history of the local games. A bronze statuette, now in the Museum für Kunst und Gewerbe, Hamburg, which has always been thought of as representing a female athlete holding a strigil, may actually depict a female gladiator, a *thraex*, a kind of gladiator who fought with a short curved dagger, a weapon that can be confused with a strigil.[14]

It seems likely that female gladiators came to the arena by a number of different routes. Some would have been slaves, coerced into the profession by their masters, the *lanistae*; others would have volunteered and received the requisite training in the gladiator schools, *ludi*; others still may have just been thrown in there as a punishment: *damnati ad gladium*, 'damned to the sword'. Women would not have faced men: rather, they would fight from chariots with the bows and arrows characteristic of the Amazons, *mulierem essedariam*, Diana and Atalanta; alternatively, as noted above, they may have been pitched against dwarfs, presumably the intention being to attempt to match, physically, like with like.

We have evidence that not all gladiatrices trained in the *ludus* like their male counterparts but in the Collegia Iuvenum – paramilitary associations set up to train free men, and women, in martial arts. Three Latin funerary inscriptions indicate female participation in these schools including one for Valeria Lacunda, who died aged 17.

The female gladiator out there with her breasts revealed would have certainly had an erotic impact on some male members of the audience, heightened by the arousing appearance of a woman in a uniform of sorts and wielding a weapon – something a typical Roman woman rarely did. Women who were trained and competent in combat were often foreign women, like

Boudica, for example, or the Amazons on the edge of empire – this exotic mystique must have stimulated further the sexual overtones of the spectacle and the libido of the male spectators. Ovid, we know, says that the games were the place to pick up a woman, and that the sight of a female leg, rarely seen outside the home, was exciting; the half-naked, weapon-wielding female gladiator would have been more exciting still.[15]

Epilogue

No Greek or Roman woman was ever conscripted into, press-ganged, volunteered or signed up to serve in the ranks of a Greek or Roman army. There was simply no official place for women in the army or the navy. However, our survey of women at war in Greece and Rome shows that women played a significant role in many aspects of battle and war in both cultures.

From the very start, Homer and the tragedians tell us how women were responsible for causing major conflict and how they interceded with male heroes in the resolution of the internecine wars and battles that followed the Trojan War. Homer was no doubt transmitting stories and legends embedded in cultures established well before his day and in the days he and others describe in the Homeric poems.

The farther we move from the centre of the two civilisations that were Greece and Rome, then the bigger, the more significant, the role that women played in fighting or prosecuting wars, and in evolving foreign policies and strategies which led to war and peace. Amazonian, Persian, Macedonian, Egyptian and Germano-Celtic women all exhibit a pugnacity quite foreign, repugnant almost, to the established norm for female conduct in Greece and Rome. We have seen how women actively assist in the defence against sieges either as combatants or as munitions manufacturers. We have seen how they form a crucial part of the baggage train providing support in everything from sewing to sex, how they loyally support proscribed husbands and how they take on invading rapists and lead armies of men in battle. They feature prominently in the Greek and Roman pantheons with martial responsibilities and as protagonists in epic and tragedy. They have leading roles in comedy and drive love poets to distraction in the war of love as soldiers of love. Their images in military settings feature prominently on Roman arches and columns and frequently on vases; and they fight in the arena as gladiators and are excited by male gladiators.

A book of this nature, with its many descriptions of the horrors of war, can easily inure the reader to the real terror, ghastliness and utter consternation and disorientation incited by battle and war – the endless atrocities routinely perpetrated over the centuries lose their ability to shock and repel. I implore the reader not to become blasé, or let the events described here be clichéd. It is vitally important we remember that every single act of war or action in battle in this book will have at the very least least one devastating, life-destroying, life changing consequence – not just for the combatants but also for non-combatants as well. Women, children and the elderly populations of fallen cities or subdued countries usually suffer terribly – physically, psychologically and socially. This will have been their fate in every one of the military and bellicose actions described in this book, be they historical, legendary or mythical.

And that is why the section on 'Women as Victims of War' is at the very heart of the book: it comes between women at war in Greece and women at war in Rome because it is pivotal to the history of warfare in both cultures and because it is, sadly, often glossed over or just absent from many of the many thousands of books and journal articles published on classical warfare in the last sixty or so years. That is why the book is dedicated to the many millions of women who have suffered in war – most often through no fault of their own – from Homeric Greece to the end of the Roman Empire, but, more significantly, from the relatively recent Second World War and the countless conflicts since then, up to and including the utterly ineffable and desperate situation afflicting innocent girls, women and their families and homelands in Syria and in other parts of the Middle East and Africa today.

War, as Homer said, may be man's work, but it is, at same time, the curse of many a woman.

Notes

Women and War in Earlier Ancient Civilisations

1. H. Thieme (1997), 'Lower Palaeolithic Hunting Spears from Germany', *Nature*, 385, 807–10.
2. R.C. Kelly, *Warless Societies and the Origin of War* (2000).
3. See H. Carreiras, *Gender and the Military: Women in the Armed Forces of Western Democracies* (2006).
4. See, for example, S. Brownmiller, *Against Our Will: Men, Women and Rape* (1975).
5. See A. Dodson, *The Complete Royal Families of Ancient Egypt* (2004).
6. See C. Graves-Brown, *Dancing for Hathor: Women in Ancient Egypt* (2010).
7. Ibid., Judges 4:6–10; 5:23–7; Song of Deborah is at 5:24.
8. See R.C. Kelly, 'The Evolution of Lethal Intergroup Violence', *Proceedings of the National Academy of Sciences*, 102, 24–9.
9. *Rigveda*, 1 and 10.
10. E.E. Vanderwerker, 'A Brief Review of the History of Amputations and Prostheses', *Inter Clinical Information Bulletin*, 15, No. 5, 15–16.
11. Diodorus Siculus, *Bibliotheca Historica*, 2, 4ff. See also Polyaenus, *Stratagemata*, 8, 26.
12. Diodorus Siculus, *Bibliotheca Historica*, 2, 7, 2.
13. Ibid., 10, 16.
14. Diodorus Siculus, *Bibliotheca Historica*, 2, 10.
15. Ammianus Marcellinus, *Res Gestae*, 14.

Chapter 1: Goddesses and War in Greek Mythology

1. Quintus Smyrnaeus, *Fall of Troy*, 8, 424.
2. Eustathius on Homer, 944.
3. Pausanias, *Description of Greece*, 4, 30, 5. Homer, *Iliad*, 5, 333, 592.
4. For which see Statius, *Thebaid.*
5. Nonnus, *Dionysiaca*, 2, 358 and 2, 475ff.
6. Hesiod, *Theogony*, 273.
7. Aeschylus, *Prometheus Bound* (trans. Weir Smyth), 788ff.
8. Pindar, *Dithyrambs*, frag. 78.
9. Hesiod, *Theogony*, 386–7; 389–94.
10. Quintus Smyrnaeus, *Fall of Troy*, 5, 25ff.
11. Hesiod, *Shield of Heracles*, 248–57.
12. See M. Egeler, 'Death, Wings, and Divine Devouring: Possible Mediterranean Affinities of Irish Battlefield Demons and Norse Valkyries', *Studia Celtica Fennica* (2008), 5–26.
13. Hesiod, *Theogony*, 226ff.
14. Hesiod, *Shield of Heracles*, 139ff.

Chapter 2: Warlike Women in Homer

1. Herodotus, *Histories*, 2, 35, 8.
2. See Apollodorus, *Bibliotheca*; Diodorus Siculus, *Bibliotheca Historica*, 4, 63, 1–3, and Plutarch, *Theseus*, 31–4.

3. Homer, *Odyssey*, 4, 277–89; Virgil, *Aeneid*, 6, 515–19.
4. Virgil, *Aeneid*, 6, 494–512.
5. Ibid., 6, 494ff.
6. Ibid., 6, 511–18.
7. Euripides, *Andromache*, 629–31.
8. Aristophanes, *Lysistrata*, 155.
9. Stesichorus, frag. 201, *PMG*.
10. *Cypria*, frag. 1; Hesiod, *Catalogues of Women and Eoiae*, frag. 204.96–101.
11. Homer, *Iliad*, 2, 688–94.
12. Ibid., 19, 261–3.
13. Ibid., 1, 110–14.
14. Ibid., 1, 10ff.
15. Ibid., 1, 320–5.
16. Ibid., 22, 431.
17. Ibid., 6, 425; 22, 470–2.
18. Ibid., 6, 450–65.
19. Ibid., 6, 485ff.; Homer, *Odyssey*, 1, 356–9; 21, 350–3.
20. Homer, *Iliad*, 6, 370–3; 6, 433–9.
21. Ibid., 6, 390.
22. Ptolemy, *Hephaeston*, 5.
23. Pausanias, *Description of Greece*, 4, 1, 1–2.
24. Suda and Hesychius of Alexandria, s.v. Ἄγραυλος; Ulpian *ad Demosth. de fals. leg.*; Plutarch, *Alcibiades*, 15; Philochorus, frag. (ed. Siebelis), p. 18.

Chapter 3: *Teichoskopeia*: A Woman's View from the Walls

1. Homer, *Iliad*, 3, 121–244.
2. Hesiod, *Shield of Heracles*, 242.
3. Servius, *ad Georgics*, 1, 18; O. Skutsch, 'The Fall of the Capitol', *JRS* (1953), 45, 77–8; E.H. Warmington, *Remains of Old Latin* (1898), p. 371.
4. Virgil, *Aeneid*, 11, 475ff. and 12, 593ff.
5. Horace, *Odes*, 3, 2, 1–14.
6. *Ciris*, 172ff. Ovid, *Metamorphoses*, 8, 21ff.
7. Euripides, *Phonissae*, 88–201; Statius, *Thebaid*, 7, 237–373; 11, 359–65.
8. Valerius Flaccus, *Argonautica*, 6, 482–506; 575–82.
9. Ibid., 6, 657–63; 6, 717–20.

Chapter 4: The Amazons

1. Pliny the Elder, *Naturalis Historia*, 6, 3, 10.
2. Herodotus, *Histories*, 6, 86; P. Walcott, 'Greek Attitudes towards Women: The Mythological Evidence', *G&R*, 31 (1984), 42.
3. Hippocrates, *Airs, Waters, Places*, 17.
4. Herodotus, *Histories*, 4, 110–17.
5. Lysias 2, 4. Strabo, *Geographia*, 5, 50.
6. Hippocrates, *Airs, Waters, Places*, 17.
7. Justinus, *Historiae Phillippicae ex Trogo Pompeio*, 2, 4.
8. Strabo, *Geographia*, 11, 503; Hellanicus, frag. 17.
9. For details see Quintus of Smyrna, *Fall of Troy*, Books 1–4.
10. Virgil, *Aeneid*, 1, 490–5. The twelve are Antibrote, Ainia, Clete, Alcibie, Antandre, Bremusa, Derimacheia, Derinoe, Harmothoe, Hippothoe, Polemusa and Thermodosa.
11. Diodorus Siculus, *Bibliotheca Historica*, 2, 46.

12. Apollodorus, *Bibliotheca*, 5, 1.
13. Propertius, *Elegies*, 3, 11.
14. Pausanias, *Description of Greece*, 10, 31, 1 and 5, 11, 2.
15. Robert Graves, *Selected Poems*, ed. Michael Longley (2013).
16. Euripides, *Hercules Furens*, 408ff.; Apollonius Rhodius, *Argonautica*, 2, 777ff. and 966ff.; Diodorus Siculus, *Bibliotheca Historica*, 4, 16; Ps.-Apollodorus, *Bibliotheca*, 2, 5, 9; Pausanias, *Description of Greece*, 5, 10, 9; Quintus Smyrnaeus, *Fall of Troy*, 6, 240ff.; Hyginus, *Fabulae*, 30.
17. Plutarch, *Theseus*.
18. Pliny the Elder, *Naturalis Historia*, 34, 75.
19. Statius, *Thebaid*, 2, 635–8.
20. Homer, *Iliad*, 2, 814; 2, 45–6; 3, 52–5; Diodorus Siculus, *Bibliotheca Historica*, 3, 54–6. See also Strabo, *Geographia*, 12, 8, 6 and Tzetzes on Lycophron, 243.
21. Stephanus of Byzantium, s.v. Myrina. See also Strabo, *Geographia*, 11, 5, 5 = 12, 3, 22.
22. Herodotus, *Histories*, 4, 110, 1–117, 1.
23. Homer, *Iliad*, 6, 186ff.; scholiast, *On Lycophron*, 17.
24. Homer, *Iliad*, 3, 189.
25. Ibid., 3, 185–9; 6, 186.
26. Diodorus Siculus, *Bibliotheca Historica* (trans. Loeb Classical Library edn, 1935), 2.45.
27. Ibid.
28. Ibid.
29. Justinus, *Epitome of Pompeius Trogus*, 2, 4.

Chapter 5: Women and War in Greek Tragedy

1. Aeschylus, *Seven Against Thebes*, 230–2.
2. Ibid., 99.
3. Ibid., 256–7.
4. Ibid., 268.
5. Ibid., 679–82.
6. Euripides, *Phoenissae*, 535f.
7. Ibid., 160ff.
8. Ibid., 1490.
9. Ibid., 1577–9.
10. Euripides, *Medea*, 263–6.

Chapter 7: Women and War in Greek History and Philosophy

1. Herodotus, *Histories* (trans. George Rawlinson), 1, 9.
2. Jordanes, *De Origine Actibusque Getarum*.
3. Herodotus, *Histories*, 1, 214; 1, 216; or, in the words of K.S. Dardi, *These Kamboj People* (2001), p. 88, is then reported to have herself cut Cyrus' head from his dead body and then grabbing it by his hair, said in a wailful and heroic avengeful crying tone: 'Cyrus I give you the fill of your blood!' (http://www.punjabi.net/forum/archive/index.php/t-927.html).
4. Herodotus, *Histories*, 4, 205ff.
5. Ibid., 7, 99, 3.
6. Pallene is the ancient name of the westernmost of the three headlands of Chalcidice in the Aegean Sea. Its modern name is Kassandra Peninsula.
7. Herodotus, *Histories*, 8, 88. Polyaenus, *Stratagemata*, 8, 53, 4.
8. Herodotus, *Histories*, 8, 93.
9. Ibid.
10. Polyaenus, *Stratagemata*, 8, 53, 4.

11. Herodotus, *Histories*, 8, 101.
12. Photius, *Bibliotheca*, Codex 190.
13. Vitruvius, *De Architectura*, 2, 8, 15. Polyaenus, *Stratagemata*, 8, 53, 4.
14. See P. Chrystal, *Women in Ancient Greece: Seclusion, Exclusion or Illusion?* (2017).
15. Pausanias, *Description of Greece*, 2, 20, 8.
16. Herodotus, *Histories*, 6, 76.
17. Tatian, *Oratio ad Graecos*, 33.
18. Herodotus, *Histories*, 9, 5, 1–2; 5, 84, 1.
19. Diodorus Siculus, *Bibliotheca Historica*, 13, 108, 2f.
20. Ibid., 13, 93f.
21. Thucydides, *History of the Peloponnesian War*, 2, 75–8.
22. Ibid., 3, 74, 2; 2, 78f.; 3, 68.
23. Athenaeus, *The Deipnosophistai*.
24. Thucydides, *History of the Peloponnesian War*, 5, 82–3. Eretria: S. Fachard, *La défense du territoire d'Eretrie. Étude de la chôra érétrienne et de ses fortifications* (2012). Evacuations: Delphi, Herodotus, *Histories*, 8, 36, 2; Troezen, Herodotus, *Histories*, 8, 41; Plataeans, Thucydides, *History of the Peloponnesian War*, 2, 6, 4; Brasidas, Thucydides, *History of the Peloponnesian War*, 4, 123, 4; Agrigentines, Diodorus Siculus, *Bibliotheca Historica*, 13, 89, 1–3; 13, 91. Scione and Mende: Thucydides, *History of the Peloponnesian War*, 5, 32, 1; 4, 130, 6; Torone: Thucydides, *History of the Peloponnesian War*, 5, 3, 2–4; Melos: Thucydides, *History of the Peloponnesian War*, 5, 116. Lyttos: Polybius, *Histories*, 4, 53, 3–54. Abydos: Polybius, *Histories*, 16, 34, 9–11. Demosthenes: *On the Embassy*, 19, 192–8.
25. Pausanias, *Description of Greece*, 8, 48, 4–5.
26. Thucydides, *History of the Peloponnesian War*, 2, 4; Aeneas Tactitus, *How to Survive under Siege*, 2, 6; Diodorus Siculus, *Bibliotheca Historica*, 12, 41.
27. Polyaenus, *Stratagemata*, 8, 69.
28. Diodorus Siculus, *Bibliotheca Historica*, 13, 55–7f.; Xenophon, *Hellenica*, 1, 1, 37.
29. Diodorus Siculus, *Bibliotheca Historica*, 13, 55, 5.
30. Pausanias, *Description of Greece*, 4, 21, 6.
31. Plutarch, *Pyrrhus*, 34, 1–4; Pausanias, *Description of Greece*, 1, 13, 8; Polyaenus, *Stratagemata*, 8, 68. Important roof tile quotation by William Barry, 'Roof Tiles and Urban Violence in the Ancient World', *GRBS*, 37 (1976), 55.
32. Polyaenus, *Stratagemata*, 8, 68, 9; Pausanias, *Description of Greece*, 4, 29, 5.
33. Pausanias, *Description of Greece*, 10, 22.
34. Plutarch, *De Virtutibus Muliebrum*, 3.
35. Ibid., 1, 3.
36. Cornelius Nepos, *On Great Generals*, 14; Aeneas Tactitus, *How to Survive under Siege*, 40, 4–5; 4, 8–11.
37. Plutarch, *Philopoemen*, 9, 5.
38. Xenophon, *Anabasis*, 5, 3, 3.
39. Ibid., 6, 1, 11–13; 4, 1, 10–15; 3, 2, 25; 4, 8, 27; 5, 4, 33; 4, 3, 18–19.
40. *Inscriptiones Graecae*, 3, 69.
41. Plutarch, *Eumenes*, 16–19.
42. Xenophon, *Oeconomicus*, 7, 22.
43. Plato, *Laws*, 7, 806a-b.
44. Plato, *Republic*, 456d-e.
45. Plato, *Laws*, 813a–814a.
46. Aristotle, *Politics*, 2, 9, 9–10.
47. Philon, *Poliorcetica*, C31.

Chapter 8: Warrior Women Catalogued

1. Aeneas Tacticus, *How to Survive under Siege*, 40, 2 and 4.
2. Frontinus, *Strategemata*, 7, 33.
3. Plutarch, *Moralia*, 242ff.
4. Thucydides, *History of the Peloponnesian War*, 2, 45. All translations adapted from Frank Cole Babbitt, Loeb edn, 1931. Plutarch, *Moralia*, 243d.
5. See also Plutarch, *Moralia*, 265b; Plutarch, *Romulus*, 1, 17f.; Polyaenus, *Stratagemata*, 8, 25, 2; Dionysius of Halicarnassus, *Roman Antiquities*, 1, 72–3.
6. See also Polyaenus, *Stratagemata*, 7, 45, 2; Justinus, *Historiae Philippicae*, 1, 6.
7. See also Xenophon, *Cyropaedia*, 8, 5, 21.
8. See also Plutarch, *Alexander*, 69 (703a).
9. Cf. Polyaenus, *Stratagemata*, 7, 50.
10. Cf. Polybius, *Histories*, 3, 14; Livy, *Ab Urbe Condita*, 21, 5. Cf. Polyaenus, *Stratagemata*, 7, 48.
11. Cf. Livy, *Ab Urbe Condita*, 38, 24; Valerius Maximus, *Factorum et Dictorum Memorabilium*, 6, 1, ext. 2; Florus, *Epitome of Roman History*, 1, 27, 6.
12. All translations adapted from the translation by R. Shepherd (1793).
13. See also Plutarch, *Camillus*, 33 and *Romulus*, 29; Varro, *De Lingua Latina*, 6, 18; Macrobius, *Saturnalia*, 1, 11, 35–40.
14. See also Athenaeus, *Deipnosophistai*, 13, 596f.; Jerome, *Beginning of the Consuls of the Romans*; Plutarch, 'On Talkativeness', *Moralia*, 505E.
15. D.L. Gera, *Warrior Women: The Anonymous* Tractatus De Mulieribus (1997).
16. Suda, *Pamphile*.
17. Julian, *Orations*, 3, 127a–c.

Chapter 9: Spartan Women: Vital Cogs in a Well-Oiled War Machine

1. See though P. Low, 'War, Death and Burial in Classical Sparta', *Omnibus*, 65 (2013), in which she cautions against taking the overtly militaristic view of Sparta at face value.
2. See Xenophon, *Constitution of the Lacedaemonians*, 3, 16.
3. For Cynisca: Xenophon, *Minor Works, Agesilaus*, 9, 1, 6; Pausanias, *Description of Greece*, 3, 5, 1. Herodotus, *Histories*, 6, 7; *L'Année philologique*, 13, 16.
4. Xenophon, *Minor Works, Agesilaus*, 9, 1, 6; 20, 1.
5. Pausanias, *Description of Greece*, 3, 8, 1–3.
6. *P. Oxy*, 268/7.
7. *IG*, 9, 2, 526, 19–20.
8. *IG*, 2, 23, 13, 9–5; 23, 13, 60; 23, 14, 50–1.
9. Plutarch, *Lacaenarum Apophthegmata*, 3, 1; Cf. Herodotus, *Histories*, 4, 51; 7, 239.
10. Plutarch, *Lacaenarum Apophthegmata*, 2, 1.
11. Ibid., 4, 2. Cf. similar sentiments spoken by a Spartan woman, quoted by Teles in Stobaeus, *Florilegium*, 108, 83.
12. Plutarch, *Lacaenarum Apophthegmata*, 5, 1; cf. *Palatine Anthology*, 7, 433.
13. Plutarch, *Lacaenarum Apophthegmata*, 6, 1; cf. the version in the *Palatine Anthology*, 7, 433.
14. Plutarch, *Lacaenarum Apophthegmata*, 6, 3.
15. Ibid., 6, 4.
16. Ibid., 6, 5.
17. Ibid., 6, 8. Cf. Teles' version in Stobaeus, *Florilegium*, 108, 83; cf. also Cicero, *Tusculanae Disputationes*, 1, 42.
18. Plutarch, *Lacaenarum Apophthegmata*, 6, 13; cf. Plutarch *Moralia*, 331b; Stobaeus, *Florilegium*, 7, 29; Cicero, *De Oratore*, 2, 61.
19. Plutarch, *Lacaenarum Apophthegmata*, 6, 14.
20. Plutarch, *Pyrrhus*, 27, 2–3; Polyaenus, *Stratagemata*, 8, 49.

21. Polyaenus, *Stratagemata*, 8, 70.
22. Plutarch, *Moralia*, 231c.
23. Lactantius, *Divine Institutes*, 1, 20, 29–32.

Chapter 10: Macedonian Women at War: Pawns and Power-Players

1. Polyaenus, *Stratagemata*, 8, 60.
2. Photius, *Bibliotheca*, cod. 92; Diodorus Siculus, *Bibliotheca Historica*, 18, 39.
3. Diodorus Siculus, *Bibliotheca Historica*, 19, 11; Justinus, *Historiae Phillippicae ex Trogo Pompeio*, 14, 5; Aelian, *Varia Historia*, 13, 36.
4. Arrian, *De Rebus Successorum Alexandri Anabasis Alexandri*, 1, 5; Photius, *Bibliotheca*; Athenaeus, *Deipnosophistai*, 13, 5; Diodorus Siculus, *Bibliotheca Historica*, 19, 52; Polyaenus, *Stratagemata*; Aelian, *Varia Historia*, 8, 36.
5. Plutarch, *Pyrrhus*, 4; Diodorus Siculus, *Bibliotheca Historica*, 19, 35; Justinus, *Epitome of Pompeius Trogus*, 14, 6.
6. Plutarch, *Demetrius*, 25; Plutarch, *Pyrrhus*, 4.
7. Plutarch, *Demetrius*, 30, 32; 53.
8. Polyaenus, *Stratagemata*, 8, 52; Justinus, *Epitome of Pompeius Trogus*, 28, 3; Pausanias, *Description of Greece*, 4, 35.
9. Diodorus Siculus, *Bibliotheca Historica*, 19, 35; Justinus, *Epitome of Pompeius Trogus*, 14, 6.
10. Diodorus Siculus, *Bibliotheca Historica*, 19, 52; Pausanias, *Description of Greece*; Strabo, *Geographia*, 7; Stephanus of Byzantium, *Ethnica*, *'Thessalonike'*.
11. The author of a romantic novel entitled *The Wonders Beyond Thule* (Τὰ ὑπὲρ Θούλην ἄπιστα – *Apista huper Thoulen*) for which Photius wrote a synopsis. Photius, *Bibliotheca*, 166.
12. Diodorus Siculus, *Bibliotheca Historica*, 18, 18.
13. Plutarch, *Pyrrhus*, 14.
14. Ibid., 22, 32, 35, 37, 38, 45; Diodorus Siculus, *Bibliotheca Historica*, 20, 93.
15. Athenaeus, *Deipnosophistai*, 6, 66.
16. Justinus, *Epitome of Pompeius Trogus*, 28, 1.
17. Ibid.; Josephus, *Against Apion*, 1, 22.
18. Plutarch, *Pyrrhus*, 2, 1.
19. Arrian, *De Rebus Successorum Alexandri Anabasis Alexandri*, 2, 11, 9; Diodorus Siculus, *Bibliotheca Historica*, 35f.; Plutarch, *Alexander*, 20, 6–21; Quintus Curtius Rufus, *Historiae Alexandri Magni*, 3, 11, 24–6.
20. Quintus Curtius Rufus, *Historiae Alexandri Magni*, 3, 12, 21; see also Plutarch, *Alexander*, 21, 3; Athenaeus, *Deipnosophistai*, 13, 603b-d; Diodorus Siculus, *Bibliotheca Historica*, 17, 38, 1; Justinus, *Epitome of Pompeius Trogus*, 11, 9.
21. See Diodorus Siculus, *Bibliotheca Historica*, 18, 2–4; Arrian, *De Rebus Successorum Alexandri Anabasis Alexandri*, 9.
22. Plutarch, *Alexander*, 77, 4; Diodorus Siculus, *Bibliotheca Historica*, 19, 11.
23. Diodorus Siculus, *Bibliotheca Historica*, 19, 9–51; Justinus, *Epitome of Pompeius Trogus*, 14, 6; Diodorus Siculus, *Bibliotheca Historica*, 19, 52, 4; Justinus, *Epitome of Pompeius Trogus*, 15, 2, 5; Diodorus Siculus, *Bibliotheca Historica*, 19, 105.
24. Polyaenus, *Stratagemata*, 8, 40; Plutarch, *Alexander*.
25. Diodorus Siculus, *Bibliotheca Historica*, 17, 72.
26. Ovid, *Remedia Amoris*, 383.
27. Dante, *Inferno*, 18, 133–6.
28. *Metz Epitome*, 40–1.
29. Quintus Curtius Rufus, *Historiae Alexandri Magni*, 8, 10, 34–6; *Metz Epitome*, 45; Justinus, *Epitome of Pompeius Trogus*, 12, 7.
30. Translation in J.W. McCrindle, *The Invasion of Alexander* (2004), p. 270.

Chapter 11: War Rape and Other Atrocities in the Classical World

1. Exodus 17:14, 16:1 Samuel 15:3.
2. See D.D. Luckenbill, *Ancient Records of Assyria and Babylonia II* (1926), p. 314.
3. Kings 8:12; 15:16.
4. Zechariah 14:2 and Isaiah 13:16; Lamentations 5:11. All King James Version.
5. Homer, *Iliad*, 6, p. 504ff; Diomedes: 11, 3993; Achilles: 18, 122–4; rape: 3, 301; 4, 238–9; Hector: 8, 165–6 and 16, 830–2; Cleopatra: 9, 594; Achilles, 20, 191–4; Priam: 22, 62–5; Andromache: 24, 729; defending women: 5, 485–6; 8, 55–7; 15, 496; 17, 220–8; 18, 265; 21, 586–8; 24, 729–32; 22, 56–7.
6. Herodotus, *Histories*, 1, 1, 3.
7. Ibid., 1, 205ff.; 7, 99; 8, 68ff.; 8, 87; 8, 101; 4, 162ff., 206ff.
8. Ibid., 1, 4, 2.
9. Ovid, *Ars Amatoria*, 1, 663–8.
10. Lewis and Short, *A Latin Dictionary ad loc*; Liddell and Scott, *Greek-English Lexicon ad loc*. See also P. Chrystal, *In Bed with the Romans: Sex and Sexuality in Ancient Rome*, ch. 9 on rape and ch. 12 on sexual vocabulary. See also Cicero, *Ad Familiares*, 14, 11 and *Ad Atticus*, 10, 8. Also M. McDonnell, *Roman Manliness:* Virtus *and the Roman Republic* (2006).
11. See A. Ziolkowski, '*Urbs Direpta* or How the Romans Sacked Cities', in J. Rich (ed.), *War and Society in the Ancient World* (1993), for an analysis of rape and plunder and 'how the Romans sacked cities'.
12. Cicero, *In Verrem*, 4, 116.
13. Sallust, *Jugurtha*, 91.
14. Tacitus, *Agricola*, 30, 4–5.
15. Tacitus, *Annals*, 14, 35.
16. Livy, *Ab Urbe Condita*, 21, 57, 13–14.
17. Ibid., 1, 9.
18. Valerius Maximus, *Factorum et Dictorum Memorabilium*, 6, 1, 1. Agrippina: Tacitus, *Annals*, 1, 69; Dido: Virgil, *Aeneid*, 1, 364; Boudica: Tacitus, *Agricola*, 16, 1; 31, 4; Tacitus *Annals*, 14, 35, 1.
19. Livy, *Ab Urbe Condita*, 1, 57.
20. Ovid, *Fasti*, 2, 720–58.
21. Livy, *Ab Urbe Condita*, 3, 44.
22. Sandra R. Joshel, 'The Body Female and the Body Politic: Livy's Lucretia and Verginia', in Amy Richlin (ed.), *Pornography and Representation in Greece and Rome* (1992).
23. Livy, *Ab Urbe Condita*, 26, 13.
24. Ibid., 29, 17, 15–16.
25. Pausanias, *Description of Greece*, 4, 16, 9–10. Livy, *Ab Urbe Condita*, 26, 13, 15; 26, 50, 7–14.
26. Alba: Livy, *Ab Urbe Condita*, 1, 29. Cicero, *In Verrem*, 2, 1, 62; 2, 1, 64; 4, 116.
27. Livy, *Ab Urbe Condita*, 2, 17, 2; 37, 2.
28. Tacitus, *Histories*, 2, 73; 4, 14; 3, 34.
29. For 'unruly' soldiers generally, see: Juvenal, *Satires*, 16; Petronius, *Satyricon*, 82; 62; cf. New Testament: Matthew 27:26–35; Mark 15:15–19; John 19:23–4; Luke 3:14; Epictetus, *Discourses*, 4, 1, 79. Vitellius: Tacitus, *Histories*, 2, 56.
30. Polybius, *Histories*, 10, 15.
31. Livy, *Ab Urbe Condita*, 23, 7.
32. Caesar, *De Bello Gallico*, 8, 44.
33. Ibid., 7, 14, 26. Tacitus, *Histories*, 18.
34. Josephus, *Bellum Judaicum*, 2, 41, 5.
35. Ibid., 2, 305–7.
36. Ibid., 2, 408; 4, 7, 2.

37. Ibid., 5, 10, 3. Translation adapted from William Whiston, *Wars of the Jews* (2016).
38. Josephus, *Bellum Judaicum*, 6, 201–13.
39. Ibid., 6, 9, 3.
40. Ammianus Marcellinus, *Res Gestae*, 18, 7.
41. *Chronicle of Pseudo-Joshua*, 76–7.

Chapter 12: Military Women in Roman Legend

1. See Livy, *Ab Urbe Condita*, 1, 3, 11; Ennius, *Annals*, I, frag. 19; Cicero, *Divinatio in Caecilium*, 1, 30; *Fabius Pictor: Fragmente der Griechischen Historiker*, 809, f. 4a; Virgil, *Aeneid*, 7, 483–99.
2. Livy, *Ab Urbe Condita*, 1, 11.
3. Ibid.
4. Propertius, *Elegies*, 4, 4.
5. Livy, *Ab Urbe Condita*, 1, 9.
6. Livy, *Ab Urbe Condita*, 2, 13; Valerius Maximus, *Factorum et Dictorum Memorabilium*, 3, 2, 2.
7. Livy, *Ab Urbe Condita*, 2, 40.

Chapter 13: Military Women in Roman History

1. Livy, *Ab Urbe Condita*, 4, 9, 4–6.
2. Ibid., 1, 13, translation adapted from B.O. Foster, Loeb edn (1919), 39–40.
3. Appian, *Civil Wars*, 4, 39–40.
4. Sallust, *In Catilinam*, 25. See J.P.V.D. Balsdon, *Roman Women* (1962), pp. 47–9 for the controversy surrounding Sallust's description of her.
5. Pliny the Younger, *Epistles*, 6, 24; Plutarch, *Brutus*, 13, 23, 6 and Dio Cassius, *Historia Romana*, 44, 13–14. See also Valerius Maximus, *Factorum et Dictorum Memorabilium*, 4, 6 and 6, 7 on brave and faithful wives.
6. Cicero, *Ad Familiares*, 5, 2, 6; Dio Cassius, *Historia Romana*, 48, 16, 3; 51, 2, 4–5.
7. Plutarch, *Cato Maior*, 8, 4; Plutarch, *Themistocles*, 18.
8. Plutarch, *Camillus*, 22, 4; Servius, *On Virgil*, 1, 720; 3, 519 and 8, 688.
9. Frontinus, *Strategemata*, 4, 1, 1; Livy, *Periochae*, 57; Appian, *Iberica*, 14, 85; Valerius Maximus, *Factorum et Dictorum Memorabilium*, 2, 7, 1; Plutarch, *Moralia*, 201b. Cicero, *In Catilinam*, 2, 23.
10. Pseudo-Quintilian, 3, 12, 3–4.
11. Propertius, *Elegies*, 4, 3, 45.
12. Tacitus, *Annals*, 3, 33–4; Livilla: *Annals*, 3, 34, 6.
13. Tacitus, *Histories*, 1, 48; Dio Cassius, *Historia Romana*, 59, 18.
14. Plutarch, *Galba*, 7ff.; Tacitus, *Histories*, 1, 5.
15. See S.E. Phang, *Marriage of Roman Soldiers (13 BC–AD 235)* (2001).
16. Alan Bowman and David Thomas, *The Vindolanda Writing Tablets II* (1994), No. 5.
17. Herodian, *Roman History*, 3, 8, 5. Tacitus, *Histories*, 3, 69.
18. Tacitus, *Annals*, 13, 35.
19. Suetonius, *Caligula*, 9.
20. Diodorus Siculus, *Bibliotheca Historica*, 32, 9.
21. Polybius, *Histories*, 1, 72, 5.
22. Polyaenus, *Stratagemata*, 8, 67.
23. *Historia Augusta: The Two Maximini*, 33; Lactantius, *Divine Institutes*, 1, 20, 27; Caesar, *De Bello Gallico*, 3, 9, 3; Frontinus, *Strategemata*, 1, 7, 3.
24. See N. McGregor, *Germany: Memories of a Nation* (2014), p. 252. The difference here is that the Prussian women were given iron jewellery in return, in emulation of the German Iron Cross.

25. Hero of Alexandria, *Treatise on Ranged Weapons*, 112.
26. Carthage: Polybius, *Histories*, 4, 56, 3.
27. Appian, *Punica*, 93.
28. Livy, *Ab Urbe Condita*, 5, 21, 10.
29. Sallust, *Jugurtha*, 67, 1.
30. Plutarch, *Lucullus*, 6.
31. Suetonius, *Julius Caesar*, 50; Plutarch, *Cato Minor*, 24, 1; Plutarch, *Brutus*, 5, 2; Appian, *Civil Wars*; Cicero, *Ad Familiares*, 12, 7; Cicero, *Ad Atticum*, 14, 21; 15, 11; 15, 12; 32. Cornelius Nepos, *Atticus*, 33.
32. Velleius Paterculus, *Historiae Romanae*, 2, 74.
33. Dio Cassius, *Historia Romana*, 47, 8, 4.
34. Plutarch, *Cicero*, 49.
35. Cicero, *Phillipics*, 13, 18.
36. Appian, *Civil Wars*, 3, 8, 51.
37. Dio Cassius, *Historia Romana*, 48, 4, 1–6; Appian, *Civil Wars*, 5, 3, 19.
38. Ibid., 5, 4, 32ff..
39. Velleius Paterculus, *Historiae Romanae*, 2, 74, 3.
40. See J. Hallet, '*Perusinae Glandes* and the Changing Image of Augustus', *AJAH*, 2 (1977), 151–71.
41. Martial, *Epigrams*, 11, 20.
42. Sallust, *In Catilinam*, 27, 3–28, 3.
43. Dio Cassius, *Historia Romana*, 47, 7, 4–5.
44. See Appian, *Civil Wars*, 5, 76; Plutarch, *Antony*, 33, 3.
45. Appian, *Civil Wars*, 5, 76, 93–5; Plutarch, *Antony*, 35, 1–4; Dio Cassius, *Historia Romana*, 48, 54, 1–5.
46. Dio Cassius, *Historia Romana*, 48, 54, 5.
47. Plutarch, *Antony*, 53, 54, 1–2; Dio Cassius, *Historia Romana*, 49, 32, 4–5 and 33, 3–4; 50, 3, 2.
48. Tacitus, *Annals*, 1, 41.
49. Dio Cassius, *Historia Romana*, 57, 5.
50. Tacitus, *Annals*, 1, 69.
51. Ibid., 2, 55, 6.
52. Josephus, *Antiquities of the Jews*, 18, 31; 143.
53. Tacitus, *Histories*, 3, 77, 6–7.
54. Claudian, *Minor Works*, 30.
55. Claudian, *Bellum Geticum*, 83–5.
56. Ibid., 623–8.
57. Ibid., 188–93.
58. E. Gibbon, *History of the Decline and Fall of the Roman Empire* (1996), ch. 31.
59. Ibid., ch. 33.2.
60. Socrates of Constantinople, *Historia Ecclesiastica*, 21.
61. Procopius, *History of the Wars*, 3, 3.
62. Procopius, *De Aedificiis*, 1, 8, 5.
63. Procopius, *History of the Wars*, 1, 24, 33–7.
64. Ibid.
65. Lynda Garland, *Byzantine Empresses: Women and Power in Byzantium AD 527–1204* (1999), p. 143.
66. Procopius, *History of the Wars*, 1, 24, 33–7.
67. Ibid.
68. Ibid.
69. Ibid., 1, 11. Translations are by G.A. Williamson in *Procopius: The Secret History* (1966).

Chapter 14: Foreign Women Fighters

1. Jane Margaret Strickland and Agnes Strickland, *Rome, Regal and Republican: A Family History of Rome* (1854), pp. 290–1.
2. Dio Cassius, *Historia Romana*, 42, 34, 3–6. Translation is by E. Cary. Caesar, *Bellum Civile*, 3, 109.
3. Plutarch, *Caesar*, 49, 1–3; Lucan, *De Bello Civili*, 10, 56–8. Cf. Dio Cassius, *Historia Romana*, 42, 34, 6–35, 1. The translation is by Rex Warner, *Plutarch, the Fall of the Roman Republic* (2006), p. 290.
4. Lucan, *De Bello Civili*, 10, 139–40.
5. Suetonius, *Julius Caesar*, 76, 3.
6. Dio Cassius, *Historia Romana*, 43, 27, 3; Cicero, *Ad Atticum*, 15, 15, 2.
7. Appian, *Civil Wars*, 2, 102, 424; Dio Cassius, *Historia Romana*, 51, 22, 3.
8. Cicero, *Ad Atticum*, 15, 15, 2; 14, 8, 1.
9. Josephus, *Antiquities of the Jews*, 15, 89.
10. Horace, *Epodes*, 9, 11–16.
11. Appian, *Civil Wars*, 4, 61, 262–3; Dio Cassius, *Historia Romana*, 47, 30, 4 and 47, 31, 5.
12. For Dido's curse see E. O'Gorman, 'Does Dido's Curse Work?', *Omnibus*, 64 (2012), 10–12; Virgil, *Aeneid*, 8, 696ff.
13. Horace, *Epodes*, 9, 11–16.
14. For themes used by the love poets and their significance see P. Chrystal, 'Differences in Attitude to Women as Reflected in the Work of Catullus, Propertius, the *Corpus Tibullianum*, Horace and Ovid', Diss., University of Southampton (1982).
15. Propertius, *Elegies*, 4, 6, 57–66.
16. Juvenal, *Satires*, 2, 109.
17. Martial, *Epigrams*, 4, 59.
18. Lucan, *De Bello Civili*, 10, 104–6; 107ff.
19. Ibid., 353ff. M. Grant, *Cleopatra* (1992), p. xvii.
20. Dio Cassius, *Historia Romana*, 60, 19–22.
21. Tacitus, *Histories*, 3, 45; Tacitus, *Annals*, 14, 32; 14, 34.
22. Suetonius, *Vespasian*, 4; Tacitus, *Agricola*, 14.
23. Tacitus, *Agricola*, 12, 13; 13, 22. Suetonius, *Nero*, 35, 4. Dio Cassius, *Historia Romana*, 61, 30. Suetonius, *Claudius*, 24, 2.
24. Tacitus, *Annals*, 12, 36; Tacitus, *Agricola*, 14–17; Tacitus, *Annals*, 14, 29–39; Dio Cassius, *Historia Romana*, 62, 1–12. 'The Roman Senate and People to Tiberius Claudius Caesar Augustus Germanicus, son of Drusus, Pontifex Maximus, Tribunician power eleven times, Consul five times, Imperator twenty-two times, Censor, Father of the Fatherland, for taking the surrender of eleven kings of the Britons defeated without any loss, and first brought barbarian peoples across the Ocean under the sway of the Roman people.', *Inscriptiones Latinae Selectae*, 216–17; cf. Suetonius, *Claudius*, 25.
25. Tacitus, *Annals*, 12, 31; 14, 31; Dio Cassius, *Historia Romana*, 62, 2.
26. Tacitus, *Agricola*, 15.
27. Dio Cassius, *Historia Romana*, 62, 7.
28. Ibid.
29. Tacitus, *Annals*, 14, 27.
30. Dio Cassius, *Historia Romana*, 62, 5.
31. Strabo, *Geographia*, 7, 2, 3.
32. Ibid.
33. Tacitus, *Germania*, 44.
34. Ibid., 8, 2.
35. Tacitus, *Histories*, 4, 61.

36. *Année Épigraphique* (1953), 25.
37. Tacitus, *Annals*, 1, 57.
38. Gibbon, *Decline and Fall of the Roman Empire*, ch. 11, pp. 128–9. For a more up-to-date assessment see also M. Bragg, *Queen Zenobia In Our Time* (http://www.bbc.co.uk/programmes/b01snjpp)
39. *Scriptores Historiae Augustae*, 107.
40. Trebellius Pollio, *Scriptores Historia Augusta*, trans. David Magie (1932), 3, 135–43.
41. *Historia Augusta: The Life of Aurelian*, 26. See Zenobia, *Encyclopedia of World Biography* (2004).
42. Virgil, *Aeneid*, 2, 559–87.
43. Zosimus, *Historia Nova*, Book I.
44. Zonaras, *The History of Zonaras*, 27.
45. Al-Tabari, *The History of al-Tabari*, Vol. IV, 124.
46. Lawrence Wright, 'Homage to Zenobia', *The New Yorker* (www.newyorker.com/magazine/2015/07/20/homage-to-zenobia).
47. Glen Warren Bowersock, *Late Antiquity: A Guide to the Postclassical World* (1999).
48. Rufinus, quoted in Anne Jensen, *God's Self-confident Daughters: Early Christianity and the Liberation of Women* (1996), pp. 93–5.

Chapter 15: Women and War in Roman Epic

1. Virgil, *Aeneid*, 11, 648.
2. Ibid., 7, 803ff.
3. Ibid., 11, 603–7.
4. Ibid., 11, 552–66.
5. Ibid., 11, 581–2.
6. Ibid., 11, 651–63.
7. Ibid., 4, 300–1; 384–7; 450–73; 483ff.
8. Ibid., 6, 456–76.
9. Ibid., 12, 593–613.
10. Ibid., 12, 139–60; 222ff.; 460–500.
11. Ibid., 7, 546ff.
12. Ovid, *Metamorphoses*, 2, 411ff.
13. Ibid., 8, 318ff.; Apollonius of Rhodes, *Argonautica*, 1, 769–73.
14. Statius, *Thebaid*, 4, 2, 48; see also 4, 267; 6, 563.
15. Ibid., 4, 337–40.
16. Lucan, *De Bello Civili*, 6, 508ff.
17. Silius Italicus, *Punica*, 2, 56ff.
18. Ibid., 2, 193–205.
19. Statius, *Thebaid*, 12, 177–86.
20. Ibid., 12, 177–9.
21. Seneca the Younger, *Consolatio ad Helviam*, 16, 5. Seneca the Younger, *Consolatio ad Marciam*, 16, 1.
22. Statius, *Thebaid*, 12, 312–13.
23. Valerius Flaccus, *Argonautica*, 2, 107ff.
24. Ibid., 191–5.
25. Statius, *Thebaid*, 5, 129ff.
26. Apollonius of Rhodes, *Argonautica*, 1, 625ff.
27. Ibid.
28. Valerius Flaccus, *Argonautica*, 2, 324–5.

29. Aeschylus, *Libation Bearers*, 594ff.
30. Apollodorus, *Bibliotheca*, 1, 9, 17.

Chapter 16: Women and War and *Militia Amoris*

1. See Propertius, *Elegies*, 1, 6; Tibullus, *Elegies*, 1, 3; Ovid, *Amores*, 1, 19; 2, 12; *Ars Amatoria*, 2, 233–42.
2. Propertius, *Elegies*, 3, 12, 1–6.
3. Ibid., 9–13.
4. Tibullus, *Elegies*, 1, 1, 51–2.
5. Ibid., 1, 3, 9–14.
6. Ibid., 1, 5, 31–6; 2, 3, 47–58.
7. Propertius, *Elegies*, 1, 8, 29–46.
8. Virgil, *Eclogues*, 10, 46, 9, now ascribed to Gallus.
9. Propertius, *Elegies*, 2, 16, 11–12.
10. Ovid, *Amores*, 1, 10, 13–16.
11. Ibid., 1, 14, 45–50.
12. Ovid, *Ars Amatoria*, 1, 171–6.
13. Propertius, *Elegies*, 4, 5, 49–58.
14. Sappho, 1, 28 . Cf. Anacreon , 46; Theognis,1285; Aeschylus, *Supplices*, 1003ff., *Libation Bearers*, 594ff.; Sophocles, *Trachiniae*, 488; *Antigone*, 781; frag. 932; Euripides, *Hercules Furens*, 299; *Hippolytus*, 525ff., 727, frags 430, 431, 1132; 19.
15. *L'Année philologique*, 5, 99, 3.

Chapter 17: Military Tendencies in Women in Seneca's *Troades*

1. Seneca the Younger, *Consolatio ad Marciam*, 1, 1; 11, 1.
2. Seneca the Younger, *Troades*, 63–6.
3. Ibid., 89–122.
4. Ibid., 68–78.
5. Ibid., 469–74.
6. Ibid., 547–8.
7. Ibid., 623–4; 630–1.
8. Ibid., 662.
9. Ibid., 676–7.
10. Ibid., 762–5.
11. Ibid., 1157–9.

Chapter 18: Military Women in the Visual Arts

1. There are two other statues of Winged Victory in the Samothrace temple complex: a Roman copy now in the Kunsthistorisches Museum, Vienna, and a third Winged Victory found in 1949, now in a museum at the Samothrace site.
2. H.W. Janson, *History of Art*, 5th edn, revised and expanded by Anthony F. Janson (1995), pp. 157–58.

Chapter 19: Woman as Gladiators

1. Tacitus, *Annals*, 15, 32.
2. Dio Cassius, *Historia Romana*, 63, 3, 1.
3. Petronius, *Satyricon*, 45.
4. Dio Cassius, *Historia Romana*, 66, 25, 1.
5. Ibid., 68, 15.
6. Martial, *De Spectaculis*, 6; 8. Statius, *Silvae*, 1, 6, 51–6; Dog: Martial, *Epigrams*, 11, 59.
7. Suetonius, *Domitian*, 6, 1. Dio Cassius, *Historia Romana*, 67, 8, 4.

8. Statius, *Silvae*, 1, 6, 53.
9. Juvenal, *Satires*, 1, 22–3; 6, 252ff.
10. Nicolaus of Damascus, *Athletica*, 4, 153.
11. Dio Cassius, *Historia Romana*, 66, 26, 7.
12. Dio Cassius, *Historia Romana*, 76, 16.
13. *CIL*, 9, 2237. *Corpus Inscriptionum Graecarum* (1828–77), 6855f.; *Archaeologische Zeitung*, (1848), 202.
14. Hostilianus, *CIL*, 9, 2237. See Manas, 'New Evidence of Female Gladiators: the Bronze Statuette at the Museum für Kunst und Gewerbe of Hamburg', *International Journal of the History of Sport*, 28 (2011), 2726–52, for the bronze.
15. Ovid, *Ars Amatoria*, 1, 156.

Bibliography

Primary Sources Cited

Aelian, *Varia Historia*

Aeneas Tacticus, *How to Survive under Siege*

Aeschylus, *Seven Against Thebes*; *Prometheus Bound*; *Libation Bearers*; *Supplices*

Al-Tabari, *The History of al-Tabari*

Ammianus Marcellinus, *Res Gestae*

Anacreon

Anon., *Tractatus De Mulieribus*

Appian, *Civil Wars*; *Iberica*; *Punica*

Apollodorus, *Bibliotheca*

Apollonius Rhodius, *Argonautica*

Aristophanes, *Lysistrata*; *Ecclesiazusae*; *Birds*

Aristotle, *Politics*

Arrian, *De Rebus Successorum Alexandri Anabasis Alexandri*

Athenaeus, *Deipnosophistai*

The Bible: Exodus; Samuel; 2 Kings; Zechariah; Isaiah; Lamentations; Judges, The Song of Deborah; Judith; Matthew; Mark; John; Luke

Caesar, *De Bello Gallico*; *Bellum Civile*

Catullus

Chronicle of Pseudo-Joshua

Cicero, *Ad Familiares*; *Ad Atticum*; *Philippics*; *Tusculanae Disputationes*; *In Verrem*; *De Oratore*; *Divinatio in Caecilium; In Catilinam*

Ciris

Claudian, *Minor Works*; *Bellum Geticum*

Ctesias, *Persica*

Cypria

Dante, *Inferno*

Demosthenes, *On the Embassy*

Dio Cassius, *Historia Romana*

Diodorus Siculus, *Bibliotheca Historica*

Dionysius of Halicarnassus, *Roman Antiquities*

Ennius, *Annals*

Epictetus, *Discourses*

Euripides, *Phoenissae*; *Bacchae*; *Andromache*; *Hippolytus*; *Hercules Furens*; *Medea*

Eustathius on Homer

Eutropius, *Abridgement of Roman History*

Fabius Pictor

Florus, *Epitome of Roman History*

Frontinus, *Strategemata*

Heliodorus, *Aethiopica*

Hellanicus
Hero of Alexandria, *Treatise on Ranged Weapons*
Herodian, *Roman History*
Herodotus, *Histories*
Hesiod, *Theogony*; *Catalogues of Women and Eoiae*; *Shield of Heracles*
Hippocrates, *Airs, Waters, Places*
Historia Augusta: The Two Maximini
Homer, *Iliad*; *Odyssey*
Horace, *Odes*; *Epodes*
Hyginus, *Fabulae*
Jerome, *Beginning of the Consuls of the Romans*
Jordanes, *De Origine Actibusque Getarum*
Josephus, *Bellum Judaicum*; *Against Apion*; *Antiquities of the Jews*
Julian, *Orations*
Justinus, *Epitome of Pompeius Trogus*
Juvenal, *Satires*
Lactantius, *Divine Institutes*
Livy, *Ab Urbe Condita*; *Periochae*; *Epitome*
Lucan, *De Bello Civili*
scholiast, *On Lycophron*
Macrobius, *Saturnalia*
Martial, *Epigrams*; *De Spectaculis*
Metz Epitome
Nepos, Cornelius, *On Great Generals*; *Atticus*
Nicolaus of Damascus, *Athletica*
Nonnus, *Dionysiaca*
Onasander, *Strategikos* (Στρατηγικός)
Orosius, *Historiae Adversus Paganos*
Ovid, *Metamorphoses*; *Fasti*; *Remedia Amoris*; *Ars Amatoria*; *Amores*
The Palatine Anthology or *Anthologia Palatina*
Pausanias, *Description of Greece*
Petronius, *Satyricon*
Philochorus
Philon, *Poliorcetica*; *Mechanike Syntaxis*
Photius, *Bibliotheca*
Pindar, *Pythian Ode*; *Dithyrambs*
Plato, *Republic*; *Laws*
Pliny the Elder, *Naturalis Historia*
Pliny the Younger, *Epistles*
Plutarch, *Alexander*; *Galba*; *Lacaenarum Apophthegmata*; *Antony*; *Moralia*; *De Virtutibus Muliebrum*; *Pyrrhus*; *Cato Maior*; *Theseus*; *Philopoemen*; *Eumenes*; *Camillus*; *Romulus*; *Demetrius*; *Brutus*; *Themistocles*; *Lucullus*; *Alcibiades*; *Cicero*; *Mulierum Virtutes*; *De Garrulitate*
Polyaenus, *Stratagemata*
Polybius, *Histories*
Procopius, *History of the Wars*; *De Aedificiis*
Propertius, *Elegies*
Pseudo-Quintilian
Ptolemy, *Hephaeston*
Quintus Curtius Rufus, *Historiae Alexandri Magni*
Quintus Smyrnaeus, *Fall of Troy*

Sallust, *In Catilinam*; *Jugurtha*; *Histories*
Sappho
Scriptores Historiae Augustae
Seneca the Younger, *Consolatio ad Marciam*; *Troades*; *Consolatio ad Helviam*
Servius, *ad Georgics*; *On Virgil*
Silius Italicus, *Punica*
Socrates of Constantinople, *Historia Ecclesiastica*
Sophocles, *Trachiniae*; *Antigone*
Statius, *Silvae*; *Thebaid*
Stephanus of Byzantium, Myrina, *Ethnica*, *'Thessalonike'*
Stesichorus
Stobaeus, *Florilegium*
Strabo, *Geographia*
Suda, *Pamphile*; Suda and Hesychius of Alexandria, s.v. Ἄγραυλος
Suetonius, *Caligula*; *Julius Caesar*; *Domitian*; *Vespasian*; *Claudius*; *Nero*
Tacitus, *Agricola*; *Germania*; *Annals*; *Histories*
Tatian, *Oratio ad Graecos*
Tertullian, *De Spectaculus*
Thucydides, *History of the Peloponnesian War*
Tibullus, *Elegies*
Trebellius Pollio, *Scriptores Historia Augusta*
Tzetzes on Lycophron
Ulpian *ad Demosth. de fals. leg.*
Valerius Flaccus, *Argonautica*
Valerius Maximus, *Factorum et Dictorum Memorabilium*
Varro, *De Lingua Latina*
Velleius Paterculus, *Historiae Romanae*
Virgil, *Aeneid*; *Eclogues*
Vitruvius, *De Architectura*
Xenophon, *Constitution of the Lacedaemonians*; *Cyropaedia*; *Anabasis*; *Minor Works*, *Agesilaus*; *Hellenica*; *Oeconomicus*; *Cyrus*
Zonaras, *The History of Zonaras*
Zosimus, *Historia Nova*, Book I

Secondary Sources

Adie, K. (2003), *Corsets to Camouflage: Women and War* (London)
Adler, E. (2008), 'Boudica's Speeches in Tacitus and Dio', *CW*, 101, 173–95
Afflerbach, H. (ed.) (2012), *How Fighting Ends: A History Of Surrender* (Oxford)
Allason-Jones, L. (2012), 'Women in Roman Britain', in S.L. James (ed.) (2012), *Companion to Women in the Ancient World* (Chichester), pp. 467–77
—— (1989), *Women in Roman Britain* (London)
Allinson, Mrs (1919), 'Berlin and Athens', *Unpopular Review* (March)
Allison, P.M. (2013), *People and Spaces in Roman Military Bases* (Cambridge)
Alonso-Nunez, J. (1987), 'Herodotus on the Far West', *AC*, 56, 243–9
Aly, A.A. (1992), 'Cleopatra and Caesar at Alexandria and Rome', in Caratelli (ed.) (1992), *Roma e l'Egitto nell'Antichità Classica* (Rome), pp. 47–61
—— (1995), *Soldier and Society in Roman Egypt: A Social History* (London)
Annas, J. (1976), 'Plato's *Republic* and Feminism', *Philosophy* 51, 307–21
Arieti, J.A. (1997), 'Rape and Livy's View of Roman History', in S. Deacy (ed.) (1997), *Rape in Antiquity* (London), pp. 209–29

—— (1980), 'Empedocles in Rome: Rape and the Roman Ethos', *Clio*, 10, 1–20
Arthur, M.B. (1973), 'Early Greece: The Origins of the Western Attitude Toward Women', *Arethusa*, 6, 7–58
Askin, K.D. (1997), *War Crimes Against Women: Prosecution in International War Crimes Tribunal* (Amsterdam)
Austin, N. (1994), *Helen of Troy and her Shameless Phantom* (Ithaca NY)
Babcock, C. (1965), 'The Early Career of Fulvia', *AJPh*, 86, 1–32
Bahrani, Z. (2001), *Women of Babylon: Gender and Representation in Mesopotamia* (London)
Baldwin, B. (1972), 'Women in Tacitus', *Prudentia*, 4, 83–101
Balsdon, J.P.V.D. (1962), *Roman Women* (London)
Barber, E.W. (1994), *Women's Work: The First 20,000 Years – Women, Cloth and Society in Early Times* (New York NY)
Barr, R. (2000), 'Woman gladiator's remains discovered', *Charleston Gazette*, 13 September, p. P4C
Barry, W. (1976), 'Roof Tiles and Urban Violence in the Ancient World', *GRBS*, 37, 55–74
Bauman, R.A. (1992), *Women and Politics in Ancient Rome* (London)
Beard, M. (2015), *SPQR* (London)
Beard, M.R. (1946), *Women as a Force in History: A Study in Tradition and Realities* (New York NY)
Becker, T.H. (1997), 'Ambiguity and the Female Warrior: Virgil's Camilla', *Electronic Antiquity*, 4 (www.scholar.lib.vt.edu/ejournals/ElAnt/V4N1/becker.html)
Benario, H.W. (2007), 'Boudica Warrior Queen', *Classical Outlook*, 82, 70–3
Bennett, F.M. (1912), *Religious Cults Associated with the Amazons* (New York NY)
Benoist, S. (2015), 'Women and *Imperium* in Rome', in J. Fabre-Serris (ed.) (2015), *Women and War in Antiquity* (Baltimore NJ), pp. 266–88
Bergmann, F.G. (1853), *Les Amazones dans l'histoire et dans la fable* (Colmar)
Bergren, A. (1979), 'Helen's Web: Time and Tableau in the Iliad', *Helios*, 7, 19–34
Bessone, F. (2015), 'Love and War: Feminine Models, Epic Roles and Gender Identity in Statius' Thebaid', in J. Fabre-Serris (ed.) (2015), *Women and War in Antiquity* (Baltimore NJ), pp. 119–37
Bisset, K.A. (1971), 'Who Were the Amazons?', *G&R*, 18, 150–1
Blanshard, A.J.L. (2011), *Classics on Screen: Ancient Greece and Rome on Film* (London)
Blok, J.H. (trans. Peter Mason) (1995), *The Early Amazons: Modern and Ancient Perspectives on a Persistent Myth* (Leiden)
Bluestone, N.H. (1997), *Women and the Ideal Society: Plato's Republic and Modern Myths of Gender* (Oxford)
Blythe, J.M. (2001), 'Women in the Military: Scholastic Arguments and Medieval Images of Female Warriors', *History of Political Thought*, 22
Bonfante, W.L (1994), 'Etruscan Women', in E. Fantham (1994), *Women in the Classical World: Image and Text* (New York NY), pp. 243–59
—— (1973), 'The Women of Etruria', *Arethusa*, 6, 91–102
von Bothmer, D. (1957), *Amazons in Greek Art* (Oxford)
Bowersock, Glen Warren (1999), *Late Antiquity: A Guide to the Postclassical World* (Harvard, NJ)
Bowman, A.K. (2003), *The Vindolanda Writing Tablets III* (London)
Bowman, A.K. and D. Thomas (1994), *The Vindolanda Writing Tablets II* (London)
Boyd, B.W. (1992), 'Virgil's Camilla and the Traditions of Catalogue and Ecphrasis', *AJPh*, 113, 213–34
Bragg, E. (2007), 'Beyond the Battlefield: Caesar on Massacres, Executions and Mutilations', *Omnibus*, 54, 15–18

Bragg, M. (2013), *Queen Zenobia In Our Time*, BBC Radio 4, Thursday, 30 May 2013 (www.bbc.co.uk/programmes/b01snjpp)
Branigan, K. (1985), *The Catuvellauni* (Stroud)
Brauer, G.C. (1970), *Judea Weeping: The Jewish Struggle Against Rome from Pompey to Masada, 63 BC to AD 73* (New York NY)
Braund, D.C. (1996), *Ruling Roman Britain: Kings, Queens, Governors, and Emperors from Julius Caesar to Agricola* (New York NY)
—— (1984), *Rome and the Friendly King: The Character of the Client Kingship* (London)
Brennan, T.C. (2012), 'Perception of Women's Power in the Late Republic: Terentia, Fulvia and the Generation of 63 BC', in S.L. James (ed.) (2012), *Companion to Women in the Ancient World* (Chichester), pp. 354–66
Brier, R. (1999), *Daily Life of the Ancient Egyptians* (New York NY)
Briguel, D. (1992), 'Les Femmes Gladiateurs', *Ktema*, 16, 47–53
de Brohun, J. (2007), 'The Gates of War (and Peace)', in Raaflaub (ed.) (2007), *War and Peace in the Ancient World* (Cambridge MA), pp. 256–78
Brosius, M. (1998), *Women in Ancient Persia: 559–331 BC* (Oxford)
Brown, C. (2011), 'The Search for Cleopatra', *National Geographic*, July, 40–63
Brown, F.S. (1985), 'ἐκτιλώσαντο: A Reading of Herodotus' Amazons', *CJ*, 80, 297–302
Brown, R.D. (1995), 'Livy's Sabine Women and the Ideal of *concordia*', *TAPA*, 125, 291–319
Brownmiller, S. (1975), *Against Our Will: Men, Women and Rape* (Harmondsworth)
Brule, P. (2003), *Women of Ancient Greece* (Edinburgh)
Brunet, S. (2014), 'Women with Swords: Female Gladiators in the Roman World', in P. Christesen (ed.), *A Companion to Sport and Spectacle in Greek and Roman Antiquity* (Chichester), pp. 478–91
Buitelaar, M. (1995), 'Widow's Worlds', in J. Bremmer (1995), *Between Poverty and the Pyre* (London), pp. 1–18
Bulst, C. (1961), 'The Revolt of Queen Boudica in AD 60: Roman Politics and the Iceni', *Historia*, 10, 496–509
Burke, P.F. (1976), 'Virgil's Amata', *Vergilius*, 22, 24–9
Cadoux, C.J. (1912), *The Early Christian Attitude towards War* (New York NY)
Campbell, B. (ed.) (2013), *Oxford Handbook of Warfare in the Classical World* (Oxford)
—— (1978), 'The Marriage of Roman Soldiers Under the Empire', *JRS*, 68, 153–66
Campbell, D.B. (2006), *Besieged: Siege Warfare in the Ancient World* (Oxford)
—— (2005), *Ancient Siege Warfare: Persians, Greeks, Carthaginians* (Oxford)
Capponi, L. (2007), '"Signed Cleopatra" – Fact or Factoid?', *Ad Familiares*, 32, 15–16
Carney, E. (2006), *Olympias: Mother of Alexander the Great* (London)
—— (2005), 'Women and Dunasteia in Caria', *AJP*, 126, 65–91
—— (2004), 'Women and Military Leadership in Macedonia', *AncW*, 35, 184–95
—— (2000), *Women and Monarchy in Macedonia* (Norman OK)
—— (1987), 'The Career of Adea Eurydice', *Historia*, 36, 496–502
Carreiras, H. (2006), *Gender and the Military: Women in the Armed Forces of Western Democracies* (New York NY)
Carreiras, H. (ed.) (2008), *Women in the Military and in Armed Conflict* (Wiesbaden)
Carson, A. (1990), 'Putting Her in Her Place: Women, Dirt and Desire', in D.M. Halperin (ed.) (1990), *Before Sexuality* (Princeton NJ), pp. 35–69
Cartledge, P. (2012), 'Surrender in Ancient Greece', in H. Afflerbach (ed.) (2012), *How Fighting Ends: A History of Surrender* (Oxford), pp. 15–28
—— (1981), 'Spartan Wives: Liberation or Licence?', *CQ*, 31, 84–105
Chaniotis, A. (2012), 'Foreign Soldiers – Native Girls? Constructing and Crossing Boundaries in Hellenistic Cities with Foreign Garrisons', Paper given at the 19th International

Congress of Historical Sciences (2000) (http://www.oslo2000.uio.no/program/papers/r5/r5-chaniotis.pdf)
Chapman, A. (2007), *The Female Principle in Plutarch's* Moralia (Dublin)
Chavalas, M.V. (2013), *Women in the Ancient Near East* (London)
Chrystal, P. (2017), *Ancient Greece in 100 Facts* (Stroud)
—— (2017), *Roman Record Keeping and Communication* (Stroud)
—— (2017), *Women in Ancient Greece: Seclusion, Exclusion or Illusion?* (Stroud)
—— (2016), *In Bed with the Ancient Greeks: Sex and Sexuality in Ancient Greece* (Stroud)
—— (2016), 'Deadlier than the Male', *Minerva*, September–October, 36–40
—— (2015), *Roman Military Disasters: Dark Days and Lost Legions* (Barnsley)
—— (2015), *Roman Women: The Women Who Influenced the History of Rome* (Stroud)
—— (2015), 'Season of the Witch', *Minerva*, September–October, 44–8
—— (2015), *Wars and Battles of the Roman Republic* (Stroud)
—— (2014), *In Bed with the Romans: Sex and Sexuality in Ancient Rome* (Stroud)
—— (2014), 'A Powerful Body of Women', *Minerva*, January–February, 10–13
—— (2014), 'Roman Women Go to War I–IV', *Omnibus*, October
—— (2014), *Women in Ancient Rome* (Stroud)
—— (1982), 'Differences in Attitude to Women as Reflected in the Work of Catullus, Propertius, the Corpus Tibullianum, Horace and Ovid', Diss., University of Southampton
Clayton, E.C. (1879), *Female Warrior* (London)
Cloud, D. (1993), 'Roman Poetry and Anti-Militarism', in J. Rich (ed.) (1993), *War and Society in the Ancient World* (London), pp. 113–38
Cloutier, G. (2013), *Andromache: Denial and Despair. The First-Year Papers. Trinity College Digital Repository* (Hartford CT), www.digitalrepository.trincoll.edu/fypapers/37
Cogan, M. (1983), 'Ripping Open Pregnant Women in Light of an Assyrian Analogue', *Journal of the American Oriental Society*, 103
Coleman, K. (2000), 'Missio at Halicarnassus', *HSCP*, 100, 487–500
Colin, G. (1935), 'Luxe Oriental et Parfums Masculins dans la Rome Alexandrine', *Revue Belge de Philologie et d'Histoire*, 33, 5–19
—— (1905), *Rome et la Grece de 200 a 146 BC avant JC* (Paris)
Cook, B.A. (ed.) (2006), *Women and War: A Historical Encyclopedia from Antiquity to the Present*, 2 vols (Oxford)
Cooper, H.M. (ed.) (1989), *Arms and the Woman: War, Gender, and Literary Representation* (Chapel Hill NC)
Copper, J.S. (1983), *The Curse of Agade* (Baltimore NJ)
Cornell, T.J. (1997), *Gender and Ethnicity in Ancient Italy* (London)
—— (1995), *The Beginnings of Rome* (London)
Courtney, E. (1988), 'Virgil's Military Catalogues and their Antecedents', *Vergilius*, 34, 3–8
Cowan, R. (2007), *For the Glory of Rome: A History of Warriors and Warfare* (London)
Crawford, H. (2004), *Sumer and Sumerians* (Cambridge)
Curto, S. (1971), *The Military Art of the Ancient Egyptians* (Milan)
Dardi, K.S. (2001), *These Kamboj People* (http://www.punjabi.net/forum/archive/index.php/t-927.html)
Davies, G. (2006), *Roman Siege Works* (Stroud)
Davies, R.W. (1989), *Service in the Roman Army* (Edinburgh)
—— (1970), 'The Roman Military Medical Service', *Sonderdruck aus dem Saalburg Jahrbuch* (Edinburgh)
Davis, P.K. (2001), *Besieged: 100 Great Sieges from Jericho to Sarajevo* (Oxford)
—— (1999), *100 Decisive Battles from Ancient Times to the Present* (Oxford)

Dawson, D. (1997), *The Origins of Western Warfare: Militarism and Morality in the Ancient World* (Boulder CO)
Deacy, S. (ed.) (1997), *Rape in Antiquity* (London)
Delbrück, H. (1920), *Warfare in Antiquity: History of the Art of War, Volume 1* (Lincoln NE)
Delia, D. (1991), 'Fulvia Reconsidered', in S.B. Pomeroy (ed.) (1991), *Women's History and Ancient History* (Charlotte NC), pp. 197–217
Desmond, A.C. (1971), *Cleopatra's Children* (New York NY)
—— (1997), 'Romans and Pirates in a Late Hellenistic Oracle from Pamphylia', *CQ*, 47, 477–81
Dewald, C. (2013), 'Women and Culture in Herodotus' Histories', in R.V. Munson (ed.) (2013), *Herodotus: Volume 2: Herodotus and the World*, pp. 151–72 (Oxford)
—— (1980), 'Biology and Politics: Women in Herodotus' Histories', *PCP*, 15, 11–18
Dillon, S. (ed.) (2006), *Representations of War in Ancient Rome* (Cambridge)
Dodson, A. (2004), *The Complete Royal Families of Ancient Egypt* (London)
Donaldson, I. (1982), *The Rapes of Lucretia: A Myth and Its Transformations* (Oxford)
Donato, G. (1989), *The Fragrant Past: Perfumes of Cleopatra and Julius Caesar* (Rome)
Dougherty, C. (1988), 'Sowing the Seeds of Violence: Rape, Women and the Land', in M. Wyke (ed.) (1988), *Parchments of Gender: Deciphering the Bodies of Antiquity* (Oxford), pp. 267–84
Dowden, K. (1997), 'The Amazons: Development and Function', *RhM*, 140, 97–128
Druett, J. (2005), *She Captains: Heroines and Hellions of the Sea* (New York NY)
Du Bois, P. (1982), *Centaurs and Amazons: Women and the Pre-History of the Great Chain of Being* (Ann Arbor MI)
Ducrey, P. (2015), 'War in the Feminine in Ancient Greece', in J. Fabre-Serris (ed.) (2015), *Women and War in Antiquity* (Baltimore NJ), pp. 181–99
Dudley, D.R. (1962), *The Rebellion of Boudica* (London)
Duke, T.T. (1955), 'Women and Pygmies in the Roman Arena', *CJ*, 50, 223–4
Eck, W. (1999), 'The Bar Kokhba Revolt: The Roman Point of View', *JRS*, 89, 76–89
Egeler, M. (2008), 'Death, Wings, and Divine Devouring: Possible Mediterranean Affinities of Irish Battlefield Demons and Norse Valkyries', *Studia Celtica Fennica*, 5–26
Eggenberger, D. (1986), *Encyclopaedia of Battles: Accounts of Over 1,560 Battles from 1479 BC to the Present* (New York NY)
Ellis, P.B. (2003), *A Brief History of the Celts* (London)
Elshtain, J.B. (1995), *Women and War* (Chicago IL)
—— (ed.) (1990), *Women, Militarism, and War* (Lanham MD)
Errington, R.M. (1972), *The Dawn of Empire* (Ithaca NY)
Evans, J.A. (2011), *The Power Game in Byzantium: Antonina and the Empress Theodora* (London)
Evans, J.K. (1991), *War, Women and Children in Ancient Rome* (London)
Ewigelben, C. (2000), 'What these Women Love is the Sword', in E. Kohne (ed.) (2000), *Gladiators and Caesars* (Berkeley CA), pp. 125–39
Fabia, P. (1903), *Titii Livii loci qui sunt de praeda belli Romana* (Lyon, pp. 305–68
Fabre-Serris, J. (ed.) (2015), *Women and War in Antiquity* (Baltimore NJ)
—— (2015), 'Women after War in Seneca's *Troades*: A Reflection on Emotions', in J. Fabre-Serris (ed.), *Women and War in Antiquity* (Baltimore NJ), pp. 100–18
Fachard, S. (2012), *La défense du territoire d'Eretrie. Étude de la chôra érétrienne et de ses fortifications* (Gollon)
Fantham, E. (1994), *Women in the Classical World: Image and Text* (New York NY)
Faulkner, N. (2004), *Apocalypse: The Great Jewish Revolt Against Rome* (Stroud)
Ferrill, A. (1997), *The Origins of War: From the Stone Age to Alexander the Great* (Boulder CO)
Ferris, I. (2009), *Hate and War: The Column of Marcus Aurelius* (Stroud)
Fielding, S. (ed.) (1757), *The Lives of Cleopatra and Octavia* (Lewisburg PA)
Fischer, N.R.E. (1976), *Social Values in Classical Athens* (London)

Fitton-Brown, A.D. (1955), 'After Cannae', *Historia*, 4, 365ff.
Fletcher, J. (2008), *Cleopatra The Great: The Woman Behind the Legend* (London)
Flory, M.B. (1998), 'The Integration of Women into the Roman Triumph', *Historia*, 47, 489–94
Flower, H.I. (ed.) (2004), *The Cambridge Companion to the Roman Republic* (Cambridge)
Foley, H. (ed.) (2008), *Reflections of Women in Antiquity* (London)
——, 'Women in Ancient Epic', in H. Foley (ed.) (2008), *A Companion to Ancient Epic* (Chichester), pp. 105–18
Ford, M.C. (2004), *The Last King: Rome's Greatest Enemy* (New York NY)
Forsythe, G. (1999), *Livy and Early Rome* (Stuttgart)
Fowler, R. (1996), 'How the *Lysistrata* Works', *EMC*, 15, 245–59
Fox, M. (1996), *Roman Historical Myths: The Regal Period in Augustan Literature* (Oxford)
Frank, T. (1927), 'Roman Historiography Before Caesar', *American Historical Review*, 32, 232–40
Fraschetti, A. (ed.) (2001), *Roman Women* (Chicago IL)
Fraser, A. (2002), *The Warrior Queens: Boadicea's Chariot* (London)
Freeman, P. (2002), *War, Women, and Druids* (Austin TX)
Fry, P.S. (1982), *Rebellion Against Rome* (Lavenham)
Fuhrer, T. (2015), '*Teichoskopia*: Female Figures Looking on Battles', in J. Fabre-Serris (ed.) (2015), *Women and War in Antiquity* (Baltimore NJ), pp. 52–70
Fuller, J.F.C. (1970), *Decisive Battles of the Western World and Their Influence Upon History: v. 1; 1* (London)
Furneaux, R. (2004), *The Roman Siege of Jerusalem* (London)
Gabba, E. (1981), 'True History and False History in Classical Antiquity', *JRS*, 61, 50–62
—— (1971), 'The Perusine War and Triumviral Italy', *HSCP*, 75, 139–60
Gabriel, R.A. (1992), *A History of Military Medicine Vol. 1: From Ancient Times to the Middle Ages* (New York NY)
Gale, M. (1997), 'Propertius 2, 7: *Militia Amoris* and the Ironies of Elegy', *JRS*, 87, 77–91
Garland, Lynda (1999), *Byzantine Empresses: Women and Power in Byzantium, AD 527–1204* (London)
Garlick, B. (ed.) (1992), *Stereotypes of Women in Power* (New York NY)
Georgiou, I.E. (2002), *Women in Herodotus' Histories* (Swansea)
Georgoudi, S. (2015), 'To Act, Not Submit: Women's Attitudes in Situations of War in Ancient Greece', in J. Fabre-Serris (ed.) (2015), *Women and War in Antiquity* (Baltimore NJ), pp. 200–13
Gera, D.L. (1997), *Warrior Women: The Anonymous* Tractatus De Mulieribus (Leiden)
Gibbon, E. (1996), *History of the Decline and Fall of the Roman Empire* (London)
Giles, R.D. (2008), 'Roman soldiers and the Roman army: A study of military life from archaeological remains', PhD Thesis, University of Pennsylvania (http://repository.upenn.edu/dissertations/AAI3328561)
Gill, C. (ed.) (1993), *Lies and Fiction in the Ancient World* (Liverpool)
Goldman, N.L. (ed.) (1982), *Female Soldiers – Combatants or Noncombatants?* (Westport CT)
Goldstein, J.S. (2003), *War and Gender: How Gender Shapes the War System and Vice Versa* (Cambridge)
Golitko, M. (2007), 'Beating ploughshares back into swords: warfare in the Linearbandkeramik', *Antiquity*, 81, 332–42
Goodman, M.D. (2007), *Rome and Jerusalem: The Clash of Ancient Civilizations* (London)
—— (2004) (ed.), *Jews in a Graeco-Roman World* (Oxford)
—— (1987), *The Ruling Class of Judaea: The Origins of the Jewish War Against Rome* (Cambridge)
Goudchaux, G.W. (2001), 'Cleopatra's Subtle Religious Strategy', in S. Walker (ed.) (2001), *Cleopatra of Egypt: From History to Myth?* (London)

Gould, J. (1980), 'Women in Classical Athens', *JHS*, 100, 38–59
Graca, M. (2001), *The Impact of War on Children* (Vancouver)
Graf, F. (1984), 'Women, War and Warlike Divinities', *Zeitschrift fuur Papyrlogie und Epigraphik*, 55, 245–54
Grant, M. (1995), *Greek and Roman Historians: Information and Misinformation* (London)
—— (1992), *Cleopatra* (New York NY)
—— (1973), *The Jews in the Roman World* (London)
Grant de Pauw, L. (2000), *Battle Cries and Lullabies: Women in War from Prehistory to the Present* (Norman OK)
Graves, Robert (2013), *Selected Poems*, ed. Michael Longley (London)
Graves-Brown, C. (2010), *Dancing for Hathor: Women in Ancient Egypt* (London)
Green, M. (1995), *Celtic Goddesses: Warriors, Virgins and Mothers* (London)
Greene, E. (1998), *The Erotics of Domination: Male Desire and the Mistress in Latin Love Poetry* (Baltimore NJ)
Greensmith, E. (2013), 'The Elle-iad: Female Empowerment in the Iliad', *Omnibus*, 66, 10–12
Griffin J. (1976), 'Augustan Poetry and the Life of Luxury', *JRS*, 66, 87–105
Groten, F.J. (1968), 'Homer's Helen', *G&R*, 15, 32–9
Grunewald, T. (2008), *Bandits in the Roman Empire: Myth and Reality* (London)
Guilaine, Jean (2005), *The Origins of War: Violence in Prehistory* (Chichester)
Gurval, R.A. (2004), 'Dying Like a Queen: the Story of Cleopatra and the Asp(s)', in R. Miles (ed.) (2004), *Constructing Identities in Late Antiquity* (London)
Gutman, A. (1996), *The Erotic in Sports* (New York NY)
Haberling, W. (1943), 'Army Prostitution and its Control', in V. Robinson (ed.) (1943), *Morals In Wartime* (New York NY), pp. 3–90
Hacker, B.C. (ed.) (2012), *A Companion to Women's Military History*.
Haggard, H.R. (1889), *Cleopatra* (London)
Haley, S.P. (1985), 'The Five Wives of Pompey the Great', *G&R*, 32, 49–59
Hallet, J. (2015), 'Fulvia: The Representation of An Elite Roman Woman Warrior', in J. Fabre-Serris (ed.) (2015), *Women and War in Antiquity* (Baltimore NJ), pp. 247–65
—— (1977), '*Perusinae Glandes* and the Changing Image of Augustus', *AJAH*, 2, 151–71
Hamblin, J.W. (2006), *Warfare in the Ancient Near East to 1600 BC* (New York NY)
Hamer, M. (2008), *Signs of Cleopatra: Reading an Icon Historically*, 2nd edn (Liverpool)
Hansen, E.V. (1971), *The Attalids of Pergamon* (New York NY)
Hanson, A. (1990), 'The Medical Writers' Woman', in D.M. Halperin (ed.) (1990), *Before Sexuality* (Princeton NJ), pp. 309–38
Hanson, W.S. (1987), *Agricola & the Conquest of North Britain* (London)
Hardwick, L. (1990), 'Ancient-Amazon Heroes: Outsiders or Women?', *G&R*, 37, 14–36
Harich-Schwarzbauer, H. (2015), 'The Femine Side of War in Claudian's Epics', in J. Fabre-Serris (ed.) (2015), *Women and War in Antiquity* (Baltimore NJ), pp. 289–302
Harris, R. (1989), 'Independent Women in Ancient Mesopotamia', in B.S. Lesko (ed.) (1989), *Women's Earliest Records* (Atlanta GA), pp. 145–56
Harris, W.V. (2004), 'The Rage of Women', in S. Braund and G.W. Most (eds) (2004), *Ancient Anger: Perspectives from Homer to Galen* (Cambridge), pp. 121–43
Harrison, T. (1997), 'Herodotus and the Ancient Greek Idea of Rape', in S. Deacy (ed.) (1997), *Rape in Antiquity* (London), pp. 185–208
Harvey, D. (1985), 'Women in Thucydides', *Arethusa*, 18, 67–90
Hassall, M. (1974), 'Rome and the Eastern Provinces at the End of the 2nd Century BC', *JRS*, 64, 195–220
Hatke, G. (2013), *Aksum and Nubia: Warfare, Commerce, and Political Fictions in Ancient Northeast Africa* (New York NY)

Hawkes, C. (ed.) (1973), *Greeks, Celts and Romans, Studies in Venture and Resistance* (London)
Haynes, D.E.L. (1923), *Ancient Tripolitania* (London)
Hays, M. (1803), *'Telesilia'. Female Biography; or Memoirs of Illustrious and Celebrated Women of all Ages and Countries*, 6 vols (London), Vol. 6, p. 424
Hayward, R. (1978), *Cleopatra's Needles* (London)
Hazewindus, M.W. (2004), *When Women Interfere. Studies in the Role of Women in Herodotus' Histories* (Amsterdam)
Heath, E.G. (1980), *Archery: A Military History* (London)
Henderson, J. (1998), *Fighting for Rome: Poets and Caesars, History and Civil War* (Cambridge)
Hendry, M. (1930), 'Three Problems in the Cleopatra Ode', *CJ*, 82, 137–1464
Herzog, C. (1978), *Battles of the Bible* (London)
Higgins, C. (2013), 'Roman Britain Under Attack', *BBC History*, August, 20–6
Hindley, C. (1994), 'Eros and Military Command in Xenophon', *CQ*, 44, 347–66
Hinds, K. (2009), *Ancient Celts* (London)
Hingley, R. (2005), *Boudica: Iron Age Warrior Queen* (London)
Hodkinson, S. (ed.) (2006), *Sparta and War* (Swansea)
Hogg, O.F. (1968), *Clubs to Cannon: Warfare and Weapons Before the Introduction of Gunpowder* (London)
Holmes, R. (2004), *Acts of War: The Behaviour of Men in Battle* (London)
Holum, K.G. (1982), *Theodosian Empresses* (Berkeley CA)
Hopkins, M. (1978), *Conquerors and Slaves* (Cambridge)
Hopwood, K. (1999), *Organised Crime in the Ancient World* (Swansea)
Howarth, N. (2008), *Cartimandua, Queen of the Brigantes* (Stroud)
Humphreys, S. (1983), *The Family, Women, and Death: Comparative Studies* (London)
Huzar, E.G. (1985), 'Mark Antony: Marriages vs Careers', *CJ*, 81, 97–111
—— (1978), *Mark Antony: A Biography* (Minneapolis, MN)
Inscriptiones Graecae
Ireland, S. (ed.) (2008), *Roman Britain: A Sourcebook*, 3rd edn (London)
Jackson, A.H. (1970), 'Some Recent Works on the Treatment of Prisoners of War in Ancient Greece', *Talanta*, 2
Jackson, L. (2012), 'Euripides in the Modern World', *Omnibus*, 66, 29–30
Jackson-Laufer, G.M. (1999), *Women Rulers throughout the Ages: An Illustrated Guide* (New York NY)
Jaeger, M. (1995), *Livy's Written Rome* (Ann Arbor, MI)
James, S. (2011), *Rome and the Sword: How Warriors and Weapons Shaped Roman History* (London)
—— (1983), 'Archaeological Evidence for Roman Incendiary Projectiles', *Saalburg Jahrbuch*, 40, 142–3
James, S.L. (1997), 'Slave-rape and Female Silence in Ovid's Love Poetry', *Helios*, 24, 60–76
James, S.L. (ed.) (2012), *Companion to Women in the Ancient World* (Chichester)
Jed, S. (1989), *Chaste Thinking: The Rape of Lucretia and the Birth of Humanism* (Bloomington IN)
Jenkins, T.E. (2006), 'Epistolary Warfare', in T.E. Jenkins (2006), *Intercepted Letters: Epistolarity and Narrative in Greek and Roman Literature* (Lanham MD), pp. 51–9
Jensen, Anne (1996), *God's Self-confident Daughters: Early Christianity and the Liberation of Women* (Westminster)
Jestice, P.G. (2006), 'Greek Women and War in Antiquity', in B.A. Cook (ed.) (2006), *Women and War: A Historical Encyclopedia from Antiquity to the Present*, 2 vols (Oxford), pp. 256–8.
Johnson, W.R. (1967), 'A Quean, a Great Queen? Cleopatra and the Politics of Misrepresentation', *Arion*, 6, 151–80
Jones, D.E. (1997), *Women Warriors: A History* (Washington DC)
Jones, H.P. (1918), 'Usages of Ancient Warfare', *Edinburgh Review* (January)

Jones, P. (2012), 'A Woman's Place in Homer' (http://www.spectator.co.uk/2012/12/a-womans-place-in-homer/)
Jones, P.J. (2006), *Cleopatra: A Sourcebook* (Norman OK)
Joshel, Sandra R. (1992), 'The Body Female and the Body Politic: Livy's Lucretia and Verginia', in Amy Richlin (ed.), *Pornography and Representation in Greece and Rome* (Oxford)
Kaplan, M. (1979), 'Agripinna semper atrox: A Study in Tacitus' Characterisation of Women', in C. Deroux (ed.) (1979), *Studies in Latin Literature and Roman History*, I, 410–17 (Brussels)
Karapanagioti, N. (2010), *Female Revenge Stories in Herodotus' Histories* (http://athensdialogues.chs.harvard.edu/cgi-bin/WebObjects/athensdialogues.woa/wa/dist?dis=93)
Keegan, J. (1993), *A History of Warfare* (London)
Keeley, L.H. (1996), *War Before Civilisation: The Myth of the Peaceful Savage* (Oxford)
Keith, A. (2015), 'Elegiac Women and Roman Warfare', in J. Fabre-Serris (ed.) (2015), *Women and War in Antiquity* (Baltimore NJ), pp. 138–56
Kelly, R.C. (2005), 'The Evolution of Lethal Intergroup Violence', *Proceedings of the National Academy of Sciences*, 102(43): 15294–8
—— (2000), *Warless Societies and the Origin of War* (Ann Arbor MI)
Kennedy, D.L. (2013), *Settlement and Soldiers in the Roman Near East* (London)
Kennedy, G. (1986), 'Helen's Web Unraveled', *Arethusa*, 19, 5–14
Kennell, N.M. (1995), *The Gymnasium of Virtue: Education and Culture in Ancient Sparta* (Chapel Hill NC)
Kern, P.B. (1999), *Ancient Siege Warfare* (London)
Klein, J. (1981), *The Royal Hymns of Shulgi, King of Ur* (Philadelphia PA)
Kleiner, D. (2005), *Cleopatra and Rome* (Cambridge MA)
Knapp, B.L. (1977), 'Virgil's Aeneid: Let us Sing of Arms and Women: Dido and Camilla', in B.L. Knapp (1977), *Women in Myth* (New York NY), Chapter 6
Knapp, R. (2011), *The Invisible Romans* (London)
Koch, J.T. (ed.) (2006), *Celtic Culture: A Historical Encyclopedia* (Santa Barbara CA)
de La Bedoyere, G. (2003), *Defying Rome: The Rebels of Roman Britain* (Stroud)
Laiou, A. (ed.) (1993), *Consent and Coercion to Sex and Marriage in Ancient and Medieval Societies* (Dunbarton Oaks, Washington DC)
Laqueur, W. (1977), *Guerrilla Warfare: A Historical & Critical Study* (London)
Larson, J. (1995), *Greek Heroine Cults* (Madison WI)
Lattimore, R. (1939), 'The Wise Adviser in Herodotus', *CP*, 34, 24–35
Law, H. (1919), 'Atrocities in Greek Warfare', *CJ*, 15, 132–47
Lazenby, J. F. (2011), *The Spartan Army* (Barnsley)
Lee, A.G. (1953), 'Ovid's Lucretia', *G&R*, 22, 107–18
Lee, J. (2004), 'For There Were Many *hetairae* in the Army: Women in Xenophon's *Anabasis*, *AncW*, 35, 45–65
Lee, W.E. (2011), *Warfare and Culture in World History* (New York NY)
Leitao, D.D. (2014), 'Sexuality in Greek and Roman Military Contexts', in T.K. Hubbard (ed.) (2014), *Companion to Greek and Roman Sexualities* (Chichester), pp. 230–43
Lendon, J.E. (2007), 'War and Society', in Sabin (2007), *The Cambridge History of Greek and Roman Warfare, Vol. 1* (Cambridge), pp. 498–516
Levene, D.S. (ed.) (2002), *Clio and the Poets: Augustan Poetry and the Traditions of Ancient Historiography* (Leiden)
Levick, B. (2014), *Faustina I and II: Imperial Women of the Golden Age* (Oxford)
—— (2007), *Julia Domna: Syrian Empress* (London)
Levithan, J. (2013), *Roman Siege Warfare* (Ann Arbor MI)
L'Hoir, F.S. (1994), 'Tacitus and Women's Usurpation of Power', *CW*, 88, 5–25

—— (1992), *The Rhetoric of Gender Terms: 'Man', 'Woman' and the Portrayal of Character in Latin Prose* (Leiden)
de Libero, L. (2012), 'Surrender in Ancient Rome', in H. Afflerbach (ed.) (2012), *How Fighting Ends: A History Of Surrender* (Oxford), pp. 29–40
Lightman, M. (2008), *A to Z of Ancient Greek and Roman Women* (New York NY)
Lissarrague, F. (2015), 'Women Arming Men: Armor and Jewelry', in J. Fabre-Serris (ed.) (2015), *Women and War in Antiquity* (Baltimore NJ), pp. 71–81
Litchfield, H.W. (1914), 'National Examples of *virtus* in Roman Literature', *HSCP*, 25, 1–71
Loman, P. (2004), 'No Woman No War: Women's Participation in Ancient Greek Warfare', *G&R*, 51, 31–54
Luce, J.V. (1963), 'Cleopatra as *Fatale Monstrum*', *CQ*, 3, 251–7
Luckenbill, D.D.(1926), *Ancient Records of Assyria and Babylonia II* (Chicago IL)
Lyne, R.O.A.M. (1983), 'Virgil and the Politics of War', *CQ*, 33, 193–203
Lynn, J.A. (2003), *Battle: A History of Combat from Ancient Greece to Modern America* (Boulder CO)
McAuslan, I. (ed.) (1996), *Women in Antiquity* (Oxford)
McCrindle, J.W. (2004), *The Invasion of Alexander* (New York NY)
McCullough, A. (2008), 'Female gladiators in imperial Rome: literary context and historical fact', *CW*, 101, 197–210
McCune, B.C. (2014), 'Lucan's Militia Amoris', *CJ*, 109, 171–98
MacDonald, S. (1987), 'Boadicea: Warrior, Mother and Myth', in S. Macdonald (1987), *Images of Women in Peace and War*, pp. 1–26 (London)
McDonnell, M. (2006), *Roman Manliness:* Virtus *and the Roman Republic* (Cambridge)
McDougall, I. (1990), 'Livy and Etruscan Women', *Ancient History Bulletin*, 4, 24–30
McGregor, N. (2014), *Germany: Memories of a Nation* (London)
MacMullen, R. (1986), 'Women's Power in the Principate', *Klio*, 68, 434–43
—— (1980), 'Women in Public in the Roman Empire', *Historia*, 29, 208–18
—— (1963), *Soldier and Civilian in the Later Roman Empire* (Cambridge MA)
Macurdy, G.H. (1927), 'Queen Eurydice and the Evidence for Woman Power in Early Macedonia', *AJPh*, 48, 201–14
Mader, D. (1989), 'Heroism and Hallucination: Cleopatra in Horace C. 1.37 and Propertius 3.11', *Grazer Beitrage*, 16, 183–201
Manas, A. (2011), 'New evidence of female gladiators: the bronze statuette at the Museum für Kunst und Gewerbe of Hamburg', *International Journal of the History of Sport*, 28, 2726–52
Marshall, A.J. (1990), 'Women on Trial Before the Roman Senate', *Echos du Monde Classique*, 34, 333–66
—— (1984–6), 'Ladies in Waiting: The Role of Women in Tacitus' *Histories*', *Ancient Society*, 15–17, 167–84
—— (1975), 'Roman Women and the Provinces', *Ancient Society*, 6, 109–29
—— (1975), 'Tacitus and the Governor's Lady, A Note on *Annals* 3, 33–34', *G&R*, 22, 11–18
Martin, S. (1997), *Private Lives and Public Personae* (www.dl.ket.org/latin2/mores/women/womenful.htm)
Matthews, V.H. (ed.) (2004), *Gender and Law in the Hebrew Bible and the Ancient Near East* (London)
Mayor, A. (2015), 'When Alexander Met Thalestris', *History Today*, 65, 10–17
—— (2014), *The Amazons: Lives and Legends of Warrior Women across the Ancient World* (Princeton NJ)
Miles, G.B. (1997), *Livy: Reconstructing Early Rome* (New York NY), Chapter 5
Miles, M. (2003), *Cleopatra and Egyptomania* (Berkeley CA)
Millar, F. (1971), 'Paul of Samosata, Zenobia and Aurelian', *JRS*, 61, 1–17

Milnor, K. (2009), 'Women in Roman Historiography', in A. Feldherr (ed.) (2009), *The Cambridge Companion to the Roman Historians* (Cambridge), 276–87
Mitchell, R. (1991), 'The Violence of Virginity in the *Aeneid*', *Arethusa*, 24 (1991), 219–38
Montagu, J.D. (2006), *Greek and Roman Warfare: Battles, Tactics and Trickery* (London)
—— (2000), *Battles of the Greek and Roman Worlds* (London)
Morkot, R.G. (2010), *The A to Z of Ancient Egyptian Warfare* (Washington DC)
—— (2003), *Historical Dictionary of Ancient Egyptian Warfare* (Washington DC)
Moses, D. (1993), 'Livy's Lucretia and the Validity of Coerced Consent in Roman Law', in A. Laiou (ed.) (1993), *Consent and Coercion to Sex and Marriage in Ancient and Medieval Societies*, (Dunbarton Oaks, Washington DC), p. 50
Munson, R. (1988), 'Artemisia in Herodotus', *CA*, 7, 91–106
Murgatroyd, P. (1975), '*Militia amoris* and the Roman Elegists', *Latomus*, 34, 59–79
Murnaghan, S. (2005), 'Women in Greek Tragedy', in R. Bushnell (ed.) (2005), *A Companion to Greek Tragedy* (Chichester), 234–50
Murray, S. (2003). 'Female gladiators of the ancient Roman world', *Journal of Combative Sport*
Murray, W. (2012), *Hybrid Warfare: Fighting Complex Opponents from the Ancient World to the Present* (Cambridge)
Mustakallio, K. (1999), 'Legendary Women and Female Groups in Livy', in P. Setala (1999), *Female Networks and the Public Sphere in Roman Society* (Rome), pp. 53–64
Nakhai, B.A. (2008), *The World of Women in the Ancient and Classical Near East* (Newcastle)
Nappi, M. (2015), 'Women and War in the *Iliad*: Rhetorical and Ethical Implications', in J. Fabre-Serris (ed.) (2015), *Women and War in Antiquity* (Baltimore NJ), pp. 34–51
Newark, T. (1989), *Women Warlords: An Illustrated History of Female Warriors* (London)
Oakley, S.P. (1997), *A Commentary on Livy Books VI–X, Volume 1 Introduction and Book VI* (Oxford)
—— (1998), *A Commentary on Livy Books VI–X, Volume II: Books VII–VII* (Oxford)
—— (1993), 'The Roman Conquest of Italy', in J. Rich (ed.) (1993), *War and Society in the Ancient World*, pp. 9–37
Ogilvie, R.M. (1976), *Early Rome and the Etruscans* (Glasgow)
—— (1965), *A Commentary on Livy Books 1–5* (Oxford)
—— (1962), 'The Maid of Ardea', *Latomus*, 21, 477–83
O'Gorman, E. (2012), 'Does Dido's Curse Work?', *Omnibus*, 64, 10–12
Omitowoju, R. (2002), *Rape and the Politics of Consent in Classical Athens* (Cambridge)
O'Neill, K. (1995), 'Propertius 4,4: Tarpeia and the Burden of Aetiology', *Hermathena*, 158, 53–60
Oost, S. (1968), *Galla Placidia Augusta* (Chicago IL)
—— (1968), 'Galla Placidia and the Law', *CP*, 63, 114
—— (1965), 'Some Problems in the History of Galla Placidia', *CP*, 60, 1–10
Oppen de Ruiter, B.F. van (2015), *Berenice II Euergetis: Essays in Early Hellenistic Queenship* (New York NY)
Ormerod, H.A. (1924), *Piracy in the Ancient World* (Liverpool)
Otis, B.A. (1938), 'A Reading of the Cleopatra Ode', *Arethusa*, 1, 48–61
Otterbein, K.F. (2004), *How War Began* (College Station TX)
Otwell, J.H. (1977), *And Sarah Laughed: the Status of Woman in the Old Testament* (Philadelphia PA)
Packman, Z.M. (1993), 'Call it Rape: A Motif in Roman Comedy and Its Suppression in English-speaking Publications, *Helios*, 20, 42–55
Panagopoulos, A. (1978), *Captives and Hostages in the Peloponnesian War* (Athens)
Pantelia, M. (2002), 'Helen and the Last Song for Hector', *TAPA*, 132, 21–7

Payen, P. (2015), 'Women's Wars, Censored Wars?', in J. Fabre-Serris (ed.) (2015), *Women and War in Antiquity* (Baltimore NJ), pp. 214–27
Peel, M. (ed.) (2004), *Rape as a Method of Torture* (London)
Pembroke, S. (1970), 'Locres et Tarente, le Rôle des Femmes dans la Fondation de Deux Colonies Grecques', *Annals ESC*, 25, 1240–70
—— (1967), 'Women in Charge: The Function of Alternatives in Early Greek Tradition and the Idea of Matriarchy', *Journal of the Warburg and Courtauld Institutes*, 30, 1–35
Pennington, R. (ed.) (2003). *Amazons to Fighter Pilots: A Biographical Dictionary of Military Women* (London)
Phang, S.E. (2008), *Roman Military Service: Ideologies of Discipline in the Late Republic and Principate* (Cambridge)
—— (2004), 'Intimate Conquests: Roman Soldiers' Slavewomen and Freedwomen', *AncW*, 35, 207–37
—— (2001), *Marriage of Roman Soldiers (13 BC–AD 235)* (Leiden)
Phipps, W.E. (1992), *Assertive Biblical Women* (Westport CT)
Pollock, S. (1999), *Ancient Mesopotamia: The Eden that Never Was* (Cambridge)
Pomeroy, S.B. (2002), *Women in Hellenistic Egypt* (New York NY)
—— (2002), *Spartan Women* (Oxford)
—— (1991), *Women's History and Ancient History* (Charlotte NC), pp. 197–217
Postgate, N. (1994), *Early Mesopotamia: Society and Economy at the Dawn of History* (London)
Postlethwaite, N. (1985), 'The Duel of Paris and Menelaos and the *Teichoskopia* in *Iliad* 3', *Anthichthon*, 19, 1–6
Potter, D.S. (1999), *Life, Death and Entertainment in the Roman Empire* (Ann Arbour MI)
Powell, A. (ed.) (1998), 'Julius Caesar and the Presentation of Massacre', in K.E. Welch (ed.) (1998), *Julius Caesar as Artful Reporter* (Swansea)
Powell, L. (2013), *Germanicus: The Magnificent Life and Mysterious Death of Rome's Most Popular General* (Barnsley)
Powers, D. (ed.) (2013), *Irregular Warfare in the Ancient World* (Chicago IL)
Price, J.J. (1992), *Jerusalem Under Siege: The Collapse of the Jewish State 66–70 CE* (Leiden)
Pringle, H. (2001), 'Gladiatrix', *Discover*, 22, 48–55
Prodanović, N.Ć. (1973), *Teuta, Queen of Illyria* (Oxford)
Pucci, P. (1993), 'Antiphonal Lament between Achilles and Briseis', *Colby Quarterly*, 29, 258–72
Quaegebeur, J. (1988), 'Cleopatra VII and the Cults of the Ptolemaic Queens', in R. Bianchi (ed.) (1988), *Cleopatra's Egypt* (New York NY), pp. 41–54
Qviller, B. (1996), 'Reconstructing the Spartan Partheniai: Many Guesses and a Few Facts', *SO*, 71, 34–41
Rankin, H.D. (1987), *The Celts and the Classical World* (London)
Rauh, N.K. (2003), *Merchants, Sailors and Pirates in the Ancient World* (Stroud)
Ray, J. (1994), 'Hatshepsut: the Female Pharaoh', *History Today*, 44, 23–9
Redfield, J. (1977), 'The Women of Sparta', *CJ*, 73, 146–61
Reinhold, M. (1933), 'The Perusine War', *CW*, 26, 180–2
Rich, J. (ed.) (1993), *War and Society in the Ancient World* (London)
Richmond, I.A. (1954), 'Queen Cartimandua', *JRS*, 44, 43–52
Ridgway, D. (ed.) (1979), *Italy Before the Romans* (Edinburgh)
Robinson, V. (ed.) (1943), *Morals In Wartime* (New York NY)
Roisman, H. (2006), 'Helen in the *Iliad*; *Causa Belli* and Victim of War', *AJP*, 127, 1–36
Roller, D.W. (2010), *Cleopatra: A Biography* (Oxford)
Roller, M.B. (2012), *Cornelia: On Making One's Name as* mater Gracchorum (www. krieger. jhu.edu/classics/wpcontent/uploads/sites/20/2013/06/Mother-of-the-Gracchi)

—— (2004), 'Exemplarity in Roman Culture: The Case of Horaius Cocles and Cloelia', *CP*, 99, 1–56

Rosaldo, M.Z. (1974), 'Women, Culture and Society', in M.Z. Rosaldo (1974), *Women, Culture and Society* (Stanford CT), pp. 17–42

Rosen, K. (1976), 'Ad Glandes Perusinas (*CIL* I 682 Sqq)', *Hermes*, 104, 123–4

Rothman, J. (2014), 'The Real Amazons', *The New Yorker*, October

Rousseau, P. (2015), 'War, Speech and the Bow are Not Women's Business, in J. Fabre-Serris (ed.) (2015), *Women and War in Antiquity* (Baltimore NJ), pp. 16–33

Ruffell, J. (1995), *Brave women warriors of Greek myth: an Amazon roster* (http://www.whoosh.org/issue12/ruffel3.html)

Saggs, H.W.F. (1989), *Civilisation Before Greece and Rome* (London)

Salmonson, J.A. (1991), *The Encyclopedia of Amazons* (London)

Samson, L. (2015), *Crossing Boundaries and Preserving Social Order: Women Who Advise Persian and Greek Leaders in Herodotus' Histories*, Classical Association of the Middle West (Monmouth IL)

Sancisi-Weerdenburg, H. (1993), 'Exit Atossa: Images of Women in Greek Historiography on Persia', in A. Cameron (1993), *Images of Women in Antiquity* (London), pp. 20–33

Santosuosso, A. (2001), *Storming the Heavens: Soldiers, Emperors and Civilians in the Roman Empire* (Boulder CO)

—— (1997), *Soldiers, Citizens, and the Symbols of War: From Classical Greece to Republican Rome, 500–167 BC* (Boulder CO)

Sartre, M. (2005), *The Middle East under Rome* (London)

Scafuro, A. (ed.) (1989), 'Studies on Roman Women Part 2', *Helios*, 16, 143–64

Schaps, D. (1982), 'The Women of Greece in Wartime', *CP*, 77, 193–213

Schatzmann, I. (1972), 'The Roman General's Authority over the Distribution of Booty', *Historia*, 21, 17–28

Schiff, S. (2010), *Cleopatra – A Life* (London)

Schulman, A.R. (1995), 'Military Organisation in Pharaonic Egypt', in J.M. Sasson (ed.) (1995), *Civilisations of the Ancient Near East Vol. 1* (New York NY), pp. 289–301

—— (1982), 'The Battle Scenes of the Middle Kingdom, *Jnl of the Society for the Study of Egyptian Antiquities*, 12, 165–82

Scodel, R. (1997), 'Teichoscopia, Catalogue, and the Female Spectator in Euripides', *Colby Quarterly*, 33, 76–93

Sealey, P.R. (1997), *The Boudican Revolt Against Rome* (London)

Sebillotte Cuchet, V. (2015), 'The Warrior Queens of Caria', in J. Fabre-Serris (ed.) (2015), *Women and War in Antiquity* (Baltimore NJ), pp. 228–46

Seevers, B. (2013), *Warfare in the Old Testament: The Organization, Weapons, and Tactics of Ancient Near Eastern Armies* (Grand Rapids MI)

Segal, C. (1971), *The Theme of the Mutilation of the Corpse in the* Iliad (Leiden)

Setala, P. (1999), *Female Networks and the Public Sphere in Roman Society* (Rome)

Seward, D. (2009), *Jerusalem's Traitor: Josephus, Masada and the Fall of Judea* (Cambridge)

Shapiro, H.A. (1983), 'Amazons, Thracians and Scythians', *GRBS*, 24, 105–14

Sharrock, A. (2015), 'Warrior Women in Roman Epic', in J. Fabre-Serris (ed.) (2015), *Women and War in Antiquity* (Baltimore NJ), pp. 157–78

Shatzman, I. (1972), 'The Roman General's Authority Over Booty', *Historia*, 21, 177–205

Shaw, I. (1991), *Egyptian Warfare and Weapons* (Oxford)

Sheppard, S.I. (2013), *The Jewish Revolt AD 66–74* (Oxford)

Sicker, M. (2000), *The Pre-Islamic Middle East* (London)

Siefert, R. (1992), 'Rape in Wars: Analytical Approaches', *Minerva – Quarterly Report on Women and the Military*, 11, 17–22

Sitwell, N.H.H. (1984), *Outside the Empire: The World the Romans Knew* (London)
Skeat, T.C. (1953), 'The Last Days of Cleopatra: A Chronological Problem', *JRS*, 43, 98–100
Skutsch, O. (1953), 'The Fall of the Capitol', *JRS*, 45, 77–8
Smallwood, E.M. (1976), *The Jews Under Roman Rule* (Leiden)
Smethurst, S.E. (1950), 'Women in Livy's History', *G&R*, 19, 80–7
Sobol, D. (1973), *The Amazons of Greek Mythology* (Cranbury NJ)
Sorek, S. (2008), *The Jews Against Rome* (New York NY)
Southern, P. (2011), *Antony and Cleopatra* (Stroud)
—— (2009), *Empress Zenobia: Palmyra's Rebel Queen* (London)
—— (1998), *Mark Antony* (Stroud)
de Souza, P. (ed.) (2011), *Piracy in the Greco-Roman World* (Cambridge)
Spalinger, A.J. (2005), *War In Ancient Egypt* (Chichester)
Spaulding, O.A. (1933), 'The Ancient Military Writers', *CJ*, 28, 657–69
Stadter, P.A. (1965), *Plutarch's Historical Methods: An Analysis of the Mulierum Virtutes* (Cambridge MA)
Stallibrass, S. (2008), *Feeding The Roman Army: The Archaeology of Production and Supply in NW Europe* (Oxford)
Stark, F. (1966), *Rome on the Euphrates* (London)
Stehle, E. (1989), *Venus, Cybele and the Sabine Women: The Roman Construction of Female Sexuality in Scafuro* (Princeton NJ), pp. 43–64
Stoneman, R. (1995), *Palmyra and its Empire: Zenobia's Revolt Against Rome* (An Arbor MI)
Strickland, Jane Margaret and Agnes Strickland (1854), *Rome, Regal and Republican: A Family History of Rome*
Suzuki, M. (1989), *Metamorphoses of Helen: Authority, Difference, and the Epic* (Ithaca NY)
Swartley, W.M (1983), *Slavery, Sabbath, War and Women: Case Issues in Biblical Interpretation* (Harrisonburg VA)
Tarn, W.W. (1965), *Octavian, Antony and Cleopatra* (Cambridge)
Taylor, C.C.W. (2012), 'The Role of Women', in R. Kamtekar (ed.) (2012), *Plato's Republic in Virtue and Happiness* (Oxford)
Thapliyal, U.P. (2010), *Warfare in Ancient India* (New Delhi)
Thieme, H. (1997), 'Lower Palaeolithic Hunting Spears from Germany', *Nature*, 385, 807–10.
Turner, B.A. (2011), *Military Defeats, Casualties of War and the Success of Rome* (London)
Turney-High, H. (1949), *Primitive War: Its Practice and Concepts* (Columbia SC)
Turton, G. (1974), *The Syrian Princesses: The Women Who Ruled Rome AD 193–235* (London)
Tyldesley, J. (1996), *Hatshepsut: The Female Pharaoh* (New York NY)
Tyrrell, W.B. (1984), *Amazons: A Study in Athenian Myth-Making* (Baltimore NJ)
Ullman, B.L. (1957), 'Cleopatra's Pearls', *CJ*, 52, 193–201
Vaiopoulos, V. (2008), *From militia patri'ae to militia amoris. Love Labour and post obitum Remuneration (Tib. 1.3)* (Naples)
Van Creveld, M. (2001), *Men, Women and War: Do Women Belong on the Front Line?* (London)
Vanderwerker, E.E (1976) 'A Brief Review of the History of Amputations and Prostheses', *Inter-Clinic Iinformation Bulletin*, 15, No. 5, 15–16
Vaughan, A.C. (1967), *Zenobia of Palmyra* (New York NY)
Veit, V. (ed.) (2001), 'The Role of Women in the Altaic World', Permanent International Altaistic Conference, 44th Meeting, Walberberg, 26–31 August 2001 (Wiesbaden)
Vesley, M.E. (1998), 'Gladiatorial training for girls in the Collegia Iuvenum of the Roman Empire', *EMC*, 62, 85–93
Vikman, E. (2005), 'Ancient Origins: Sexual Violence in Warfare, Part I', *Anthropology & Medicine*, 12 (1), 21–31

Virlouvet, C. (1993), 'Fulvia the Woman of Passion', in A. Fraschetti (ed.) (2001), *Roman Women* (Chicago IL), pp. 66–81
Vivante, B. (2006), *Daughters of Gaia: Women in the Ancient Mediterranean World* (Westport CT)
Volkmann, H. (1958), *Cleopatra: A Study in Politics and Propaganda* (London)
Waite, J. (2007), *Boudica's Last Stand* (Stroud)
Walcott, P. (1991), 'On Widows and their Reputation in Antiquity', *SO*, 66, 5–26
—— (1984), 'Greek Attitudes towards Women: The Mythological Evidence', *G&R*, 31, 37–47
—— (1978), 'Herodotus on Rape', *Arethusa*, 11, 137–47
Walters, J. (1997), 'Soldiers and Whores in Pseudo-Quintilian Declamation', in T.J. Cornell (1997), *Gender and Ethnicity in Ancient Italy* (London), pp. 109–14
Warmington, E.H. (1898) *Remains of Old Latin* (London)
Warner, Rex (2006), *Plutarch, the Fall of the Roman Republic* (London)
Wasinski, V.M. (2004), 'Women, War and Rape: The Hidden Casualties of Conflict', Diss., University of Leeds
Watts, D. (2008), *Boudica's Heirs: Women in Early Britain* (London)
Webster, G. (1978), *Boudica: The British Revolt Against Rome, AD 60* (London)
Weiden Boyd, B. (1992), 'Virgil's Camilla and the Traditions of Catalogue and Ecphrasis', *AJP*, 113, 213–34
Weigall, A. (1914), *The Life and Times of Cleopatra Queen of Egypt* (London)
—— (1931), *The Life and Times of Marc Antony* (New York NY)
Weir, A.J. (2007), 'A Study of Fulvia', Diss., Queen's University, Kingston, Ontario
Welch, K.E. (ed.) (1998), *Julius Caesar as Artful Reporter* (Swansea)
—— (1995), 'Antony, Fulvia and the Ghost of Clodius in 47 BC', *G&R*, 42, 182–201
Welch, T. (2012), 'Perspectives On and Of Livy's Tarpeia', *Journal on Gender Studies in Antiquity*, 2, 1–31
Wellesley, K. (1969), 'Propertius' Tarpeia Poem (IV, 4)', *Acta Classica*, 93–103
Wells, C.M. (1995), 'The Daughters of the Regiment: Sisters and Wives in the Roman Army', in W. Groenman-van-Waateringe (ed.), *Proceedings of the XVIth International Congress of Roman Frontier Studies* (1995), pp. 571–4
West, G.S. (1985), 'Chloreus and Camilla', *Vergilius*, 31, 23–5
—— (1975), 'Women in Virgil's *Aeneid*', Diss, UCLA
Wheeler, E.L. (1998), *Strategem and the Vocabulary of Military Trickery* (Leiden)
Wheelwright, J. (1989), *Amazons and Military Maids* (London)
Whiston, W. (2016), *Wars of the Jewsby Josephus* (London)
Whitehead, D. (2002), *Aineias Tacticus: How to Survive Under Siege*, 2nd edn (Bristol)
Wicker, K.O. (1978), 'Mulierum Virtutes', in H.D. Betz (ed.) (1978), *Plutarch's Ethical Writings and Early Christian Literature* (Leiden), pp. 106–34
Wiedemann, T.E.J. (1983), 'Thucydides, Women and the Limits of Rational Analysis', in I. McAuslan (1996), *Women in Antiquity* (Oxford), pp. 83–90
Wilden, A. (1987), *Man and Woman, War and Peace: the Strategist's Companion* (London)
Wildman, B.J. (1908), 'Juno in the *Aeneid*', *CW*, 2, 26–9
Wilhelm, H. (1987), 'Venus, Diana, Dido and Camilla in the *Aeneid*', *Vergilius*, 33, 225–6
Wilkes, J.J. (1992), *The Illyrians* (Chichester)
Williams, D. (1998), *Romans and Barbarians* (London)
Williamson, G.A. (1966), *Procopius: The Secret History* (London)
Winsbury, R. (2010), *Zenobia of Palmyra: History, Myth and the Neo-Classical Imagination* (London)
Wintjes, J. (2012), '"Keep the Women out of the Camp!" Women and Military Institutions in the Classical World', in B. Hacke (ed.) (2012), *A Companion to Women's Military History* (Leiden), pp. 17–60

Wiseman, T.P. (1998), 'Roman Republic, Year One', *G&R*, 45, 19–26
—— (1993), 'Lying Historians – Seven Types of Mendacity', in C. Gill (ed.), *Lies and Fiction in the Ancient World* (Liverpool)
—— (1987), *The Credibility of the Roman Annalists in Roman Studies: Literary and Historical* (Liverpool), pp. 293–6
—— (1983), 'The Wife and Children of Romulus', *CQ*, 33, 445–52
Witt, R.E. (1971), *Isis in the Graeco-Roman World* (London)
Wrangham, R. (1996), *Demonic Males: Apes and the Origins of Human Violence* (Boston MA)
Wright, L. (2015), 'Homage to Zenobia', *The New Yorker*, July
Wyke, M. (2009), '*Meretrix Regina*: Augustan Cleopatras', in J. Edmondsson (ed.) (2009), *Augustus* (Edinburgh), pp. 334–80
—— (1992), 'Augustan Cleopatras: Female Power and Poetic Authority', in A. Powell (ed.) (1992), *Roman Poetry and Propaganda in the Age of Augustus* (London), pp. 98–104
—— (ed.) (1988), *Parchments of Gender: Deciphering the Bodies of Antiquity* (Oxford)
Yadin, Y. (1963), *The Art of Warfare in Biblical Lands* (Jerusalem)
Zahran, Y. (2013), *Zenobia: Queen of the Desert*, 2nd edn (London)
Zaidman, L.B. (2015), 'Women and War: From the Theban Cycle to Greek Tragedy', in J. Fabre-Serris (ed.) (2015), *Women and War in Antiquity* (Baltimore NJ), pp. 82–99
Ziegler, K. (1937), 'Tomyris', *RE*, vi, A2, 1702–4
Ziolkowski, A. (1993), '*Urbs Direpta* or How the Romans Sacked Cities', in J. Rich (ed.), *War and Society in the Ancient World* (London), pp. 69–91
Zoll, A. (2002), *Gladiatrix: The true story of history's unknown woman warrior* (New York NY)

Websites

www.romanarmy.net – the website of the Roman Military Research Society
www.roman-empire.net/army/army/html
oxbow@oxbowbooks.com – booksellers and publishers specialising in classics and archaeology
www.hellenicbookservice.com
www.jact.org – Joint Association of Classical Teachers, publishers of *Omnibus*
www.friends-classics.demon.co.uk – publishers of *Ad Familiares*
www.scholar.lib.vt.edu/stats/ejournals/ElAnt-current.html – the website for Virginia Tech's *Electronic Antiquity*
www.paulchrystal.com – author's website
www.womenforwomen.org.uk – helping women survivors of war rebuild their lives
www.reading.ac.uk/Ure/ – the Ure Museum of Greek Archaeology at Reading University
www.rescue.org/sites/default/files/resource-file/IRC_WomenInSyria_Report_WEB.pdf
http://syriainstitute.org/siege-watch/

General Index

Abducting/kidnapping of women, xiii, xv, 9, 23, 29, 33, 50–1, 105
Aeneas Tacticus, *Poliokretika* or *How to Survive under Siege*, 60
Aeschylus: *Seven Against Thebes*, 28–9
Libation Bearers, 192
Agricola, 116
Alba, sack of, 113–14
Amalekites, 102
Amazonomachy, 21, 204
Amida, siege of, 119–20
Amputations, 116
Apollonius Rhodius, 191–2
Aristomenes, 113
Aristophanes: *Lysistrata*, 36–7
Birds, 37
Aristotle, 59
Army wives, 10, 12, 131, 145–6, 148, 189, 203
Atrocities committed against women, 11, 13, 15, 22, 29, 33, 52–3, 56, 88, 102–103, 105, 109, 116, 119–20, 189
Atrocities committed by women, 26, 30, 35, 39–41, 45–6, 57, 96, 137, 152–3, 190–1
Attic War, 23

Babylon, xix, xxi
Baggage train, women part of, 95, 129–30
Bible, The, xvi–xvii, 102–103
Black Hole of Calcutta, 115
Books on military strategy and war, 60–1
Bragging soldier, 37
Bravery, and women, xvi–xvii, 8, 15, 20, 26–7, 41–2, 46–8, 53, 55, 59, 65, 72–3, 127
Brutalisation of women, 18, 34

Camp followers, 49, 54–6, 129–30, 133
Cannibalism, xv, 116–20
Capua, 115, 126
Casualties of war, women as, xv, xvii, 13, 22, 33–4, 52–3, 81, 101–20
Catalogues, warrior women in, 60–81
Catullus, 197
Cemetery 117, xvii
Children as war victims, xv–xvii, 102, 108, 116–19, 203
Child soldiers, 135
Cicero, 97, 109, 114, 129–30, 137–9
Cities, destruction of, 3, 13, 28, 32–3
Cremona, sack of, 114–15
Cross-dressing, 15, 20, 46–7, 54, 61, 63–4, 67, 109, 165, 205

Dante: *Divine Comedy*, 97
Decapitation, xv–xviii
Diplomats, women as, 4, 63, 69–70, 93, 123, 128–9, 139–40
Displacement of women, xv, 1, 28, 33, 113–14
Donations by women for the war effort, 125, 133–4
Dryden: *Alexander's Feast*, 98

Egyptian women at war, xvi
Elephants, xix
Ennius, 17
Enslavement of women, xv, 11, 13, 28–30, 33, 49, 50, 52, 56–7, 77–8, 101, 108, 195, 199
Ethnic cleansing, 102
Euripides: *Andromache*, 30
Hecuba, 29
Iphigenia in Aulis, 31–2
Medea, 35
Meleager, 37
The Phoenissae, 33
The Suppliant Women, 30
The Trojan Women, 32
Evacuation of women and children, 48–50, 86

Feminisation of women, 183, 187, 189, 191, 192, 206

Frontinus: *Stratagemata*, 60–1
Funeral/burial rites, and women, 30–4, 190

Gallus, 194–5
Garrisons, women in, 57, 61, 130, 145–7
Genocide and women, xv, 102
Gladiators, women, 205–10
Goddesses of war, xxi, 3–6
Graves, Robert, 22–3
Guile, and women, xvii, xix, 10, 81

Heliodorus, 116, 193
Herodotus, 38–48, 105–107
Hesiod, 11, 16
Historia Regum Britanniae, xvii–xviii
Homer, 7–15, 103, 105
Horace, 17

Impact of war on women, 16–18, 28, 36–7
Imprisonment of women, xviii, 33, 35, 95
Infanticide/filicide, 27, 32, 50, 83, 85, 103, 108, 200
Ingenuity in military women, xviii, 41–2, 53, 61, 63–4, 68, 72, 81
Injuries, battlefield and women, xvii, 3
Intelligence (military), use of by women, xvii, 9, 16, 17, 26–7, 69, 148, 187–9
Iraq War 2003, 28, 116
ISIS or Daesh, 180–1

Jerusalem, siege of, 116–19
Jordanes, 24
Julius Caesar, 116, 127, 158–9
Julius Caesar: *Bellum Gallicum* and *Bellum Civile*, 60
Juvenal, 206–208

Killing of women by kin, 50

LBK Culture, xvii
Leir, King, xvii–xviii
Lex Oppia, 129
Livy, 110–11, 113, 115
Lucan, 187–9

Magic and warlike women, 71–2, 152–3
Marcus Aurelius, Column of, 115
Mark Antony, 162
Marlowe: *Doctor Faustus*, 98
Marriage, 57, 131
Martial, 206
Masculinisation of women, xix, 15, 20, 26, 34, 41–4, 66, 111, 127–8, 135, 143, 145, 190, 200, 206
Medicine, battlefield, xvii
Mercy shown to women and children, 50
Miles/militia amoris, 187, 194–8
Miles gloriosus see Bragging soldier
Misogyny, 29, 36
Mothers, 82, 187, 200, 203

Oikos, women's role in, 13–14, 192–3, 203
Onasander: *Strategikos*, 60
Ovid, 17, 97, 106, 187, 195, 207, 210

Palmyra, 179–81
Pawns, women as, xviii, 9, 12, 89–98, 149–50
Petronius, 207
Pirates, 50–1
Philo Mechanicus, 59
Philon of Byzantium *see* Philo Mechanicus
Picasso: *Guernica*, 120
Plato, 58–9
Plutarch: *Mulierum Virtutes*, *On the Bravery of Women*, 61–5
Lacaenarum Apophthegmata, *Famous Sayings of Spartan Women*, 84–5
Polyaenus: *Stratagemata*, 61, 65–78
Polybius: *Histories* (especially camps), 60, 115
Propaganda, 120, 203
Propertius, 194ff
Prostitutes in baggage trains, 133
Prostitution/concubinage – women forced into, 11, 101, 120
Psychological warfare, 65, 135

Quintilian, 115

Rape, xv, 28–30, 32–3, 51–2, 65, 78, 101–20, 167, 187
Rejection of women as militarists, 3, 7, 13–14, 23, 29, 58
Rigveda, xvii
Role reversal, 7–8, 14–15, 20, 26–7, 34–8, 41–4, 47, 51, 55, 143
Royal women, xvii, 38–40, 74, 81
Rule of women, opposition to, xviii, 7, 38

Sallust, 109, 115, 128, 135
Scipio Africanus, 113, 156

Seneca: *Troades*, 199
Sexual abuse of women, 54, 56, 101, 109
Sexuality and women's use of, 11, 17–20, 27, 32, 36–7, 62–3, 70–1, 106, 152–3, 159, 207
Sexual mutilation, xv, 103, 105
Sexual slurs against military women, xix–xx, 23, 36, 39–40, 137, 145–6, 152–3, 207
Shelley, Mary: *Frankenstein*, 189
Siege warfare, women's role in xix, 28–9, 33, 48–54, 56, 58–9, 76–7, 83, 135
Siege warfare, women as victims of, 52, 95, 96, 101–20
Silius Italicus, 189
Sinope, siege of, 134
Slavery *see* enslavement
Song of Deborah, xvi
Sophocles: *Antigone*, 34–5
Sparta, 63–4, 81–8
Statius, 18, 187, 189, 191, 206
Stoning to death of women, 11, 15, 96
Strategists, women as, xix, 4, 8, 14, 32, 41–6
Stripping naked of women, 56
Suicide of women, xviii, 10, 15, 31, 33–5, 44–5, 51, 56, 70–1, 77–8, 80, 86, 91, 93, 111, 128, 131, 147, 154, 156, 162, 185
Syracuse, sack of, 109, 114
Syria 2017, ii, 33, 102, 120, 180–1

Tacitus, 110–11, 114
Tacticians *see* Strategists
Talheim Death Pit, xv
Teichoskopeia, 16–18, 33, 53, 200
Thucydides, 38, 48–9, 61, 108
Tibullus, 194ff
Tile-slinging by women, 49, 51–4, 135
Titus, Arch of, 120
Torture and women, 69, 72–3, 77–8, 96, 102, 144
Tractatus de Mulieribus, 61, 79–81
Trajan's Column, 120
Trauma *see* Injuries
Troy, siege and fall of, 3–4, 8, 22, 31–3, 62

Valerius Flaccus, 18, 190–2
Vegetius: *De Re Militari*, 61
Vercingetorix, 114
Verres, 114
Victims of war, women as, xv, 11, 56–7, 101–20, 187, 203
Vindolanda, 131
Virgil, 10, 17, 80
Vitellius, 114

Warmongering in women/goddesses, 3, 15, 186
Widows, war, 8, 33, 130
Witchcraft, allegations of against military women, xx, 90, 187–9, 193
Women acting as men, xix, 13, 34, 41–4, 66, 87
Women and acceptance of war, 14, 18, 81–8
Women and opposition to the military, 29
Women and war, origins of, xiv
Women as a weapon of war, 183
Women as builders of cities, xix, 21, 24, 26–7
Women as *casus belli*, xv, 8–11, 31–2, 183, 189–90
Women as killers, xvi–xvii, 30, 48, 53, 57, 67, 69–75, 79–80, 83, 96, 161
Women as mourners, 8, 30–2, 190
Women as military advisors, 4, 8, 13, 41–4, 97, 123
Women as regents, xix, 90, 93–4, 149, 184
Women as spoils of war, 8, 11–12, 33, 52, 55, 68, 101, 104, 107
Women, devotion to husbands, 127
Women, duplicitous, 10, 17–18, 23, 53–4, 63–4, 72, 123–4
Women Intelligent & Brave in War see Tractatus de Mulieribus
Women in the arms industry/war effort, 54, 133–4, 135
Women on accompanied postings, 57
Women, powerful, xvii, 33–4, 75–6, 89–98, 136
Women rated as equal to men in war, 26, 41–2, 47–8
Women sacrificed, 31–3

Xenophon, 101–102

Yahweh, 102

Zosimus: *Historia Nova*, 61

Index of Women

Acanthis, 195
Ada I of Caria, 96–7
Aelia Galla Placidia, 149–51
Aelia Galla (Propertius), 194
Aello, 24
Aglauros, 15
Agrippina the Elder, xix, 94, 111, 133, 141–3
Agripinna the Younger, 94, 144–5
Ague, 7
Ahhotep I, xvi, 158
Ahhotep II, xvi
Albia Dominica, 147
Alecto, 186
Amage, 75–6
Amanikhatashan, 157
Amanirenas, 157
Amata, 185
Amazons, the, 7–8, 19–27, 183–4, 189–90, 197, 204, 206, 209
Ambrones, the, 173–4
Andromache, 8, 12–14, 30, 32, 97, 104, 199–200
Andromache, Amazon, 24
Antigone, 18, 33
Antiope, 24–5
Antonina, 152–3
Aphrodite, 3, 9, 16
Arcotatus, 86–7
Aretaphila of Cyrene, 69–70
Argia, 189
Arsinoe, 76
Arsinoe III Philopator, 154–5
Artemis, 31–2, 187
Artemisia I of Caria, 7, 41–4, 80–1
Artemisia II of Caria, 45–6
Asbyte, 189
Atalanta, 187
Athena, 3–4, 7, 76–7, 204
Atossa, 80, 108–109
Audata, 89
Bessa, the old woman of, 193
Boudica, 81, 110–11, 145, 165–71
Bracari women, The, 157
Briseis, 8, 11

Caecilia Metella, 130
Caeria of Illyria, 90
Callisto, 187
Camilla, 22, 183–4
Camma, 70–1
Candace of Meroe, 156
Cartimandua, 165–6
Cassandra, 30, 32
Celtic women, 63, 80–1
Caphene, 78
Charicleia, 193
Cheilonis, 67
Chiomara, 65
Chrysame, 71–2
Chryseis, 8, 11–12
Cimbrian women, the, 171–3
Claudia Severa, 131
Cleopatra, daughter of Olympias, 94–5
Cleopatra II, 157
Cleopatra VII, 157–64, 185
Cleophis, 98
Cloelia, 111, 124–5
Cluvia Facula, 126
Clytemnestra, 12, 31–2
Cordelia, Queen, xvii–xviii
Cornelia, wife of Gaius Calvisius Sabinus, 131
Cratesipolis, 76
Cyana *see* Hydna of Scione
Cynane, 89–90
Cynisca, 83
Cynthia, 194, 197
Cyrene, 7

Deborah, xvi
Deidameia, 74

Deidamia, 91–2
Deidamia = Laodamia *see* Laodamia
Delia, 194, 197–8
Dido, 111, 184–6
Domitia Decidiana, 132
Dripetrua, 125–6
Drypetis, 95–6

Enyo, 3
Epicharis, 77
Epipole, 15
Eppia, 207
Erictho, 187–9
Eris, 3
Eryxo, 71
Etazeta, 93–4
Etruscan women, 63–4
Europa, 106
Eurydice, wife of Creon, 35
Eurydice I of Macedon, 89, 91
Eurydice II of Macedon, 90–1
Evadne, 31

Faustina the Younger, 145–6
Fulvia, 138–9
Fulvia Flacca Bambula, 137–8
Furies, the, 186

German women, 174–7
Gorgo, 84
Graiae, 3
Gwendolen, Queen, xvii
Gygaie, 109

Harpe, 189
Hatchepsut, xvi, 158
Hecuba, 8, 14, 29–30, 32–3, 199, 204
Helen of Troy, 8–11, 14–16, 31–2, 106–107, 198
Hera, 3–5, 23, 205
Hersilia, 123
Hippo, 154
Hippodameia, 205
Hippolyta, 20–1, 23, 184
Hortensia, 127
Hydna of Scione, 47–8
Hypsicratea, 165
Hypsipile, 191–2

Io, 33, 106
Iphigenia, 31–2
Iris, 16
Ismene, 35

Jael the Kenite, xvi
Joan of Arc, xix, 15
Jocasta, 33–4, 189
Judith, xvi–xvii
Julia Domna, 146–7
Junia Tertia, or Tertulla, 136
Juno, 18, 185–6
Juturna, 185–6

Kambojas, The, 98
Kandakes of Kush, The, 156–7

Lampsace, 69
Laodice, 16, 73
Laodamia, 92
Lavinia, 184, 186, 198
Leaena, 72–3
Lemnian women, 190–1
Livia, 94, 130
Lucretia, 103–104, 111, 190
Lycoris, 194–5
Lysippe, 25
Lysistrata, 7, 36–7

Mania, 74
Marpesia, 24
Marpessa, 51
Mavia, 181
Medea, 18, 34, 106, 187
Menalippe, 24, 25
Messalina, 94
Messene, 15
Mevia, 206
Mucia Tertia, 128–9
Munatia Plancina, 143
Myrina, 24
Mysta, 77

Nefertiti, 158
Nereis, 92
Nitocris of Babylon, 80
Nike, 3, 6, 204
Nitetis, 66–7
Nyssia of Lydia, 38–9

Octavia the Younger, 139–41
Olympias, 90–2, 94
Onomaris, 80–1

Pamphile of Epidaurus, 79
Pantariste, 25
Parysatis, 95
Paulina Busa, 135–6
Penelope, 8–9, 14, 194
Penthesilea, 7, 8, 20–2, 26, 184, 204
Persian women, 62–3
Pheretima, 40–1
Phila, 91–2
Philocomasium, 37
Philotis, 67
Phocian women, 51, 78, 107–108
Phthia, 93
Pieria, 68
Polycleia, 72
Polycrete, 68–9
Polyxena, 29–30, 32, 199–200
Pomponia, 137
Pomponia Graecina, 132
Porcia Catonis, 128, 136
Praecia, 136
The Priestess of Athena, 76–7

Rhea Silvia, 123
Rhodugune of Parthia, 66
Roxana, 91, 95–6

Sabine women, 103–104, 110–11, 124, 198
Salmantica, The Women of, 64
Samsi, xx
Scylla, 17–18
Semiramis, xviii, 66
Sempronia, 127–8
Serena, 148
Servilia Caepionis, 136–7
Shammuramat, xviii
Shanakdakheto, 156
Sophonisba, 156
Stateira, 95–6
Stratonice of Macedonia, 93

Tanusia, 140
Tarpeia, 17, 123–4
Tegea, women of, 154
Telesilla of Argos, 46–7
Teuta, 155–6
Thais, 97–8
Thalestris, 25
Thargelia, 80
Thasos, women of, 134
Theiosso (Dido), 80
Theodora, 151–2
Thessalonike, 92
Thetis, 3–5, 203–204
Timessa, 51
Timocleia, 96
Timycha, 87
Tirgatao, 74–5
Tomyris of the Massagetae, 39–40
Triaria, 145
Trojan women, 62
Turia, 139
Tutela, 67

Valasca, 25
Verginia, 103–104, 112, 190
Verulana Gratilla, 145
Vestal Virgins, 145
Vestia Oppia, 126
Veturia, 125
Vishpala, xvii
Vitruvia, 147
Volumnia, 125

Youtab, 154

Zarinaea, 79–80
Zenobia, 177–81